Also in the Variorum Collected Studies Series:

NATHAN SIVIN
Science in Ancient China:
Researches and Reflections

JOHN HEDLEY BROOKE
Thinking about Matter: Studies in the History of
Chemical Philosophy

MAURICE CROSLAND
Studies in the Culture of Science in France and Britain since the
Enlightenment

ROBERT FOX
Science, Industry and the Social Order in Post-Revolutionary France

ROY M. MacLEOD
Public Science and Public Policy in Victorian Science

A. RUPERT HALL
Science and Society: Historical Essays on the Relations of Science,
Technology and Medicine

TREVOR H. LEVERE
Chemists and Chemistry in Nature and Society, 1770-1878

A.I. SABRA
Optics, Astronomy and Logic: Studies in Arabic Science and Philosophy

JULIO SAMSÓ
Islamic Astronomy and Medieval Spain

SILVIO A. BEDINI
Science and Instruments in 17th-Century Italy

DAVID A. KING
Astronomy in the Service of Islam

ALLEN G. DEBUS
Chemistry, Alchemy and the New Philosophy, 1550-1700

R.W. HOME
Electricity and Experimental Physics in 18th-Century Europe

WALTER PAGEL
From Paracelsus to Van Helmont: Studies in Renaissance
Medicine and Science

VARIORUM COLLECTED STUDIES SERIES

Medicine, Philosophy and
Religion in Ancient China

Professor Nathan Sivin

Nathan Sivin

Medicine, Philosophy and Religion in Ancient China
Researches and reflections

Ashgate
VARIORUM

Aldershot · Brookfield USA · Singapore · Sydney

This edition copyright © 1995 by Nathan Sivin

Published in the Variorum Collected Studies Series by

Ashgate Publishing Limited
Gower House, Croft Road,
Aldershot, Hampshire GU11 3HR
Great Britain

Ashgate Publishing Company
Old Post Road,
Brookfield, Vermont 05036–9704
USA

ISBN 0–86078–493–2

British Library CIP Data
   Sivin, Nathan
   Medicine, Philosophy and Religion in Ancient China
   Researches and Reflections
   (Variorum Collected Studies Series;CS512)
   1. Title II. Series
   931

US Library of Congress CIP Data
   Sivin, Nathan
   Medicine, Philosophy and Religion in Ancient China:
   Researches and Reflections / Nathan Sivin.
   p. cm. – (Variorum Collected Studies Series: CS512)
   includes index.   (cloth;alk.paper)
   ISBN 0-86078-493-2
   1. Medicine, Chinese–History. 2. Science–China–History. 3. Taoism.
   I. Title. II. Series: Collected Studies: CS512
   R601.S58 1995   95-19571
   610´.951–dc20   CIP

This book is printed on acid free paper

Reprinted 1999

Printed in Great Britain by Biddles Limited, Guildford and King's Lynn

VARIORUM COLLECTED STUDIES SERIES CS512

CONTENTS

Introduction		vii-x
Acknowledgements		x
Publications of Nathan Sivin		xi-xvii
I	Comparing Greek and Chinese Philosophy and Science *First publication*	1-11
II	Emotional Counter-therapy *First publication*	1-19
III	The First Neo-Confucianism: An introduction to Yang Hsiung's 'Canon of Supreme Mystery' *(T'ai hsuan ching,* ca. 4 B.C.) With Michael Nylan. *Chinese Ideas about Nature and Society. Studies in Honour of Derk Bodde, ed. C. Le Blanc and S. Blader.* Hong Kong University Press, 1987, revised	1-42
IV	The Myth of the Naturalists *First publication*	1-33
V	On the Limits of Empirical Knowledge in Chinese and Western Science *Rationality in Question. On Eastern and Western Views of Rationality,* ed. S. Biderman & B. Scharfstein. Leiden: E. J. Brill, 1989	165-190

VI	On the Word 'Taoist' as a Source of Perplexity. With Special Reference to the Relations of Science and Religion in Traditional China *History of Religions 17. University of Chicago Press, 1978.*	303-330
VII	Taoism and Science *First publication*	1-72
VIII	Research on the History of Chinese Alchemy *Alchemy Revisited. Proceedings of the International Conference on the History of Alchemy at the University of Groningen 17-19 April 1989, ed. Z. R. W. M. von Martels. E. J. Brill, 1990*	3-20
IX	An Introductory Bibliography of Traditional Chinese Medicine: Books and Articles in Western Languages *Revised version of part of 'An introductory bibliography of traditional Chinese science: books and articles in Western languages,' Chinese Science: Explorations of an Ancient Tradition, ed. S. Nakayama and N. Sivin. Cambridge, MA: MIT Press, 1973, and of supplement in Chinese Science 7. Philadelphia, 1986*	1-15
Index		1-6

This book contains xviii + 274 pages

INTRODUCTION

This volume and its companion, *Science in Ancient China*, contain a selection of essays published between 1966 (the year I received my doctorate) and the present. *Medicine, Philosophy, and Religion in Ancient China* reprints three papers that appeared between 1978 and 1990, provides revised versions of one essay and a bibliography, and presents four new papers. I chose them because they are likely to be useful to people interested in Chinese culture.

My main response to being reminded of that long expanse of time is surprise. It has passed like a moment, with no sense of changing preoccupations, of staleness, of old issues done with. This is, I suppose, a variant on the old but true idea that, no matter what the course is called, every teacher has one thing to teach, namely his own curiosity and what it has turned up.

What motivated these studies and others is precisely that: an intense curiosity about how people in a civilization very different from that of Europe have gone about understanding Nature (whether external or represented in their own bodies) and defining their relation to it. Historical research on East Asian science has played more than a small role in the growing awareness that Europe realized only a few of the possibilities of science and technology. Studying other cultures historically is at least as serviceable a way as philosophic speculation to learn about ways of thinking that Western traditions never evolved.

If there is anything unconventional about my work, it is due to a dilettante's equal familiarity with the advancing frontiers of the history of science and medicine on one hand and Chinese studies on the other. Most people on one side think of the other as it was a generation or more ago. Fence-straddlers are no more objective than anyone else, but at least we are seldom tempted to treat the conventional wisdom on either side reverently. That is especially true of these two disciplines, because their styles are so different.

The main difference between technological history in 1966 and 1995 is the growing sense among students of history that Nature and the body are not objective entities, but were differently imagined, invented and reinvented in diverse cultures and eras. These years have spanned two major fads in the history of science: the study of ideas while ignoring their settings, and the study of social circumstances while ignoring what scientists thought about and did. There have been several minor ones as well: the Strong Programme, retrospective technology assessment, cultural studies, the attempt within anthropology to deny the validity of all generalizations, the postmodern realization that all is vanity, etc. Fads leave me uneasy, because when they pass their adherents vegetate, but they have yielded useful tools and insights. When I hear a colleague in the course of a lecture breaking the news yet again that science is constructed by the shared activity of human beings rather than being preordained by the character of physical reality, I feel irresistibly impelled to leave the room. The point is well taken (at least when taken sparingly), but the "rather than" robs it of most of its value. I realize that either/or stances are essential to the success of fads, but dealing with the complexity of events in the real world calls for all the subtlety and interconnection that we can grasp. That complexity is precisely what makes the practice of history zestful.

Classical sinology, still obsessed with texts and tending to treat them as sacred objects, has barely noticed the fruitfulness of new methodologies, and has rolled unseeing like a juggernaut over wave after wave of intrepid social constructionists, literary deconstructionists, discourse analysts, and other representatives of the academic millennium. That sublime oblivion leaves me equally uneasy.

Nevertheless, scholarship is very much a common enterprise, and new light often shines from old lamps. I greatly admire, and eagerly draw on, the quite fresh understanding of early Chinese thought that a number of young specialists have created since ca. 1990 using philological methods that have not changed a great deal in the past thousand years. I have probably profited as much from those trotting along on various hobbyhorses as from those who are not aware that they exist. I have directly addressed such disparate convictions about scholarship in Vol. I, Chap. VIII.

As in most research that aims at more than adding details to what is already understood, the key is finding the questions that are likely to be fruitful. No system, no readymade method, generates such questions. They often arise from innovative research in one of the seven or eight disciplines that I try with limited success to follow. They are as likely to jump out of a chat with someone who doesn't seem to see the point, or to simply pop up in the shower, as to crystallize from rumination. When one of them promises to shed light on the large issues, I simply choose the period and area of Chinese enterprise that is most likely to yield new understanding.

In that way, led by serendipitous questions, I have immersed myself at one time or another in all of the main dynasties and most of the traditional sciences. This volume reflects that promiscuity of interest. The four new articles deal with issues that have concerned me for a long time. "The Myth of the Naturalists" (Chap. IV) makes a case that I decided ca. 1963 needed to be made, and about which I had an invigorating correspondence ten years later with A.C. Graham (we met in Kyoto and quickly undertood a collaboration). Chap. VII, "Science and Taoism," was drafted in 1979 for the Third International Conference on Taoism. In an odd sense it was a collaboration with Joseph Needham, whose view of the historical connection between the two endeavors, although much deeper than those of his predecessors, had never been paid the compliment of a broad critical assessment. I wrote the paper in his research institute, we discussed it while it was aborning, and he was commentator when it was discussed at Unterägeri.

One of the things that made the three Taoism conferences so fruitful was that there was no pressure to produce a conference volume, so that the work that shaped the new field was able to mature in its own time. Over the years that followed, the study of Taoism as a historical entity accelerated so that there are now many scholars familiar with the remarkable literature in the Taoist Canon. The view that prompted generalizations about a special connection between science and religion depended on the understanding of the religion that these new studies rendered obsolete. The idea that Taoism (any Taoism) generated the Chinese sciences can no longer be taken seriously. A coherent new view of the alignment has taken some time to emerge. We now have a more defensible view of what the Taoist religious movements were about, and of the senses in which we can talk about Taoist philosophy. While addressing the issue of science I have tried to give a clear account of this new view.

The other two new papers did not gestate for so long. Chap. II was inspired by research for a book review a mere five years ago. It turned up a trove of material related to early ideas about therapy for emotional disorders that called for understanding without distraction from modern psychiatric theory. Since present-day viewpoints pervade almost everything published on the topic, I decided, despite my own innocence of the field, to carry out an exercise in what one might call non-destructive analysis. As any applied scientist knows, destructive analysis can be extremely powerful. Perhaps those who read this essay will agree that there is room for both.

The first essay, on comparison of Greek and Chinese science, is simply a preview of my current research. It summarizes the methodology that makes this work, for my collaborator and myself, absorbing.

I do not represent these as attempts at definitive statements. Laymen may speak of this or that book as definitive, but scholars know that the best we can do is strive tentatively to understand. My strivings, because I have chosen breadth at the inevitable expense of depth, are particularly tentative. A great cosmologist once said to me that he would know his field was seriously deficient in vitality if his work were not obsolescent within five years of the time it was published. I would be delighted if the velocity of the fields that interest me were as great.

I have somewhat simplified the discussions below by sometimes using the word "science" in general contexts to stand for mathematics, the natural sciences, medicine, and technology. In a more rigorous argument the considerable historic differences between these fields would have to be considered.

When citing Chinese sources in the revised essays in this volume, I have given

Chinese characters in the form of the original publication data, whether traditional or simplified. When I refer to the period between 1635 and 1644, I use "1635–1644"; "1635/1644" refers to some unknown moment within that interval. Translations are my own except where otherwise indicated.

Finally, since many of these essays have carried their own dedications, it would be excessive to give this volume a formal dedication. But the associations of the ensemble are too rich to pass over without acknowledgement. Consider this book and its companion an offering to my teachers, colleagues, and students, past, present, and future.

Acknowledgements

I am grateful to the following publishers for permission to publish original or revised papers in this volume: Hong Kong University Press (Chap. III), E. J. Brill (Chaps. V & VIII), the MIT Press (Chap. IX), and the University of Chicago Press (Chap. VI).

I would like to thank the Harvard-Yenching Library and the University of Pennsylvania Libraries for permission to reproduce illustations from books in their collections. I am grateful for support of several kinds from the School of Arts and Sciences at the University of Pennsylvania, including its Educational Technology Services, and to the Needham Research Institute, St. John's College, Cambridge, the Institute for Research in the Humanities, Kyoto University, the Institute for the History of Natural Science of the Chinese Academy of Sciences, and the China Institute for the History of Medicine and Medical Literature of the China Academy of Traditional Chinese Medicine (both in Beijing), for hospitality over many years.

I have acknowledged in individual chapters the invaluable criticisms and suggestions of many colleagues. Among my most concrete debts are those to Mary Beyer, who has served resourcefully as my secretary for many years, and to my wife Carole, who is always broadening my horizons. I am also grateful to Charles Cox for his editorial suggestions, and John Smedley and Ruth Peters for their very competent oversight of publication.

Correction

In Vol. I, the publication date of the original version of Chap. IV should be 1975. The item should also be marked "revised."

PUBLISHED WRITINGS OF NATHAN SIVIN

BOOKS

1968. *Chinese Alchemy: Preliminary Studies.* Harvard Monographs in the History of Science, 1. Cambridge, MA: Harvard University Press. Chinese translation, Taipei: National Translation Bureau, 1973.

1969. *Cosmos and Computation in Early Chinese Mathematical Astronomy.* Leiden: E. J. Brill. Separate book version of 1969 essay (see below).

1972. *Chinese Science: Explorations of an Ancient Tradition.* MIT East Asian Science Series, 2. Edited by Shigeru Nakayama & N.S. Cambridge, MA: MIT Press. Includes an introduction and three articles by N.S., listed below.

1977. *Science and Technology in East Asia. Articles from Isis, 1913-1975.* Selected and edited by N.S. New York: Science History Publications. Includes an introduction and an article by N.S., listed below.

1979. *Astronomy in Contemporary China. A Trip Report of the American Astronomy Delegation.* By ten members of the Delegation. CSCPRC Reports, 7. Washington, DC: National Academy of Sciences. Includes several contributions by N.S., including a chapter on the history of astronomy.

1980. *Science and Civilisation in China.* Vol. 5, Part 4. *Chemical Discovery.* By Joseph Needham, Lu Gwei-djen, Ho Ping-yu, & N.S. Cambridge, England: At the University Press. Includes a section by N.S. on the theoretical background of laboratory alchemy.

1984. *Chûgoku no Kopernikusu* 中國の コプルニクス (Copernicus in China), trans. Nakayama Shigeru 中山茂 & Ushiyama Teruyo 牛山耀代. Selected essays by N.S., 1. Tokyo: Shisakusha.

1985. *Chûgoku no renkinjutsu to ijutsu* 中國の錬金術と醫術 (Chinese alchemy and medicine), trans. Nakayama & Ushiyama. Idem, 2.

1987. *Traditional Medicine in Contemporary China. A Partial Translation of Revised Outline of Chinese Medicine (1972) with an Introductory Study on Change in Present-day and Early Medicine.* Science, Medicine and Technology

---

Following citations of essays, (J) indicates that a Japanese translation of the essay appears in the 1984 or 1985 book. (R) indicates that the essay was reprinted, and (U) that it was revised and updated, in this volume or its companion volume *Science in Ancient China*.

in East Asia, 2. Ann Arbor: University of Michigan, Center for Chinese Studies.

1988. *Contemporary Atlas of China.* Boston: Houghton Mifflin. Consulting Editor.

1989. *Science and Medicine in Twentieth-Century China: Research and Education,* ed. John Z. Bowers, William J. Hess, & N.S. Science, Medicine, and Technology in East Asia, 3. Ann Arbor: Center for Chinese Studies, University of Michigan.

1995. *Science in Ancient China. Researches and Reflections.* Variorum Collected Studies Series. Aldershot, Hants: Variorum.

1995. *Medicine, Philosophy and Religion in Ancient China. Researches and Reflections.* Idem.

ESSAYS

1963. William Lewis (1708-1781) as a Chemist. *Chymia, 8:* 63-88.

1965. On 'China's Opposition to Western Science during Late Ming and Early Ch'ing.' *Isis, 56:* 201-205.

1966. A Simple Method for Mental Conversion of a Year Expressed in Cyclical Characters to the Corresponding Year in the Western Calendar. *Japanese Studies in the History of Science, 4:* 132-134.

1966. Chinese Conceptions of Time. *The Earlham Review, 1:* 82-92. (J, R)

1967. A Seventh-Century Chinese Medical Case History. *Bulletin of the History of Medicine, 41:* 267-273.

1967. On the Reconstruction of Chinese Alchemy. *Japanese Studies in the History of Science, 6:* 60-86.

1968. Chinese Alchemy as a Science (Abstract). *Transactions of the International Conference of Orientalists in Japan, 13:* 117-129.

1969. Cosmos and Computation in Early Chinese Mathematical Astronomy. *T'oung Pao* (Leiden), *55:* 1-73. (R)

1969. On the *Pao p'u tzu nei p'ien* and the Life of Ko Hung. *Isis, 60:* 388-390.

1970. Chinese Alchemy as a Science. In *Nothing Concealed. Essays in Honor of Liu Yü-yün,* ed. Frederic Wakeman, Jr., pp. 35-50. Taipei: CMRASC, Inc.

1973. A Systematic Approach to the Mohist Optical Propositions, with A. C. Graham. In Sivin & Nakayama 1973: 105-152.

1973. Man as a Medicine. Pharmacological and Ritual Aspects of Drugs Derived from the Human Body, with W. C. Cooper. In *ibid.,* pp. 203-272.

1973. An Introductory Bibliography of Traditional Chinese Science. Sources in Western Languages. In *ibid.,* pp. 279-314. (U)

1973. Copernicus in China. *Studia Copernicana* (Warsaw), *6:* 63-122. In special series Colloquia Copernicana, 2. (J, U)

1973. The Celestial Elder's Canon of the Spirit Lights. *Io*, 4: 232–238.

1973. Li Shih-chen (1518–1593). In *Dictionary of Scientific Biography*, VIII, 390–398. New York: Charles Scribner's Sons.

1974. Next Steps in Learning about Science from the Chinese Experience. *Proceedings, XIVth International Congress of the History of Science* (Tokyo and Kyoto, 19–27 August 1974), I, 10–18. (J)

1974. Kagakushi kenkyû ni okeru saikin no keikô 科學史研究における最近傾向 (Recent trends in understanding the evolution of science). *Chûgoku no kagaku to bunmei* 中國の科學と文明, December, 1. 2: 2–4.

1975. Shen Kua (1031–1095). In *Dictionary of Scientific Biography*, XII, 369–393. New York: Charles Scribner's Sons. Reprinted as "Shen Kua: A Preliminary Assessment of his Scientific Thought and Achievements," *Sung Studies Newsletter*, 1977 (publ. 1978), 13: 331–56. (J, U)

1975. Introduction and Directory of Chinese Science. *Chinese Science*, 1: 1–51.

1976. Wang Hsi-shan (1628–1682). In *Dictionary of Scientific Biography*, XIV, 159–168. New York: Charles Scribner's Sons. Technical and interpretive, with bibliography. (J, U)

1976. Wang Hsi-shan, with Chaoying Fang. In *Dictionary of Ming Biography, 1368–1644*, ed. L. Carrington Goodrich, II, 1379–1382. New York: Columbia University Press. Mainly biographical. (U)

1976. Éloge. Giorgio de Santillana. *Isis*, 67: 439–443.

1976. Chinese Alchemy and the Manipulation of Time. *Isis*, 67: 513–527. Reprinted in Sivin 1977: 108–122. (J)

1977. Chûgoku dentô no gireiteki iryô ni tsuite 中國傳統の儀禮的醫療について (Ritual therapy in Chinese tradition). In *Dôkyô no sôgôteki kenkyû* 道教の綜合的研究 (General studies of Taoism), ed. Sakai Tadao 酒井忠夫, pp. 97–140. Tokyo: Kokusho kankôkai.

1977. Further Comments on the Use of Statistics in the Study of Han Dynasty Portents, with Hans Bielenstein. *Journal of the American Oriental Society*, 97: 185–187.

1977. Social Relations of Curing in Traditional China: Preliminary Considerations. *Nihon ishigaku zasshi* 日本醫史學雜誌, 23: 505–532. Reprinted in *History of Traditional Medicine. Proceedings of the 1st and 2nd International Symposia on the Comparative History of Medicine—East and West*, ed. Teizo Ogawa, pp. 21–46. Osaka: The Taniguchi Foundation, Division of Medical History, 1986. (J)

1978. On the Word Taoism as a Source of Perplexity. With Special Reference to the Relations of Science and Religion in Traditional China. *History of Reli-*

gions, 17: 303-330. (J, R)
1978. Current Research on the History of Science in the People's Republic of China. *Chinese Science,* 3: 39-58.
1979. Report on the Third International Conference on Taoist Studies. *Bulletin, Society for the Study of Chinese Religions,* Fall, 7: 1-23.
1980 (publ. 1981). Science in China's Past. In *Science in Contemporary China,* ed. Leo A. Orleans, pp. 1-29. Stanford University Press.
1980. The Roanoke Conference—Critical Issues in the History of Technology: Roanoke, Virginia, August 14-18, 1978. Concluding Remarks on the Conference. *Technology and Culture,* 21: 621-632.
1981. Why the Scientific Revolution Did Not Take Place in China—Or Didn't It? The Edward H. Hume Lecture, Yale University, *Chinese Science,* 1982, 5: 45-66. Reprinted in *Transformation and Tradition in the Sciences,* (ed. Everett Mendelsohn; Cambridge University Press, 1984), pp. 531-554, and in *The Environmentalist,* 1985, 5. 1: 39-50. Earlier draft published in *Explorations in the History of Science and Technology in China* (ed. Li Guohao, Zhang Mengwen, & Cao Tianqin; Shanghai Chinese Classics Publishing House, 1982), pp. 89-106. Chinese trans. by Wen-jen Chün 闻人军 & Yü Kang 余岗 in *K'o-hsueh yü che-hsueh* 科学与哲学, 1984, 1: 5-43; better trans. in *Chung-kuo k'o-chi-shih t'an-so* 中國科技史探索 (Explorations in the History of Science and Technology in China; ed. Li Kuo-hao 李國豪, Chang Meng-wen 張孟聞, & Ts'ao T'ien-ch'in 曹天欽; Shanghai: Shang-hai Ku-chi Ch'u-pan-she, 1986), pp. 97-114; Japanese trans. by Yano Michio 矢野道雄 in *Tôyô no kagaku to gijutsu. Yabuuchi Kiyoshi sensei sôju kinen rombunshû* 東洋の科學と技術。藪内清先生頌壽記念文集 (Science and Skills in Asia. *Festschrift* for Professor Yabuuchi Kiyoshi; Kyoto: Dôhôsha, 1982), pp. 252-280. (R)
1983. Chinesische Wissenschaft: Ein Vergleich der Ansätze von Max Weber und Joseph Needham. In *Max Webers Studie über Konfuzianismus und Taoismus. Interpretation und Kritik,* ed. Wolfgang Schluchter, pp. 342-362. Frankfurt: Suhrkamp.
1983. Dragons and Toads. The Chinese Seismoscope of A.D. 132, with André Wegener Sleeswyk. *Chinese Science,* 6: 1-19.
1983. A Directory of Scholars in East Asia Engaged in Research on Traditional Chinese Science, with Shigeru Nakayama. *Chinese Science,* 6: 33-58.
1984. Reflections on 'Nature on Trial.' In *Methodology, Metaphysics and the History of Science. In Memory of Benjamin Nelson,* ed. Robert S. Cohen & Marx Wartofsky, pp. 323-330. Boston Studies in the Philosophy of Science, 84. Dordrecht: D. Reidel Publishing Company.
1985. Max Weber, Joseph Needham, Benjamin Nelson: The Question of Chinese

Science (revised version of the essay in Schluchter 1983). In *Civilizations East and West. A Memorial Volume for Benjamin Nelson,* ed. E. Victor Walter et al., pp. 37–49. Atlantic Highlands, NJ: Humanities Press.

1985. Chung-kuo k'o-hsueh chi-shu i-hsueh shih chan-wang 中国科学技术医学史展望 (The Prospects of the History of Chinese Science, Technology, and Medicine). *Ta-tzu-jan t'an-so* 大自然探索 (Explorations of Nature), *4*. 11: 11–16. Transcript of a plenary lecture delivered in Chinese at the Third International Conference on the History of Chinese Science, Beijing, 21 August 1984.

1985. Chinese Science, Medicine, and Technology. *Proceedings,* The Conference on Technology and World History, Aspen, Colorado, 14–15 June, pp. 38–46. Aspen: The Conference.

1986. Traditional Chinese Medicine and the United States: Past, Present, and Future. *Bulletin,* The American Academy of Arts and Sciences, May, *39*. 8: 15–26.

1986. On the Limits of Empirical Knowledge in the Traditional Chinese Sciences. In *Time, Science and Society in China and the West. The Study of Time V,* ed. J. T. Fraser et al., pp. 151–169. Amherst: The University of Massachusetts Press.

1987. The First Neo-Confucianism. An Introduction to the "Canon of Supreme Mystery" (*T'ai hsuan ching,* ca. 4 B.C.), with Michael Nylan. In *Chinese Ideas about Nature and Society. Studies in Honour of Derk Bodde,* ed. Charles le Blanc & Susan Blader, pp. 41–99. Hong Kong University Press. (U)

1987. A Supplementary Bibliography of Traditional Chinese Science. Introductory Books and Articles in Western Languages. *Chinese Science,* 7: 33–41.

1988. Science and Medicine in Imperial China. The State of the Field. *Journal of Asian Studies, 47.* 1: 41–90. For a Japanese summary with commentary by Nakayama Shigeru see *Chûgoku kagakushi kokusai kaigi: 1987 Kyôto Symposium hôkokusho* 中國科學史國際會議。1987 京都 Symposium 報告書 (International Conference on the History of Science in China. 1987 Kyoto Symposium Proceedings), ed. Yamada Keiji 山田慶兒 & Tanaka Tan 田中淡 (Kyoto: Kyôto Daigaku Jimbun Kagaku Kenkyûshô, 1992), pp. 155–162.

1989. Chinese Archeoastronomy: Between Two Worlds. In *World Archeoastronomy. Selected Papers from the Second Oxford International Conference on Archeoastronomy Held at Merida, Yucatan, Mexico, 13–17 January 1986,* ed. Anthony R. Aveni, pp. 55–64. Cambridge University Press.

1989. On the Limits of Empirical Knowledge in Chinese and Western Science. In *Rationality in Question. On Eastern and Western Views of Rationality,* ed. Shlomo Biderman & Ben-Ami Scharfstein, pp. 165–189. Leiden: E. J. Brill.

Revision of 1986 essay. (R)

1989. A Cornucopia of Reference Works for the History of Chinese Medicine. *Chinese Science,* 9: 29–52.

1989. Alchemy: Chinese Alchemy. In *The Encyclopedia of Religion,* s.v. University of Chicago Press. Reprinted in *Hidden Truths. Magic, Alchemy, and the Occult* (Religion, History, and Culture), ed. Lawrence E. Sullivan, pp. 253–260. Idem.

1989. East Asia, Ancient. In *International Encyclopedia of Communications,* II, 71–78. New York: Oxford University Press.

1990. Science and Medicine in Chinese History. In *Heritage of China. Contemporary Perspectives on Chinese Civilization,* ed. Paul S. Ropp, pp. 164–196. University of California Press. Italian and Chinese translations. (R)

1990. Research on the History of Chinese Alchemy. In *Alchemy Revisited. Proceedings of the International Conference on the History of Alchemy at the University of Groningen. 17–19 April 1989,* ed. Z.R.W.M. von Martels, pp. 3–20. Leiden: E. J. Brill. (R)

1990. Traditional Medicine in Contemporary China and Reflections on the Situation in the People's Republic of China, 1987. *American Journal of Acupuncture,* 18. 4: 325–343.

1991. Saiyô kagaku to Tôyô kagaku no hikaku 西洋科學と東洋科學の比較 (Comparing Eastern and Western science). *Kokusai kôryu* 國際交流, 56: 24–29.

1991. Change and Continuity in Early Cosmology. In *Chûgoku kodai kagaku shiron. Zoku* 中國古代科學史論。續 (On the History of Ancient Chinese Science, 2), pp. 3–43. Kyoto: Institute for Research in Humanities.

1991. Over the Borders: Technical History, Philosophy, and the Social Sciences. *Chinese Science,* 10: 69–80. (R)

1993. *Huang ti nei ching.* In *Early Chinese Texts. A Bibliographical Guide,* ed. Michael Loewe, pp. 196–215. Early China Special Monograph Series, 2. Berkeley: The Society for the Study of Early China and the Institute of East Asian Studies, University of California.

1994. Introduction to East Asian Cartography, with Gari Ledyard. In *The History of Cartography,* ed. J. B. Harley & David Woodward, II, pt. 1, 23–35. Cartography in the Traditional East Asian Societies. University of Chicago Press.

1995. State, Cosmos, and Body in the Last Three Centuries B.C. *Harvard Journal of Asiatic Studies,* 55. 1: 5–37.

1995. Text and Experience in Classical Chinese Medicine. In *Knowledge and the Scholarly Medical Traditions,* ed. Don G. Bates, pp. 177–204. Cambridge University Press.

ESSAY REVIEWS

1966. Blofeld, John, trans. *The Book of Change* (London, 1965). *Harvard Journal of Asiatic Studies, 26:* 290–298.

1968. Needham, Joseph, et al. *Science and Civilisation in China*, IV, pt. 2 (Cambridge, England, 1965). *Journal of Asian Studies, 27:* 859–864.

1972. Needham, Joseph, et al. *Science and Civilisation in China*, IV, pt. 3 (Cambridge, 1971). *Scientific American,* January, pp. 113–118; *T'oung Pao,* 1971, *57:* 306–320.

1973 (publ. 1974). Teng Ssu-yü; Knight Biggerstaff. *An Annotated Bibliography of Selected Chinese Reference Works* (3rd ed., Cambridge, MA, 1971). *Isis, 64:* 534–538.

1978 (publ. 1979). Elvin, Mark. *The Pattern of the Chinese Past* (Stanford, 1973). *Harvard Journal of Asiatic Studies, 38:* 449–480.

1981. Needham, Joseph, et al. *Science and Civilisation in China*, V, pt. 3 (Cambridge, 1976). *Harvard Journal of Asiatic Studies, 41:* 219–235.

1981. Some Important Publications on Early Chinese Astronomy from China and Japan, 1978–1980. *Archaeoastronomy, 4.* 1: 26–31.

1990. Unschuld, Paul U. *Medicine in China* (3 vols., Berkeley, 1985–1986). *Isis, 81:* 722–731.

1991. Science, Religion, and Boundary Maintenance. *Contemporary Sociology, 20.* 4: 526–530. Review of four books on Merton's sociology of science.

1991. New reference books on Chinese medicine and on astronomy. *Chinese Science,* 10: 53–56.

1992. Ruminations on the Tao and its Disputers. *Philosophy East and West, 42.* 1: 21–29. Memoir of A. C. Graham and comments on *Disputers of the Tao.*

1992. Simon, Denis Fred; Merle Goldman, editors). *Science and Technology in Post-Mao China* (Cambridge, MA, 1989). *Minerva, 30.* 3: 432–439.

I

COMPARING GREEK AND CHINESE PHILOSOPHY AND SCIENCE

Introduction

Most people are aware by now that we can find science, technology, and medicine in every culture, but the outcome of decades of research on non-European traditions has not been, as we might have hoped, a narrative of world science.[1] What we find instead is a set of stories that remain largely unrelated until they end, one by one, with bodies of knowledge and practice pulverized by the impact of the West. Some bits survive, more or less, such as traditional Chinese medicine. So far as the rest of the world is concerned, they are exotic curiosities rather than parts of an understanding that hang together.

The history of science ought to be more than a collection of cultural ghettos. That is why people curious about the evolution of science as a general human phenomenon have always invested such high hopes in comparative studies.

For three hundred years historians have been forming judgments about the relations of European and other scientific traditions, and the main similarities and differences between them. But if we ask what we have learned from these three centuries of comparative studies, the answer is embarrassingly little.

Certainly no educated person today would claim that the Chinese or the Japanese had no aptitude for science, or that Islamic science depended entirely on borrowings from Greece. Every undergraduate who has taken an introductory course in the history of science knows that the centers of innovation in late-fourteenth-century mathematical astronomy were not Paris and Oxford but Seoul and Baghdad. That is not the result of comparative study, but of research by specialists on one civilization at a time.

Explicitly comparative study has called attention to a number of parallels between civilizations, and to regular contacts between their technical traditions. But has it uncovered new possibilities of human thought and practice? Has it led

---

1. I use "Greece" as shorthand for a wide domain whose elites wrote Greek. It included much of the Mediterranean littoral of the Middle East. In the Hellenistic period it extended to Northern India, Egypt, and other parts of north Africa. I am grateful to Judith Zeitlin, Benjamin Schwartz, and others for an exuberant discussion when I offered this paper at the Harvard Pre-Modern China Seminar.

to substantial understanding of worldwide science? Has it helped us to a deeper comprehension of Japanese medicine, or Greek epistemology, or Indian mathematics? A certain number of facts and dates has accumulated, but the conclusions drawn in the many published comparisons seldom affect our own understanding.

Current Work

Disappointment will continue to outbalance hope as long as we insist on comparing things out of context one at a time, whether they are concepts, values, machines, or groups of people. I cannot offer a ready-made formula for a more optimistic approach, but let me describe my current explorations.

They began four years ago when I met G.E.R. Lloyd, a historian of classical philosophy and science who for some time had been as worried as I had been about the need for comparisons that made sense. He had been following recent and innovative scholarship on China, as I had done with respect to Europe. We began a series of exploratory studies that we hoped would point us in revelatory directions. After three years of exchanging what turned out to be a thick stack of drafts, and meeting in the summers to argue about them, we decided that this experiment was worth continuing. It was yielding new questions that suggested new ways to look at the traditions we studied.

We decided to begin a more or less systematic collaborative study of Chinese and Greek natural philosophy and science. At present we are rereading the primary literature, led by new questions to find new things in it. We have begun drafting and rewriting parts of what will take several years to become a finished book. Our preliminary studies, not at all restricted to science, have led us to define what seem to be several useful approaches to comparison:

1. An advantage of cooperation is that it gives us approximately equal depth in the study of the two cultures. We do not want to add to the accumulation of comparisons that put together a substantial understanding of one with superficial generalities about the other. Mastery of the languages and the primary sources of both traditions is hardly optional.

2. We do not choose what to compare by the criteria of today's science, but by what turn out to be important similarities or contrasts in Greece and China. We look for superficial similarities with different significances, or different means that lead to analogous ends. We are trying to learn how different cultural circumstances push ideas and institutions in different directions.

That leads us to draw on aspects of Greek thought that have no counterpart in China, at least in the period that interests us. An obvious example is Greek element theories, which claim that things are made up of minute ultimate parts that usually do not look like the parts that are big enough for us to see. Element theories build on the idea that reality is hidden, and direct experience is in some ultimate sense not real. That fundamental claim, which we usually refer to as appearance vs. reality, has no counterpart in China. Exploring the circumstances in which it developed in Greece has drawn our attention to Chinese circumstances that leave no scope for such an idea.

Equally interesting, of course, are important Chinese ideas that were not found in Europe, for instance *cheng ming* 正名, which used to be translated "the rectification of names." This is actually quite a diverse group of doctrines about what should be done when names and the things that they designate do not correspond. The variations point to different social interests and political agendas. Greeks entertained vaguely similar ideas, but what we find interesting is how peripheral they were there.

Much effort has been wasted by comparativists straining to find logic in early Chinese philosophy, but no one has yet come to grips with the complementarity of Greek logic and Chinese semantics. Semantics, after all, is what the the people that historians lump together as *ming-chia* 名家 mostly discussed.[2]

More obvious comparisons are possible when concepts or assumptions turn up in both cultures. Because their settings are different, they are bound to have very different implications. Topics worth exploring range from patronage as a form of livelihood (which had very different implications in the two cultures)[3] to the idea that the state and the human body are both miniature replicas of Nature. I will summarize below our approach to this idea of macrocosm and microcosms, to make it clear that what we are doing is more than a conventional exercise in the history of ideas.

3. There is obviously some risk in picking activities to compare from dissimilar periods, where their settings and their significances may be very different. We have looked for a single period in which both cultures seem to be passing through more or less analogous transitions and in which the literatures are comparable. There is no particular reason that there should be such a period. But we believe that there is one.

It happens that between roughly 300 B.C. and A.D. 200 both China and Greece decisively moved away from a free-for-all in philosophic innovation. By the end of that time both were obsessed instead with preserving what had already been done.

---

2. Logic has to do with the forms of thought or its expression, and semantics with the signification and meaning of words.

3. See Sivin 1995 and, more briefly, Chap. IV below.

This was also the period in which the sciences separated out of natural philosophy and established their own institutions and distinct literatures.

Medicine did so earlier in Greece. In the fifth century B.C., when philosophy was just developing its repertory, medicine was already a flourishing intellectual as well as practical enterprise. Still, in the Hellenistic period, from about 300 B.C. on, in medicine as well as in physics, people focussed more and more attention on the past, on what had been inherited from the ancients.

Those five hundred years give us a large and more or less commensurable literature in which to explore basic scientific issues that had not been pressing earlier. In natural science and mathematics we can confront Eudoxus, Euclid, Hipparchus and Ptolemy with the Mathematical Methods in Nine Chapters *(Chiu chang suan shu* 九章算術*)*, the Arithmetical Canon of the Chou Gnomon *(Chou pi suan ching* 周髀算經*)*, and the astronomical treatises of the first two Standard Histories. In medicine we can set Galen and his predecessors side by side with Chinese classics from the Mawangdui manuscripts through the Canon of Eighty-one Problems *(Huang-ti pa-shih-i nan ching* 黃帝八十一難經*)*.

The period we are studying is especially significant in a sense that we did not realize when we began. It was an era of fundamental political change at both ends of Eurasia. The Hellenistic period marked the end of the independent Greek city state as a significant unit. The social order that had evolved thinkers such as Plato and Aristotle ceased to exist. The larger world that Alexander's empire created, the world in which Rome rose and fatally overextended itself by the year 200, was bound to redefine the scope of learning and science.

In China too, changes in society and politics paralleled a change in thought and work. The same half-millennium began with rapidly shifting constellations of power as the local potentates of the Chou dynasty wiped each other out. Over those five centuries, the first universal and centrally organized imperial state first stabilized itself, then disintegrated, and finally collapsed.

Before 300, ambitious intellectuals competed in a widespread system of princely patronage. As the Han state succeeded, they became civil servants in a single bureaucracy and eventually adherents of a single state orthodoxy. But that order faded. By A.D. 200 a position in the civil service could be as dangerous as it could be profitable, and the official philosophy was no longer being taken seriously. That leaves us with the stimulating challenge to determine how these political changes are related to the emergence of natural philosophy and science. The period is the same, and to a large extent the actors are the same.

4. My colleague and I are not comparing things or concepts, but processes, evolving activity. Like most historians of science today, we have no use for the

I

COMPARING GREEK AND CHINESE PHILOSOPHY

idea that science is one thing and its context another. We are looking at ideas, their use, and the social process that created and elaborated them, as a single phenomenon. How physicians or astronomers earned a living, how thinkers grouped themselves, in what ways they publicly disagreed, and what political significance they claimed for cosmology are just as revealing as concepts, forms of proof, and patterns of thought.

From this point of view it does not make sense to ask whether social change was the cause of scientific change, or whether philosophy changed politics. We see these as part of a single manifold of history. Our work is understanding what makes them one process. A prime example is the close linkages from the third century on between natural process, political process, and the vital processes of the human body.

There has been a great deal of writing on the development of cosmology in China, in connection with natural philosophy and the individual sciences. The yin-yang, Five Phases, and *ch'i* concepts were eventually combined to provide the framework for a new theory of resonance between Heaven and Earth on the one hand and the political realm on the other, with the body a second microcosm. A number of scholars including myself would put the final synthesis in the first century B.C., near the end of the Western Han. The document that consummated the mature synthesis of yin-yang, Five Phases, and *ch'i* now appears to be the Inner Canon of the Yellow Lord *(Huang-ti nei ching* 黃帝內經*)*.

Thinkers from the third century on did not associate concepts of the state with cosmology and science because the empirical data forced them to do so. Rather, as I show in Chap. IV, the yin-yang and Five Phases concepts were moral and political from the start. In the long sweep of Chinese thought they remained at the same time moral, political, and physical.

Ideas of *ch'i*, yin-yang, and so on permeated all three domains because they were too politically important to pass by. What made them important is that the same people were using them to claim that the operations of the unified and centralized Ch'in-Han state were based on Nature's processes. They were arguing that resisting a state built on the unity of Nature was bound to fail as surely as resisting the four seasons was bound to fail. They were also trying to persuade an emperor not bound by constitution or law to limit his own freedom. They argued that he and his officials could keep cosmos and state aligned only by modeling government on the regularities and disciplines that governed heaven and earth.[4]

Let us look at the book that first systematically developed such parallels, Lü

---

4. Queen 1991, Sivin 1995. Part of the argument here summarizes the latter.

Pu-wei's 呂不韋 Springs and Autumns *(Lü shih ch'un-ch'iu* 呂氏春秋*)*. It begins with the idea that interfering with the circulation in the human body leads to illness. It goes on first to draw out parallels in Nature, and then to point out their meaning for the state. This book was written in a chaotic time (ca. 239) when there had been no Chou king for a generation, and when no one could do more than hope that China would soon be unified:

> Human beings have 360 joints, nine body openings, and five yin and six yang systems of function. In the flesh tightness is desirable; in the blood vessels *(hsueh-mo* 血脈*)* free flow is desirable; in the sinews and bones solidity is desirable; in the operations of the heart and mind harmony is desirable; in the essential *ch'i* regular motion is desirable. When [these desiderata] are realized, illness has nowhere to abide, and there is nothing from which pathology can develop. When illness lasts and pathology develops, it is because the essential *ch'i* has become static.
>
> Analogously, water when stagnant becomes foul; a tree when [the circulation of its *ch'i* is] stagnant becomes worm-eaten; grasses when [the circulation of their *ch'i* is] stagnant become withered (?).[5]
>
> States too have their stases. When the ruler's virtue does not flow freely [i.e., if he does not appoint good officials to keep him and his subjects in touch], and the wishes of his people do not reach him, this is the stasis of a state. When the stasis of a state lasts for a long time, a hundred pathologies arise in concert, and a myriad catastrophes swarm in. The cruelty of those above and those below toward each other arises from this. The reason that the sage kings valued heroic retainers and faithful ministers is that they dared to speak directly, breaking through such stases.[6]

This union of politics, physics, and medicine was not temporary. It made its way into what you might call "the first Neo-Confucianism" (see Chap. IV). This promiscuous doctrine, much of it contrary to the teachings of Confucius, became conventional among the office-holding elite from the late second century on. Here is a sample from Abundant Dew on the Spring and Autumn Annals *(Ch'un-ch'iu fan lu* 春秋繁露, 156/130 B.C.), which designed the Han Confucian state orthodoxy:[7] "the king models himself on heaven. He models himself on its seasons and

---

5. *K'uai* 蕢 appears corrupt. None of the commentators offers a plausible reconstruction, nor can I.

6. *Lü shih ch'un-ch'iu,* 20, *Lan* 8. 5: 1373.

7. *Ch'un-ch'iu fan lu, 11:* 9b; the meaning of the last sentence remains uncertain. Cf. Queen 1991: 326–327. I accept the recommendations of the editor Su Yu 蘇輿 for reading this corrupt passage; he suggests that *shih* 施 and *chih* 治 are scribal errors for *fa* 法, and that *fa* appears for *chih*.

consummates them. He models himself on its commands and circulates them among all men. He models himself on its constant categories and uses them when initiating affairs. He models himself on its Way and thereby makes order emerge. He models himself on its will and thus commits [his realm] to benevolence."

The political focus continued into the literature of the distinct sciences that emerged in the first century B.C. It is common in the ancient books that came to be considered the founding canons of the various sciences. We can see it, for instance, in the dialogues between emperors and ministers in the medical Inner Canon and the cosmological Mathematical Canon. It recurs where we would most expect to find it, namely in technical writings for and by emperors, at least down to the medical Canon of Sagely Benefaction *(Sheng chi ching* 聖濟經) of 1118.

This passage from the Basic Questions *(Su wen* 素問), a version of the Inner Canon, is not about macrocosm and microcosm, but about the correspondence of the two microcosms, state and body. It begins as the Yellow Lord tells his minister Ch'i-po 歧伯 that he wants to hear "about the relative authority of the twelve systems of functions associated with the internal organs, about which is higher in rank and which lower." A key to medicine, in other words, is the political hierarchy of the visceral systems, seen as departments in a civil service. Ch'i-po replies:

> The cardiac system is the office of the monarch; consciousness issues from it. The pulmonary system is the office of the minister-mentors; oversight and supervision issue from it. The hepatic system is the office of the General; planning issues from it. The gall bladder system is the office of the rectifiers; decisions issue from it ... [and so on for twelve systems of body functions associated with internal organs]. It will not do for these twelve offices to lose their coordination.
>
> If the ruler is enlightened, all below him are secure. If one nourishes one's vital forces in accordance with this, one will live long and pass through life without peril. If one governs all under Heaven in accordance with this, it will be greatly prosperous.
>
> If the ruler is unenlightened, the twelve offices will be endangered, so that the thoroughfares of circulation will be closed off and movement will not be free. The body *(hsing* 形) will be seriously injured. If one nourishes one's vitalities in accordance with this, the result will be calamity. One who governs all under Heaven in accordance with this will imperil his patrimony. Take care! Take care!⁸

There is no need to pause over the choice of civil service posts, which relate each set of functions to the ensemble of vital processes. Ch'i-po's point is that

---

8. *Huang-ti nei ching su wen*, 8.1–8.2. There is no counterpart in the surviving portion

these systems make up the body's internal bureaucracy, and that the heart (which thinks, wills, and feels) coordinates their activities in the same way as the emperor keeps his civil service working together. Executive virtue is equally crucial in both spheres. Health is defined in terms of an ultimately political ideal.

These examples show that government depends on patterns that also hold for the cosmos and the body, and that the functional systems of the body make up a bureaucracy. To complete the schema, since the state is a little cosmos, the cosmos must be a civil service writ large. This is clear enough from the first of the astronomical treatises in the Standard Histories, which is in fact entitled The Book of Celestial Offices ("T'ien kuan shu 史記天官書"). Each of the constellations it enumerates turns out to be a department staffed by stars.[9]

Comparisons
Let me summarize a few of the conclusions we have drawn from comparing the relations of macrocosms and microcosms in China with those we find in Greece:

1. Greek culture in the period that concerns us encouraged disagreement and disputation in natural philosophy and science as in every other field; in China the emphasis remained on consensus.

2. For Greeks, whatever other purposes it served, oral disagreement was a tool of competition. Without sinecures or even secure employment, philosophers were teachers. They depended on skill in debate for livelihood and fame. They tended to argue face to face and to expect the public to decide, just as it decided in the assembly or at trials. Even when Greeks agreed that something was the case, they seldom agreed on why or how.

In China people who lived by their knowledge, with few exceptions, expected rulers to support them—as "guests" *(ke* 客) in the local courts of the Warring States and as imperial officials in the Han. They presented their ideas much of the time not to colleagues but rather to their patrons, who expected advice but did not have to act on it, or even to reply to it. This relationship hardly made for lively exchanges, and few are recorded. Disagreements with other scholars were, except for a testy few who tended to have unsatisfactory careers, unimportant by comparison. Open attacks were usually written—and one-sided. Patrons faced with political decisions encouraged parleys on concrete questions, but seldom showed patience for anything resembling intellectual debate.

---

of *Huang-ti nei ching tai su* 太素.
9. *Shih chi, chüan* 27.

I

COMPARING GREEK AND CHINESE PHILOSOPHY 9

On the whole, the Chinese valued consensus as much as the Greeks valued dispute. In China the relationships of masters and disciples were based on the ritual transmission of written texts.[10] A teacher and his disciples formed an internally cohesive community that avoided attacking other communities. Quarrels were not likely to be productive when teachers aspired above all to official employment for their pupils, and when parents measured success by the same criterion.

3. Like those of China, Greek macrocosms and microcosms reflected political ideals. For the Chinese those ideals remained unifying and centripetal. From the Han on, in a social system that valued civil service above every other career, philosophers who wanted to be politically engaged, or simply respectable, understood the risk of favoring alternatives to the current dispensation of power.

In Greece, with its diverse city-states, constitutions, and political tastes, some people saw the cosmos as a single order, some as an ensemble of quite distinct orders, some as a balance of opposed powers, some as a state of strife. There was no shared ideal to build on.

4. Chinese rulers formed a loyal civil service elite by building on symbols and rituals that literati valued. But this process of alignment bound both. As an element of the new ideology, rulers to varying extents accepted limitations on their power.

The interests of Greek rulers were on the whole irrelevant to the thinkers who developed diverse cosmic metaphors for the state. Philosophers had no voice in the decisions of power-holders. Because intellectuals were not constrained by reasons of state, and because their public roles were played out in disagreement rather than consensus, their stances reflected a great range of contradictory definitions of state as well as of cosmos.

Conclusion

Because of circumstances like the ones I have summed up for this single example, the two cultures differed greatly in what one said and how one said it. The comparison throws into relief the reasons that the discourse of Chinese science was so rarely confrontational. Natural philosophers in the Western Han accepted the state's view of itself because, unlike their Greek counterparts, they formed and maintained it. The same people shaped a view not merely of the political microcosm, but of the cosmos and the body. We intend to explore further the manifold in which the three were united, particularly to understand why it remained desirable to keep them linked as the separate sciences emerged. This sketch of one of our themes suggests that the pieces of the puzzle are beginning to fall into place.

References

Early Chinese Sources

Chiu chang suan shu 九章算術 (Mathematical Methods in Nine Chapters). Anonymous, early first century B.C. In Ch'ien Pao-ts'ung 1963.

Chou pi suan ching 周髀算經 (Arithmetical Canon of the Chou Gnomon). Anonymous, 50 B.C./A.D. 100. In idem.

Ch'un-ch'iu fan lu 春秋繁露 (Abundant Dew on the Spring and Autumn Annals). Tung Chung-shu 董仲舒, parts written 156/130 B.C. In *Ch'un-ch'iu fan lu i cheng* 春秋繁露義證.

Huang-ti nei ching 黃帝內經 (Inner Canon of the Yellow Lord). Anonymous, probably first century B.C. The *Su wen* 素問 (Basic questions), ed. by Wang Ping 王冰, 762, is cited from Jen Ying-ch'iu 1986 by *p'ien* 篇, *chang* 章, line, and page number. The only usable ed. of *T'ai su* 太素 (Grand basis), ed. by Yang Shang-shan 楊上善, 656/683?, is Kosoto 1981.

Huang-ti pa-shih-i nan ching 黃帝八十一難經 (Canon of Eighty-one Problems in the Inner Canon of the Yellow Lord). Anonymous, probably second century A.D. In *Nan ching pen i* 難經本義, reprint, Taipei, 1976.

Lü shih ch'un-ch'iu 呂氏春秋 (Springs and Autumns of Master Lü). Compiled under patronage of Lü Pu-wei 呂不韋, ca. 239 B.C. In Ch'en Ch'i-yu 1984.

Sheng chi ching 聖濟經 (Canon of Sagely Benefaction). Compiled by imperial order, issued 1118. In *Chen pen i shu chi ch'eng* 珍本醫書集成, vol. IX.

Shih chi 史記 (Records of the Grand Astrologer). Ssu-ma T'an 司馬談 and Ssu-ma Ch'ien 司馬遷, completed 100/90 B.C. Chung Hwa Book Co., 1974, ed.

Modern Sources

Ch'en Ch'i-yu 陳奇猷. 1984. *Lü shih ch'un-ch'iu chiao-shih* 呂氏春秋校釋 (Critical Edition of Springs and Autumns of Master Lü). 4 vols. Shanghai: Hsueh-lin Ch'u-pan-she.

Ch'ien Pao-ts'ung 錢寶琮, editor. 1963. *Suan ching shih shu* 算經十書 (Ten Mathematical Canons). Peking: Chung Hwa Book Co. Critical ed. of Sung texts.

---

10. For detailed discussions see Nakayama 1984 and Sivin 1995.

Kosoto Hiroshi 小曾戸洋, editor-in-chief. 1981. *Tôyô igaku zempon sôsho* 東洋醫學善本叢書 (Collected Rare Books on Oriental Medicine). Osaka: Tôyô igaku kenkyûkai.

Nakayama, Shigeru. 1984. *Academic and Scientific Traditions in China, Japan, and the West*, trans. Jerry Dusenberry. University of Tokyo Press. English version of *Rekishi toshite no gakumon* 歴史としての學問 (Academia as history; Tokyo: Chûo Koron, 1974).

Queen, Sarah. 1991. From Chronicle to Canon: The Hermeneutics of the *Spring and Autumn Annals* according to Tung Chung-shu. Ph.D. diss., History and East Asian Languages, Harvard University. Revised version in press.

Sivin, Nathan. 1995. State, Cosmos, and Body in the Last Three Centuries B.C. *Harvard Journal of Asiatic Studies*, 55. 1: 5-37.

Sivin, Nathan. 1995. Text and Experience in Classical Chinese Medicine. In *Knowledge and the Scholarly Medical Traditions*, ed. Don G. Bates, pp. 177-204. Cambridge University Press.

II

EMOTIONAL COUNTER-THERAPY

Introduction

Psychiatry is an invention of nineteenth-century Europe. We can hardly expect to find it in China before modern times. Nor do we find one of its key assumptions, the dichotomy of mind and body. Chinese were perfectly able to distinguish these two when they wanted to do so. Physicians were much more interested, however, in their underlying integrity and interaction. In everyday practice, they assumed that physical abnormalities would have consequences for thought and feeling, and vice versa.

This essay explores abstract thought about disorders due to excessive emotions. It is meant as an experiment in analyzing early Chinese discussions without being distracted by the largely irrelevant notions of Western psychiatry. The latter are essential in communicating ancient ideas to a modern readership, but are also ubiquitous in analyzing original sources. This habit usually encourages a disregard for the intentions of practitioners, and results in parochial conclusions. My goal, like that of most other students of ancient medicine, is to weave particular comprehensions of the past into the fabric of all experience, scientific and merely human, antique and modern. I wish to suggest through an example that understanding is more likely to add to the larger pattern if it has not depended at too early a stage on ideas imported from another time or place. Since this report is not meant solely for specialists, I do not hesitate to use modern terminology to make one or another point clear, but my concern remains reconstructing what a physician in the late sixteenth century meant by what he said.

Emotions and Material Change

Excessive, undisciplined emotion was a cause of disease in the early classics. Physicians did not understand this in any way remotely resembling modern psychodynamics, however. Emotion was an integral aspect of the body's most basic functions. Health was a balance in the dynamics of body processes, and illness was an imbalance. The circulation of *ch'i*, partly inborn and partly extracted from food and drink, maintained not only the material body but its mental and spiritual processes. Feelings caused and were caused by material change.

Abnormal emotion could affect *ch'i* functions, as in a tractate on the causes of pain in the Basic Questions of the Inner Canon of the Yellow Lord: "I know that all medical disorders arise from the *ch'i*. Anger *(nu* 怒*)* makes the *ch'i* rise; joy *(hsi*

喜) relaxes it; sorrow *(pei* 悲) dissipates it; fear *(k'ung* 恐) makes it go down; cold contracts it; heat makes it leak out; fright *(ching* 驚) makes its motion chaotic; exhaustion consumes it; worry *(ssu* 思) congeals it." The results are physical; for instance, "Anger reverses the flow of the *ch'i*. When it is extreme, [the patient] will vomit blood or void 'rice-in-liquid' diarrhea."

Since emotion was part of the body's system, causation ran both ways. Another text discusses imbalances that involve *shen* 神, the finest part of *ch'i*, responsible for consciousness and allied activities: "If there is an excess of *shen*, [the patient] will laugh unceasingly; if a deficit, the result is sorrow."[1]

Before moving on to consider therapeutic manipulation of the emotions, let us examine more closely the subtle interaction of body processes, emotions, and drugs in late classical medicine. This example comes from a work on women's disorders attributed to Fu Shan 傅山 (1607–1684), a great painter and physician:[2]

A woman who cannot bear a child because she harbors hate for someone over a long period: People think [she is infertile] because Heaven detests her. They do not realize that it is due to a stasis of her hepatic *ch'i*. For a woman to bear a child it is essential that [the *ch'i* in] her cardiac circulation vessels flows freely and smoothly, that in her splenetic vessels be unhurried and concordant, and that in her renal-genital vessels be vigorous and large, so that it perceptibly drums against the finger [reading the pulse]. Only in this case can one speak of the "pulse of happiness," signalling that there is no stasis in the vessels of the three visceral systems, so that she can give birth.

The treatment is a drug formula that brings together seven ingredients commonly used to promote fertility. The text goes on to explain its activity:

If [the formula is] cooked in water and taken for a month, the static *ch'i* will be freed. Once the stasis is unblocked, the "*ch'i* of happiness" will fill her belly, and it will be possible for her feelings of jealousy to change once and for all. Without further ado the couple will be on good terms [again], and it will be only a matter of time before she becomes pregnant. What is remarkable about this medicine is that by clearing the stasis of hepatic *ch'i* it releases the

---

1. *Huang-ti nei ching su wen* 黃帝內經素問, 39.4, 62.2.1 (pp. 113, 168). I wish to record my gratitude to Bridie Andrews, Horacio Fabrega, Charlotte Furth, and Marta Hanson for a number of useful suggestions. Dr. Fabrega's insights have been extremely helpful because of my own lack of qualification to make psychiatric assessments.

Most readers will be aware that there is nothing intrinsically Chinese about emotional counter-therapy. An amusing American instance appears in Clemens 1959: 9–10. I am grateful to Prof. Furth for drawing this anecdote to my attention.

2. *Nü k'o hsien fang* 女科仙方, 2: 333. Italics are mine. The vertical Conception and horizontal Belt tracts are circulation channels that intersect below the navel.

II

EMOTIONAL COUNTER-THERAPY

constraint in the splenetic *ch'i,* and the cardiac and renal-genital *ch'i* are relaxed along with it. [The circulation in] the area of the loins and navel becomes free and, in the Conception and Belt tracts *(jen tai* 任帶), unimpeded. There is no need to open the gate [of the womb with other therapy to allow conception] of the fetus, for the womb opens spontaneously. *This is not a specific therapy for jealousy.*

This discussion is not about jealousy of an imaginary rival. The patient has someone to hate. Fu does not, however, advise that the husband be counseled to change his behavior. It is likely that if he were willing to recognize his wife's needs, he would have done so before the physician was called in. By this stage, as the emotional relationship between husband and wife has deteriorated, it is unlikely that their sexual relations are frequent enough to lead to conception. In any case, the author holds the wife responsible for inability to provide the husband's family with offspring, the worst calamity that can strike any marriage. But a doctor has the power to transform her from an object of cosmic detestation to a patient with a curable disorder.

Fu finds the wife's unrestrained feeling responsible for the blockage of circulation that makes conception impossible. The account of etiology coheres with the general understanding of how medical disorders arise. They generally begin with some impediment to the life-giving and life-sustaining flow of *ch'i,* and are overcome by unblocking it. Therapy traces backward along the pathways of causation. Doing away with the consequences of jealousy allows human relations, otherwise stymied, to clear the inappropriate emotion away.

That is not what the medicine does. Given a correct diagnosis, the medicine can enable the wife to restrain her feelings. Fu does not use any mechanical metaphor to explain its success. The healer is not wielding technology, but bringing about the conditions, at the same time physical, social, and emotional, in which spontaneous change in a human relationship can take place. This change is a necessary condition of the "cure," a pregnancy.

There was nothing superficial about the holism that this typical example illustrates. It left no need for a special category of mental or emotional illness.

Emotion and Counter-emotion

We find such a category emerging only in late imperial China, as part of a great general elaboration of ideas about illness and therapy. The rubric was rarely used, although accounts of medical disorders commonly paid attention to what modern readers would call emotional symptoms. An unusual source permits us to glimpse certain overarching themes in thought about emotional dysfunctions and their treatment.

II

The same source also casts light on an issue often discussed in writings on the history of Chinese medicine. Scholars not trained in traditional medicine have often treated yin-yang, the Five Phases, and related concepts as a matter of abstract theory that has nothing to do with, or may even get in the way of, the application of empirical knowledge in clinical practice. Asian historians, more familiar with practice, do not make this error. Here as in other studies, examples demonstrate that physicians used these concepts to structure therapy, even in circumstances where knowledge of the patient's human relations is as important as knowledge of the body. Medical doctrine connects the physician's experience with continued study of the classical literature, a link never broken in ancient times and still prominent today in the training of traditional doctors. The cases discussed below also show physicians enforcing conventional social roles, among them those of women, in imperial China.³

Collections of therapeutic formulas and methods (the word *fang* 方 includes both) are among the earliest surviving medical books. The Inner Canon explained the rationales behind the choice of treatments. Researches on Medical Formulas (*I fang k'ao* 醫方考) of Wu K'un 吳崑 was written in 1584, by which time enormous numbers of remedies had accumulated within a number of therapeutic traditions and fashions. It was the first monograph in a new genre that set out the principles underlying particular remedies in a systematic way. Its originality lies in applying classical doctrines to standard therapies and case records to explain why one combines certain drugs or chooses certain actions. As Wu's preface puts it,

> I have taken more than seven hundred formulas and methods from excellent physicians of antiquity, measured them by the standard of the medical canons, considered them in the light of my own perceptions, and formed [my comments] from therapeutic experience to reveal the subtle meanings [of the formulas and methods].... My method, I should say, was research on drugs, research on the manifestations presented [by patients], research on nomenclature, research in historical records, research on using formulas and methods in a way responsive to changes in the disorder (*pien-t'ung* 變通), research on what succeeds and what fails, and research on the reasons that [formulas and methods] have the effects that they do.

Among the interesting features of this book is a collection of anecdotes headed "Emotions" *(ch'ing-chih* 情志). This group, assembled to illustrate the use of emotions to treat disorders caused by emotion, is the subject matter of this essay.

Wu's was not the first, nor was it the last, grouping of disorders that predomi-

---

3. See Farquhar 1994 and Sivin 1987 for the relations of doctrine and practice, and the publications of Charlotte Furth, esp. 1986, on gender in medicine.

II

EMOTIONAL COUNTER-THERAPY

nantly involve thought and feeling. The classical rubric was "Wind Disorders" *(feng* 風), a broad grouping based on one of the six external causes of disease. It appears in the early classics. On the Causes and Symptoms of Medical Disorders *(Chu ping yuan hou lun* 諸病源候論), the great etiological treatise of 610, itemizes its contents. Wind includes an assortment of disorders that today would be considered emotional and psychological or neurological, from manic-depressive syndrome to hemiplegia and tics. It excludes others in which the symptoms are also primarily emotional. We can find other sources, from medieval times on, that discuss emotions as causes of disease, and even use emotional manipulation to treat them).[4] Most of those from the Ming and Ch'ing, conforming to the trend away from ritual approaches, prescribe only drugs. The massive imperial encyclopedia of 1726 *(Ku-chin t'u-shu chi cheng* 古今圖書集成) includes an expanded chapter on "Emotions," incorporating Wu's cases and many pertinent citations from earlier classics; its discussion largely focusses on drug therapy. A slightly later palace compilation, the textbook Golden Mirror of the Medical Tradition *(I tsung chin chien* 醫宗金鑑, 1742), abandoned this classification for one that included seven basic emotions rather than Wu's five. It was entitled "spiritual disorders" *(or* "disorders of consciousness," *shen ping* 神病). The only therapies recommended were a pill and a powder.[5]

In order to understand Wu's starting point, it is necessary to peruse the main classical source for medical correspondences in the Inner Canon of the Yellow Lord. The fifth chapter of the Basic Questions spins a great web of associations for each of the systems of functions associated with the five yin organs. It begins "The east gives rise to wind; wind gives rise to the phase Wood; Wood gives rise to sour [among the five flavors]; sour gives rise to the hepatic functions," and eventually, for each phase, reaching the emotions: "among the emotions [Wood] corresponds to anger; anger damages the hepatic functions; sorrow overcomes anger," and so on. We can capture the pertinent part of this source in a table, to which I have added a fifth column inferred from the fourth:

---

4. E.g., Chu Chen-heng 朱震亨, one of whose cases is cited below (p. 10). For instances that use emotional counter-therapy see his medical case records, *Chu Tan-hsi i an* 朱丹溪醫案, pp. 26–29.

5. *I fang k'ao,* 3: 221–225; *Chu ping yuan hou lun,* ch. 1–2; *Ku-chin t'u-shu chi ch'eng,* ch. 321; *I tsung chin chien,* 41: 1077–1079. For another Ming book that discusses the same doctrine from the Inner Canon as *I fang k'ao,* but prescribes drugs, see *Chien-ming i kou* 簡明醫彀, 4: 202. My attention was first drawn to Wu's book by an interesting discussion of one of his anecdotes in Ng 1990: 40–41 (which misapprehends the source).

CYCLIC PHASE	SYSTEM OF SOMATIC FUNCTIONS	E = EMOTION THAT HARMS	EMOTION THAT OVERCOMES E	EMOTION THAT E OVERCOMES
Wood	Hepatic	Anger	Sorrow	Worry
Fire	Cardiac	Joy	Apprehension	Sorrow
Earth	Splenetic	Worry	Anger	Apprehension
Metal	Pulmonary	Sorrow	Joy	Anger
Water	Renal-genital	Apprehension	Worry	Joy
Wood	Hepatic	Anger	Sorrow	Worry

It can be put more concisely in the abstract language of Five Phases sequences. The order of the phases and functional systems is that of unfolding natural processes, usually called the Mutual Production series. Thus all five columns read downward in that order. Each phase, on the other hand, overcomes the second one after it. The order in which one emotion can be used to overcome another (anger-worry-apprehension-joy-sorrow-anger, corresponding to Wood-Earth-Water-Fire-Metal and back to Wood) is the Mutual Conquest series, which in medicine applies both to pathological changes and to therapeutic measures designed to overcome the source of disorder.[6]

Translation

To each account in the translation that follows I have appended explanatory notes in smaller type. Wu seldom identifies his sources, but I have compared his accounts to the originals when they could be found. A general analysis follows the translation.

* * *

Introduction: When emotion is overwhelmingly excessive, no drug can cure [the resulting disorder]; it must be overcome by emotion. Thus it is said that "anger damages the hepatic system, but sorrow overcomes anger; joy damages the cardiac system, but apprehension overcomes joy; worry damages the splenetic system, but anger overcomes worry; sorrow damages the pulmonary system, but joy overcomes sorrow; apprehension damages the renal system, but worry overcomes apprehension." A single saying from the Inner Canon, and a hundred generations have venerated it! These are immaterial medicines.

---

6. *Huang-ti nei ching su wen*, 5.3. On Five Phases sequences see Needham et al. 1954– : II, 253–265, and Sivin 1987: 70–80, esp. Fig. 2, p. 79.

II

EMOTIONAL COUNTER-THERAPY

Wu's Chinese terms do not translate unambiguously into English. *Yu* 憂, interchangeable here with *pei*, means not only sorrow generally, but also concern for others. *Ssu* is literally "thought," but is often used in medical discourse for excessive emotion that leads to illness. It implies preoccupation, or even obsession. I translate *ssu* as "worry," "longing," and "thinking" in different contexts below. *K'ung*, generally apprehension or fear, sometimes refers to being startled, as in item 5.

I will study eleven instances from remarkable physicians. The enlightened may make larger inferences from them, and the art will be theirs.

Not all the anecdotes involve physicians, a point to which I will return.

1. Wen Chih 文摯 was a man of Ch'i. King Wei 威 of Ch'i was ill, and had an emissary sent to summon Wen Chih. After Wen Chih arrived, he said to the heir apparent "The king will recover if he gets angry. But if he tries to have me killed on the spot, what can I do?" The heir apparent replied "Don't worry. I will save you." Wen Chih failed to be punctual when he visited the king. When he entered, he stepped onto the sitting-mat without taking off his shoes. The king, furious, had his retainers drag Wen Chih away, and was about to have him boiled to death. The heir apparent, in the nick of time, kowtowed and asked the king to spare him. The king's anger passed, and he pardoned Wen Chih. Because of this he recovered.

The reason this happened was that the king's illness was due to worry, so Wen Chih used anger to overcome it.

This episode, which (if it is historic) took place ca. 300 B.C., was told many times with variant details. In the Springs and Autumns of Master Lü,[7] two generations after the time of the story, Wen Chih had already become more than human: "The heir apparent and the queen desperately argued for him, but failed. Despite them, Wen Chih was boiled alive in a large vessel. After he had been cooked for three days and three nights, his expression had not even changed. He said 'if you really want to kill me, then why don't you put a lid on to stop [the circulation of] yin and yang *ch'i*?' The king had the vessel covered, whereupon Wen Chih died." The story was meant to point a moral: "Faithful service may be easy in an orderly time, but it can be impossible in a chaotic one."

---

7. *Lü shih ch'un ch'iu* 呂氏春秋, chi 紀 11, *p'ien* 篇 2, p. 578. This text calls the king's illness *wei* 痏. Commentators and lexicographers have done no more than guess at the meaning of this rare word. Wu is obviously drawing on one of the many late variants of the story. They also specify, as Lü's text does not, that the patient was King Min, implying the date I have tentatively given. I do not know why Wu gives the ruler as King Wei, who reigned about 60 years earlier than Min.

2. According to the Records of the Wei Kingdom, there was a Commandery Governor who fell ill. Hua T'o 華佗 believed that if the man became very angry he would recover. Hua accepted his goods [i.e., gifts in lieu of a fee] but did not perform any services. Before long Hua absconded, leaving an abusive letter. The Governor, naturally incensed, sent people to follow and kill him. The Governor's son, who knew all about it, ordered the official not to follow him. The Governor glowered for some time, and then vomited several cupfuls of black blood, after which he recovered.

The reason this happened is that the Governor's illness was also rooted in worry. The Canon says "worry makes the *ch'i* congeal." Congelation of *ch'i* is the root of yin obstruction(?). Therefore Hua T'o simply used an outburst of rage to impair the yin and restore a balanced condition.[8]

3. According to Master Shao's Record of Things Heard and Seen, when a prefectural Army Supervisor became ill from grief and worry, his son arranged for Ho Yun 赫允 to treat him. Ho Yun told the son "According to the correct method, he must be greatly alarmed, and then he will recover." The Vice Governor at the time, the Censor Li Sung-ch'ing 李宋卿, was a martinet. The Army Supervisor inwardly feared him. Ho and his son asked that Sung-ch'ing visit [the Army Supervisor], inquire into his faults, and castigate him. The Army Supervisor, terror-struck, broke out in sweat. He recovered from his illness.

It would seem that as a result of grief and worry the *ch'i* congeals, but that as a result of fright or terror the *ch'i* floats up. When that happens it can no longer congeal. This too is an instance of one emotion overcoming another.

> The story comes from a twelfth-century collection of jottings. The original, in which the physician's name is Hao 郝, not Ho, gives a number of anecdotes that illustrate his skill in avoiding the unnecessary use of drugs. Hao flourished in the mid-eleventh century.[9]

4. Chao Chih-tse 趙知則 was a native of T'ai-yuan (present Shensi province), who became ill from joy. Master Ch'ao 巢, treating him, read his pulse, and was amazed. He went out to get medicine, but in the end did not give him any. Chao,

---

8. This anecdote is quoted fairly accurately from the Standard Histories, partly from *San kuo chih, 29:* 801, and partly from *Hou Han shu, 82B:* 2738. The quotation is from *Su wen* 39.4. I take *i* in *yin i* 陰翳 in its only pertinent sense, "darkening," or, less literally, "obstruction." It occurs in *Su wen* 71.2.2.2 (p. 224), one of the spurious phase energetics chapters. Since that chapter is much later than the two biographies, the sense of the text remains uncertain.

9. *Ho-nan Shao shih wen-chien hou lu* 河南邵氏聞見後錄, *29:* 226.

EMOTIONAL COUNTER-THERAPY

grief-stricken and weeping, took leave of his family, telling them that he was not long for this world. Ch'ao knew that he was about to recover, and sent someone to comfort him. When asked why [he acted as he did, Ch'ao] replied by quoting the passage in the Basic Questions about apprehension overcoming joy. He may be called "one who has gone through the Profound Barrier."

> The last sentence means that the practitioner has learned hidden doctrines; "Profound Barrier" (*hsuan kuan* 玄關) is sometimes used for esoteric Buddhist teachings. Mr. Chao assumed that the doctor did not administer the promised medicine because he had determined that the illness was incurable.

5. When a woman was in the final stages of labor, the midwife, who intended to offer her some hot water [to drink], gave her some deerhorn marrow fat by mistake. Deerhorn fat is stuff with which women dress their temple-hair. She disgorged it, and her tongue kept sticking out. For several days after she bore the baby she could not withdraw it into her mouth. She did not respond to medicine. Chen Li-yen 甄立言 was the last [among the several physicians to be called and] to arrive. He smeared vermilion on her tongue, and then had her re-enact the delivery. He had two women support her, and had someone hide on the other side of the wall and repeatedly beat an earthenware vessel. At the crucial point [the drummer] made a great noise by dashing [the vessel] to the ground. When the sound was heard the woman's tongue went in.

The reason for this is that when one is startled *(k'ung)* the *ch'i* goes downward. Chen Li-yen first used vermilion to smear on her tongue because he feared that the affrighted *ch'i* might enter her cardiac system, and therefore took this preventive measure.

> Chen Li-yen and his brother Ch'üan 權 were famous doctors and court officials of the early seventh century. The point of this story, as reference to the Five Phases schema makes obvious, is that the joy of childbirth was responsible for the woman's disproportionate response to the taste of the fat.
>
> Vermilion is mercuric sulfide, generally artificial. In Chinese medicine it acts on the cardiac functions to calm the spirits and perform various fortifying functions. It is so nearly insoluble that its poisonous effect is negligible.

6. When Grand Councilor Han 韓 took sick, there had been no rain [for some time]. He changed doctors ten times, but without effect. Tso Yu-hsin 左友信 was the last to arrive. After reading the patient's pulse, he counted off days on his fingers, said "on such-and-such a day it will rain," and left. Han, suspicious, said "Can this mean that my illness is not treatable? Why did he talk about rain and not get round to giving me medicine?" On the designated night it did actually rain. Han, overjoyed, got up and walked about his courtyard. By dawn his illness seemed to have fallen away.

He had Tso summoned and asked him about it. Tso replied "The illnesses of the highest officials are due to sorrow and concern *(yu)*. I covertly judged that an official so high must be a faithful and benevolent person. Since there has been a prolonged drought, you must have been concerned on behalf of the people. If drought was your worry, it must be that rain would be your cure. Since deduction [indicated that the circumstances] were indeed what they should be, why should your cure wait on medicine?"

This is another instance of the Inner Canon's joy overcoming sorrow.

7. After a girl was betrothed, her husband-to-be went away on business for two years without returning. Because of this she did not eat, and miserably took to her bed as though lovesick. She had no other symptoms, but lay there all day facing the wall. Her father invited Chu Chen-heng to treat her, and told him how it had come about. After Chu read her pulse he said to her father "This is an instance of *ch'i* congealed because of longing. It cannot be cured with medicines alone. But if something made her joyous it might clear up. Lacking that, let us make her angry." Without further ado [the father?] slapped her face and accused her of being involved with someone else. At that she became very angry, weeping and sobbing for six hours. When [Chu] let [her father] explain [what had happened], she asked for food. The reason for this is that with sorrow her *ch'i* dissipated, but the anger overcame her worry. Chu said to her father "Although her illness has remitted, only joy will cure it." [The father] arranged for her fiancé to return. After this happened, as predicted, her illness did not recur.

> Chu Chen-heng 朱震亨 (1281–1358), here called by his literary name Tan-hsi Weng 丹溪翁, "Old Man of Cinnabar Creek," was the most eminent physician of his time. *Hsiao* 消, which I translate "dissipate," is the antonym of *chieh* 結, "congeal." In the version in a collection of Chu's cases, which has the couple newly married rather than engaged, the deception continued a bit further: "They untruthfully told her that there had been a letter from the husband saying that he would return at any moment. Three months later her husband did return, and she recovered."[10]

8. A district yamen runner arrested a criminal and, having chained him about the neck, was taking him to the yamen. While they were enroute, the criminal jumped into the river and died. His family charged that the runner had tried to get money out of him, and had coerced him until he killed himself. The runner was just able to escape being charged, but could not avoid losing all his property. His sorrow and resentment made him ill, so that he was like a drunkard or an idiot. He talked absurdities, and was no longer conscious of what was going on

---

10. *Chu Tan-hsi i an,* p. 28.

about him. His superior asked Tai Nien-jen 戴念仁 to examine him. Tai said that since this was an instance of illness from losing wealth, he was certain to recover by gaining wealth. [The superior?] had three tin ingots cast to look like silver, and had them placed in advance in a muddy ditch. He waited for the runner to arrive, and then, pretending he had dropped a key there, ordered the runner to retrieve it. When he pulled out the three tin ingots, the superior said "It's silver! But I have no use for such unrighteous gain. I'll give it to you." The runner held it, scrutinized it, and would not put it down. His illness got better within a day (or by the day, *jih yü* 日愈).

This too is joy overcoming sorrow.

I have not identified Tai Nien-jen, or Han Shih-liang in the next anecdote.

9. A young woman and her mother deeply loved each other. After the woman was married her mother died. The young woman thought *(ssu)* about her unceasingly. Her vitality was depleted. She was barely alive, always sleeping. No medicine had any effect. Her husband asked Han Shih-liang 韓世良 to treat her. Han said "She got this disorder through thinking [about her mother]. It is not an easy one for medicines to cure. It must be treated by art." [The husband?] bribed a female medium and taught her something secret.

One day the husband said to his wife "You think about your mother like this, but you don't know whether under the earth she is thinking of you, do you? I have to go away [on a trip]. Why don't you ask a medium to do a seance for you?" The young woman, pleased, agreed. She called in the medium. They burnt incense and worshipped, and the mother's spirit came down into the medium. In sound and silence she became just like the mother when she was alive.

The young woman burst out in great sobs. The mother scolded her: "Don't cry! Your destiny got the better of mine, so I died earlier than you. My dying is your fault. Now that I am in the Yin Administration [i.e., purgatory], I want to pay you back. Your illness that has left you barely alive is really my work. When I was alive we were mother and daughter, but in death I have become your enemy." When she finished speaking, the young woman's expression changed. She became furious. She railed at her mother: "It is because of [longing for] my mother that I became ill, but it is she who has been hurting me. Why should it make me happy to think of her?" From then on her illness got better.

This too is curing by emotion.

    This curious story comes from a collection of medical case records of the early sixteenth century.[11] The speech the husband put into the mouth of the

---

11. *Shih shan i an* 石山醫案, *3:* 17b–18a.

medium implies that the daughter was originally fated to die early, but she was able to draw on the vitality of her mother to outlive her.

10. T'an Chih 譚植 by habit was reticent. He was an Assistant in the Shao prefectural government. One day when he was at a banquet with his colleagues, among the dishes was quite a large Chinese radish (*lo-po* 蘿蔔), which everyone admired. T'an said "But there are radishes as big as a man!" Everyone laughed, confident that this was untrue. T'an was abashed, and blamed himself: "They have not seen one that big. When I told them that, I should have expected that they would consider my words wild, and laugh at them." As a result, he was full of sorrow and resentment, and could not eat for days.

His son Huang 煌 was well-read and understanding. He realized that because his father by habit did not speak lightly, his humiliation had made him ill. If he were to recover, it was essential to make known the truth of what he had said. Huang arranged for someone to go to his home and take a radish as big as a man to the office. When there was another banquet he pressed his father, despite his sickness, to attend. After the wine had made several rounds, a cart brought the radish to the dinner. Everyone was amazed. The father was overjoyed. He recovered that very day.

This too is an instance of the Inner Canon's joy overcoming sorrow.

In this story and the next no physician is involved.

11. Ho Chieh 何解 was a man of Ch'en-liu (present Kaifeng, Honan). One day he was at a drinking party for Yueh Kuang 樂廣, the Governor of Ho-nan, at the house of Chao Hsiu-wu 趙修武. After several rounds of drinking, he suddenly noticed what seemed to be a tiny snake in the bottom of his wine-cup. But when he took the wine into his mouth, he was not aware of anything in it. Still he thought of this again and again, and became suspicious. As the days passed he felt pains in his heart. He thought of the little snake growing larger and eating his vital organs. Medical treatment could not help him.

Some time later he again attended a drinking party at the Chao house. As soon as he picked up his cup he again saw a little snake. Setting down his cup and taking a good look around him, he noticed that a compound bow was hanging from a beam of the house. It turned out that the bow was indistinctly reflected in the bottom of the cup. Because of this his suspicion was quelled, and his medical disorder was no more.

In this case he became ill because of his feelings of suspicion. It was essential to quell them in order for him to recover. For all that time, treatment with drugs produced no effect.

The mention of the well-known wit Yueh (252-304) as Governor puts this story near the end of the third century.

II

EMOTIONAL COUNTER-THERAPY

General Comment

This is quite a mixed bag of anecdotes. It will be well to summarize the cases in a table, comparing the statuses of "healers" and "patients," and looking at the character of the emotional manipulation in each instance. One may question whether all of the instances are medical in character. Only the first nine of the eleven involve physicians. For those that do not specify a precipitating emotion, so that readers have to depend on Wu's analysis, I have put the cause in parentheses.

STATUS OF HEALER	STATUS OF PATIENT	CAUSE	MANIPULATION
1. Physician	King	(Worry)	Anger: insults, aided by son who appeases patient
2. Eminent physician, official family	High official	(Worry)	Anger: insults, aided by son who appeases patient
3. Physician	High official	Worry, grief	Anger: intimidates with help of son
4. Physician	Unknown, probably elite	Joy	Apprehension: deceives to frighten
5. Physician, high official	Woman, status unknown	(Joy) [or shock during parturition?]	Apprehension: re-enacts childbirth, stages event to startle
6. Physician	Very high official	Sorrow, concern	Joy: uses ability to predict rain, explains when summoned
7. Physician, eminent	Woman, fiancée of businessman	Longing for betrothed	Anger moderates lovesickness
8. Physician	Minor functionary of local government	Sorrow, resentment	Joy: aided by district magistrate, deceives by appeal to greed
9. Physician	Woman, obsessively grieving for dead mother	Longing [or grief?]	Anger: stages deceptive mediumistic ritual to destroy emotional bond to mother
10. Son of patient	Medium-level local official	Sorrow [or obsession due to loss of face?]	Joy: demonstrates at banquet that his father spoke the truth
11. Patient	Member of elite	Worry	Unclear: discovers that his fears were illusory

Though these stories are superficially similar, closer analysis reveals an unexpectedly diverse range of issues, and an intriguing variety of responses to a common subject. The shared theme is manipulation of emotions, sometimes to precipitate a catharsis, and sometimes to break a pattern of behavior responsible for some abnormality. This theme was apparently understood with enough latitude to encompass all eleven cases.

We might next ask in what sense the "therapy" was meant to change attitudes of the "patient."

In the two cases that do not involve doctors, the adjustment was *ad hoc.* In case 10, the filial son corrected the drunken disbelief of his father's colleagues in order to restore his father's dignity. He did not alter the flaw in his father's self-esteem that made him hesitate to express his opinions. The canons of filial piety did not contemplate children improving their parents. In fact, as the story is phrased, the father's diffidence is not a flaw, but a judicious habit. Still, modern readers will expect that, if faced again with derision, the father would respond similarly.

In case 11, the "patient" deluded himself when he relaxed the emotional discipline expected of a conventional gentleman, with pathological results. The "recovery" came from his own reassertion of that discipline. This may or may not affect his suggestibility in the future. In any case, how the "disorder" fits the five-emotion schema is not at all clear. It is called suspicion, and thus would be a kind of worry. Nothing in the "therapy" resembles the anger that overcomes worry, or for that matter any of the other four emotions.

Most of the physicians' therapies aimed only to cure distressing manifestations. If there were behavioral or what Wu would recognize as emotional symptoms, he does not mention them. The doctor is not striving for a long-term change of attitude or perception, and there is no reason to expect one. This is true of cases 1–6 and 8. In three of these the cause and manifestations of the disorder are not even specified (1, 2, 6; in the last of these only the diagnosis reveals the cause). In two the cause is recorded in entirely schematic terms (3, 4).

Two cases (5, 8) are decidedly odd. Case 5 arises from the shock of eating something unpalatable at the normally joyous moment of parturition. In case 8, the social status of the patient is exceptionally low. The doctor appealed to the greed of the magistrate's subordinate in order to relieve his misery. Sooner or later, one must expect, the bars of tin would be revealed for what they were. The deception may have gotten the "patient" over his shock, but one is left wondering whether finding it out will occasion another shock. Generally speaking, yamen runners were considered (by their superiors, at least) to be a species without higher instincts that one could appeal to.

In four of these cases (1–3, 6), the status of the patients was higher than that of their physicians. In case 4 the status of the patient is not mentioned; in case 5 it is unknown, because the status of elite women was simply that of their husbands. We are not even told whether the women who supported the nameless patient as she re-enacted childbirth were servants.

II

EMOTIONAL COUNTER-THERAPY

The deception in cases 7 and 9 is impressively gross. The first involves physical as well as mental abuse. The account does not say who slapped and insulted the patient, but it is difficult to believe that it was the doctor rather than the father. In the second, either the husband or the physician enlists the services of a popular religious operative, whose occupation doctors generally despised. One might compare the woman's treatment with an analogous case in Chu Chen-heng's compilation, in which Lo Chih-t'i 羅知悌 (ca. 1243–1327) treated a wandering monk who had had a breakdown because of longing for his mother. Lo's "therapy" consisted of giving the mendicant monk a place to stay, administering medicine, feeding him rich food, calming him verbally, and giving him money to pay for his further travels. The contrast is striking.[12]

In the same two cases, the "therapy" also aimed at changing attitudes. The first woman's father, and the second woman's husband, would have been motivated not only by concern for suffering, but by the universal expectation that women's emotions should not interfere with their service to the household.

In case 7 it had been arranged that the woman would leave her parental household. A bride was expected to shift her emotional ties to her husband and his parents. It was, however, unfilial to appear impatient for this transition. Chu Chen-heng encouraged the father to resolve the issue by summoning the fiancé back, and perhaps fixing an early date for the wedding. But the woman had to be taught a lesson first, namely that lovesickness implies an illegitimate affair, not engagement to be married. That, I take it, was the point of the accusation.

The woman in case 9 had already left the house of her parents. Although she might visit her patrilineal home, she belonged to her new family, no longer to the old one. Her unceasing thoughts of her mother, which decidedly interfered with her activities in her marital household, would, from the conventional point of view, amount to disloyalty. This implication did not have to be explained in the anecdote, for every sixteenth-century reader would have been aware of it.

This case fits the neat scheme of correspondences that Wu K'un is seeking to illustrate, for it asserts that anger overcomes longing. But more fundamentally and a great deal less neatly, the counter-therapy in this instance used betrayal—the husband's trick to convince his wife that her beloved mother was tricking her—to end the bride's "betrayal"—her failure to serve her new family without reservation. From the husband's point of view, no doubt, this was simply the best way to restore the wife's health and happiness. He was expected to see it not from her point of view, or for that matter his own, but from the overriding viewpoint of his family line. Each Chinese family depended on the importation of women from other families for its own continuation.

Conclusion

These anecdotes illustrate several aspects of inappropriate emotion as a problem for therapy in classical medicine.

First, Wu K'un, a sophisticated physician, did not draw a sharp line between therapy performed by doctors and emotional manipulation carried out by laymen, so long as the goal was to relieve suffering brought on by undisciplined emotion.

Second, we see from cases 7 and 9 that more than therapy was involved. The boundaries between social and medical deviance—between doing wrong, and not doing right because you are ill—were determined by values abroad in the society, and policed by physicians among others.

When people could not control their own emotions, those about them normally saw the outcome as a social problem, to be resolved by suasion, negotiation, and sanction. If such efforts within the family failed, an option was to see the failure as a medical problem and call on the expertise of the physician. Here as elsewhere in medicine, the therapy of first resort was consistently material, mainly the use of drugs to restore the balance of *ch'i* that inappropriate feelings had disturbed. If that failed, certain doctors with deeper understanding could apply "immaterial" therapy to resolve more refractory imbalances. This transformation of social failings into medical problems is what sociologists have recently called "medicalization" (as when inability or unwillingness to sit still in classrooms where immobility is valued becomes hyperkinesia). Wu reminds us that medicalization is neither modern nor Western.

This led inevitably to a tension between the physician's sensitivity to the medical situation seen whole and the expectation (on his part as well as that of others) that adjustment to one's social role is part of health. Wu speaks of being "responsive to changes in the disorder." Doctors need a great deal of clinical resourcefulness in order to respond to dynamic pathological and physiological processes. Success also depended, the sources remind us, on flexibly interpreting the constellation of somatic, emotional, and social dysfunctions that from the physician's viewpoint *are* the disease. The therapist must choose, when confronted with a withdrawn young woman, to encourage her to change her situation, or to deceive her so that she will accept it docilely. No one familiar with the complexity of life in an ancient Chinese family will find it easy to judge which course will be in her long-term interest. Nor would a physician have found it easy to decide whether her interest should override that of his male client.

Third, therapeutic deceptions were applied not only to patients of lower status

---

12. *Chu Tan-hsi i an*, p. 29.

EMOTIONAL COUNTER-THERAPY

than the practitioner, but even to those in high places. When the patient's rank was much higher, the healer might be in considerable danger. The fury he elicited might even end his life. He would have to rely on intercession by someone close to the powerful patient (1, 2), or on his own persuasiveness (6).

Fourth and from some points of view most important, the appearance of "emotional disorders" as a classification in this and a very few other books does not amount to a separation of physical and emotional illness. Within this category there is still no split between mind and body or between what moderns would call somatic and psychological symptoms. The latter are also important in more traditional categories.

Finally, Wu's section on emotional counter-therapy is not represented as a theoretical innovation. It is, rather, a practical expedient, a loose rubric for discussing an approach to therapy that may be useful for certain disorders of emotional origin.

These concrete instances carry a number of assumptions interesting from the psychiatric point of view. The interactions of one human being with others lead to emotion, which Chinese were expected to control. Sometimes its intensity overcomes one's resources of discipline. The result may be disabling illness. The physician, with his repertory of drugs and other technical means, is expected to redress the metabolic and circulatory imbalances that allow illness, whatever the cause, to spread and deepen. Some dysfunctions of emotional origin are so serious that conventional means will fail. This residuum can yield only to powerful counterbalancing emotions, which an exceptional physician (usually called in as a last resort) will know how to manipulate. Wu's cases assert that in doing so a guide is available, namely the familiar Mutual Overcoming sequence of the Five Phases. In order to make this point, Wu was willing, within limits, to rewrite details of his sources (1, 7). We have also seen him providing diagnoses when his sources lack them (1, 2, 5). He even included an anecdote (11) that is not obviously related to his Five Phases schema, and did not explain it.

These are a few basic insights about the emotions as seen in Chinese medicine. Many other writings that shed light on such topics are waiting to be explored.

II

References

Early Chinese Sources

Chien-ming i kou 簡明醫彀 (The Bowspan of Medicine). Sun Chih-hung 孫志宏, 1629. Beijing: Jen-min Wei-sheng Ch'u-pan-she. A handbook for travellers and those living in isolated places.

Chu ping yuan hou lun 諸病源候論 (On the Causes and Symptoms of Medical Disorders). Ch'ao Yuan-fang 巢元方, 610. Beijing: Jen-min Wei-sheng Ch'u-pan-she.

Chu Tan-hsi i an 朱丹溪醫案 (Medical Case Records of Chu Chen-heng). Posthumous, compiler and date unknown. In Hsu & Yao 1933.

Ho-nan Shao shih wen-chien hou lu 河南邵氏聞見後錄 (Sequel to Things Heard and Seen by Mr. Shao of Honan). Shao Po 邵博 (1057–1134), posthumously compiled 1157. Beijing: Chung Hwa Book Co., 1983. This was a sequel to Shao's father's *Wen chien lu*.

Huang-ti nei ching 黃帝內經 (Inner Canon of the Yellow Lord). Anonymous, probably first century B.C. The *Su wen* 素問 (Basic questions), ed. by Wang Ping 王冰, 762, is cited from Jen Ying-ch'iu 1986 by *p'ien* 篇, *chang* 章, line, and page number.

I fang k'ao 醫方考 (Researches on Medical Formulas). Wu K'un 吳崑, 1584. Beijing: Jen-min Wei-sheng Ch'u-pan-she, 1990.

I tsung chin chien 醫宗金鑑 (Golden Mirror of the Medical Tradition). Wu Ch'ien 吳謙 et al. Beijing: Jen-min Wei-sheng Ch'u-pan-she, 1963.

Ku-chin t'u-shu chi cheng 古今圖書集成 (Comprehensive Anthology of Sources Old and New). Chiang T'ing-hsi 蔣廷錫 et al., presented to the throne 1726. First ed.

Lü shih ch'un-ch'iu 呂氏春秋 (The Springs and Autumns of Master Lü). Compiled under patronage of Lü Pu-wei 呂不韋, ca. 239 B.C. In Ch'en Ch'i-yu 1984.

Nü k'o hsien fang 女科仙方 (Methods of the Immortals for Women's Disorders). Attributed to Fu Shan 傅山, ca. 1670?, publ. 1827? *Chung-kuo i-hsueh ta ch'eng hsu pien* 中国医学大成续编, IX.

Shao-shih wen-chien hou lu 邵氏聞見後錄. See *Ho-nan Shao shih wen-chien hou lu*.

Shih shan i an 石山醫案 (Medical Case Records of Wang Chi 汪機). Ed. Ch'en Ch'üeh 陳桷, 1519. In *Ssu k'u ch'üan shu* 四庫全書.

Modern Sources

Clemens, Samuel L. 1959. *The Autobiography of Mark Twain*, ed. Charles Neider. New York: Harper.

Chen Ch'i-yu 陳奇猷. 1984. *Lü shih ch'un-ch'iu chiao-shih* 呂氏春秋校釋 (Critical Edition of Springs and Autumns of Master Lü). 4 vols. Shanghai: Hsueh-lin Ch'u-pan-she.

Farquhar, Judith. 1994. *Knowing Practice. The Clinical Encounter of Chinese Medicine.* Boulder: Westview Press.

Furth, Charlotte. 1986. Blood, Body and Gender. Medical Images of the Female Condition in China 1600–1850. *Chinese Science*, 7: 43–66.

Hsu Heng-chih 徐衡之 & Yao Jo-ch'in 姚若琴. 1933. *Sung Yuan Ming Ch'ing ming i lei an* 宋元明清名醫類案 (Classified Medical Case Records of Celebrated Physicians of the Sung, Yuan, Ming, and Ch'ing Periods). Reprint, Taipei: Hsuan-feng Ch'u-pan-she, n.d.

Jen Ying-ch'iu 任應秋, editor-in-chief. 1986. *Huang-ti nei ching chang-chü so-yin* 黃帝內經章句索引 (Phrase Index to the Inner Canon of the Yellow Lord). Beijing: Jen-min Wei-sheng Ch'u-pan-she. Includes all phrases and technical terms in the *Ling shu* and *Su wen*, with a good text in old-style characters divided logically into sections.

Ng, Vivien. *Madness in Late Imperial China. From Illness to Deviance.* Norman: University of Oklahoma Press.

Sivin, Nathan. 1987. *Traditional Medicine in Contemporary China. A Partial Translation of Revised Outline of Chinese Medicine (1972), with an Introductory Study on Change in Present-day and Early Medicine* (Science, Medicine and Technology in East Asia, 2). Ann Arbor: University of Michigan, Center for Chinese Studies.

III

THE FIRST NEO-CONFUCIANISM
AN INTRODUCTION TO
YANG HSIUNG'S "CANON OF SUPREME MYSTERY"
(T'AI HSUAN CHING, ca. 4 B.C.)
太玄經

Michael Nylan and Nathan Sivin

Introduction

Confucius, Mencius, and Hsun-tzu were humanists; they believed achieving the good life was a matter of human interests and values. One's relation to the gods or to the cosmos was not a comparably urgent problem. Nevertheless, by 100 B.C. the first stable Chinese empire was supporting its claims to legitimacy with a Confucianism that, by a process not at all self-evident, had come to give the relation of man and Nature a place as conspicuous as that of man and man.

As new philosophic syntheses emerged from the late third century on, some of them aimed to form an orthodoxy (see Chap. I). The process can only be described as the first Neo-Confucianism, at least as great a shift in new directions as that of the Sung. The various systems drew on every contemporary current of thought, and wove them together so inextricably that it makes no sense to speak of Taoists or Legalists as specialized groups after the late second century. In these attempts at orthodoxy a single underlying pattern governed orderly change, whether in Nature, in the realm of social and political relationships, or in personal experience. Self-cultivation aimed to encompass all three of these spheres. Guided by the classics, its goal was sagehood. Only the power of sagely example could overcome social disorder and create a stable field for relationships. The monarch, as holder of the mandate bestowed by the natural order, was entitled *ex officio* to the dignity of a sage. It was the task of his advisors to guide and maintain him in sagehood. Such was the rationale of this state-centered Neo-Confucianism.

The genesis and original character of the Book of Changes *(Chou i* 周易 or *I ching* 易經) remain enigmas despite more than two millennia of intense study. By the first century B.C. the book had become not only a Confucian canon, its teaching sponsored by the state, but an infallible guide to foresight and self-discovery. A strong influence on this integration of cosmic and humanistic Confucianism was its Great Commentary *(Hsi tz'u ta chuan* 繫辭大傳), probably of the third or second century. One of its central problems was that of timely conscientious action. Its method combined emulation of the sages with numerology and yin-yang cyclic analysis. Tung Chung-shu 董仲舒 (ca. 179–ca. 104 B.C.) shaped these themes into a theory of monarchic order justified by resonance with the order of Nature. Out of its symbols and ideology evolved the book that completed the elaboration of yin-yang and Five Phases cosmology, the Inner Canon of the Yellow Lord *(Huang-ti nei ching* 黃帝內經, probably first century B.C.).

Han Neo-Confucianism culminated in another sense in Yang Hsiung's 揚雄 (53 B.C.–A.D. 18) Canon of Supreme Mystery *(T'ai hsuan ching)*. The book is a remarkable contribution to the tradition of the Changes, both as philosophy and as literature. Yang retained the metaphysical depth and psychological subtlety of the Changes in a work systematically constructed and poetically lucid. To let the reader judge whether Yang's Canon deserves more attention than it has had, we will discuss what justified it and how it is related to the Book of Changes; we will then explain and illustrate with excerpts what ideas it reflects, how it is organized, and how it uses language and imagery.

The Mystery was the most influential among the many meant to remedy inconsistencies in the Changes and to add to the old discourse current ideas about the cosmic order, the sagely life, and the beauty and precision that can be drawn from words.[1] Until the thirteenth century Yang's writings were considered central to the orthodox search for universal pattern, and thereafter were forgotten. The ruin of his reputation (see p. 10) has left the Mystery unread. Most modern histories of Chinese philosophy do not even mention it.

In referring to the Canon of the Supreme Mystery as "the Mystery" we follow the practice of Chinese authors from Yang Hsiung on. In citing it briefly as

---

1. For references to literature on the Book of Changes see Hellmut Wilhelm, *The Book of Changes in the Western Tradition. A Selective Bibliography* (Parerga, 2; Seattle, 1975). Later books in the *Chou i* tradition are listed (intermixed with treatises on divination) in *ch*. 108–110 of *Ssu k'u ch'üan shu tsung mu t'i yao* 四庫全書總目提要, and discussion with special reference to the *T'ai hsuan ching* in Suzuki Yoshijirô 鈴目由次郎, *Taigen'eki no kenkyû* 太玄易の研究 (Tokyo, 1964), pp. 26–41. Suzuki provides a complete Japanese translation of the Mystery, including Yang's commentaries. This translation is based on the views of early Chinese commentators rather than on a fresh reading of the text, but the book provides a systematic introduction. Suzuki's popular *Taigenkyô* 太玄經 (Chûgoku koten shinsho, 56; Tokyo, 1972) omits the Fathomings and Yang's commentaries, but the translation is sufficiently revised to be worth consulting. We are grateful for advice and help from Derk Bodde, David Cowhig, Michael Hearn, David Knechtges, Bernard Solomon, and Hellmut Wilhelm.

THE FIRST NEO-CONFUCIANISM 3

"*Hsuan*玄" they echo the common use of "*I* 易" ("Changes") for the *Chou i*.² "The Changes" and "the Mystery" preserve an important ambiguity in the early commentaries on the Changes and in Yang's commentaries on his own book. Certain commentaries in both groups are meant to be read on two levels at once. On one level they describe the processes of change that constitute the Way. At the same time, statements in these commentaries about "change" or "mystery" often describe the *I* or the *Hsuan* itself. Each book, its annotators assure us, comprises the diversity of the Way, as well as its unchanging mystery. Here is a typical example from the Great Commentary to the Changes:

> The Master said: The Changes in their perfection!
> Through the Changes the Sages exalted their virtue and broadened their achievement
> Exalted in wisdom, humble in ritual
> Exalted to emulate heaven
> Humble to exemplify earth
> Heaven and earth determining relations
> The Changes active between them
> Letting natures fully develop
> Preserving what exists:
> Gate of the Way and of Right.³

The Book of Changes in the Han

The archaic Book of Changes is a jumble of omens, rhymed proverbs, riddles and paradoxes, and snatches of song and story, drawn from popular lore and archaic traditions of divination. The so-called Ten Wings *(shih i* 十翼) remade the archaic text into a Confucian canon. They were actually a group of seven anonymous interpretations and commentaries, six of them from the third and second centuries B.C. To accomplish this revision, they ignored the simple and direct senses of most of these constituents.⁴ This was partly because significances had been lost as Chi-

---

2. Yang's disciple Hou Pa 侯芭 apparently elevated the book to status of a canon *(ching)*. On Hou see *Han shu pu chu* 漢書補注 (hereafter HS; Basic Sinological Series 國學基本叢書 ed., hereafter BSS), *87B:* 5135.

3. *Chou i yin-te* 周易引得 (Peiping, 1935), 40/*Hsi tz'u*, A/5, end. For an analogous example from the Mystery, see p. 17 below.

4. Among the many studies that established this view of the pre-Confucian Changes, the most important are those by Li Ching-ch'ih 李鏡池 and others in *Ku shih pien* 古史辨, III, pt. 1 (Peiping, 1931). Li's writings on the Changes have been gathered in *Chou i t'an yuan* 周易探源 (Beijing, 1978). Also important though faulty are Kao Heng's 高亨

nese culture evolved, partly because the language changed, and mostly because the Wings read into the original text philosophical and social concepts that did not exist when the old text was written down late in the ninth century B.C.

Confucians by the middle of the Han were convinced that *Chou i* had originated in a set of cosmic emblems invented by the legendary sovereign Fu-hsi 伏羲 and elaborated by King Wen 文王 at the beginning of the Chou era, and that Confucius had diligently studied it. Thus it must coherently express a perfectly formed vision of sagehood. The most influential interpretation of the Changes, the Great Commentary, described it as an ordered account of the sage in society and in the universe, perfectly attuned to the springs of change that led without fail to timely and appropriate action. The scripture was a perfect mirror of the relations it described, its compilation an example of sagely action. But the order and system outlined in the Great Commentary obviously did not lie on the surface of the archaic text. Its *non sequiturs,* rustic frivolities, and archaic puzzles were never-quite-surmountable obstacles to the recovery of an inner meaning that had nothing to do with peasant lore or fortune-telling.

Frustrating though the diversity of the text was bound to be, the determined search for order found clear underlying principles in the sequence of the hexagrams and their internal structure. The sixty-four emblems, although not in a

---

attempts to reconstruct the original meaning in *Chou i ku ching chin chu* 周易古經今注 (Shanghai, 1947) and *Chou i ta chuan chin chu* 周易大傳今注 (Jinan, 1979), an expanded version. Gerhard Schmitt has studied several hexagrams with philological rigor in *Sprüche der "Wandlungen" auf ihrem geistesgeschichtlichen Hintergrund* (Deutsche Akademie der Wissenschaften zu Berlin. Institut für Orientforschung. Veröffentlichung, 76; Berlin, 1970). The problem of pre-philosophical meaning was introduced to the West by Arthur Waley in "The Book of Changes," *Bulletin of the Museum of Far Eastern Antiquities,* Stockholm, 5 (1933), 121–142, drawing on the *Ku shih pien* studies, and then independently of both by Iulian K. Schutskii in a Russian dissertation completed in 1935, trans. into English as *Researches on the I Ching* (trans. W. L. MacDonald et al.; Bollingen Series, 62.2; Princeton, 1979). An important study of the early meaning of the Changes is Richard A. Kunst, "The Original *Yijing*: A Text, Phonetic Transcription, Translation, and Indexes, with Sample Glosses," Ph.D. diss., Oriental Studies, University of California, Berkeley, 1985. For a concise essay by Edward L. Shaughnessy on dating and the bibliography of the Changes see Michael Loewe (ed.), *Early Chinese Texts. A Bibliographical Guide* (Early China Special Monograph Series, 2; Berkeley, 1993 [publ. 1994]), pp. 216–228. Willard Peterson has studied the Great Commentary in "Making Connections: 'Commentary on the Attached Verbalizations' of the *Book of Change,*" *Harvard Journal of Asiatic Studies,* 42 (1982). 1: 67–116. Bent Nielsen, "The *Qian zuo du* 乾鑿度. A Late Han Dynasty (202 BC–AD 229) Study of the Book of Changes, Yi jing 易經" (Ph.D. diss., Sinology, University of Copenhagen, 1995), studies a book of the first century A.D.(?). Liao Ming-ch'un 廖名春 et al., *Chou i yen-chiu shih* 周易研究史 (Changsha, 1991), is a history of *I* studies.

III

regular order as they occur in the text, are at least in pairs, one member of which is the other turned upside down. The eight symmetrical hexagrams that inversion leaves unchanged are paired by changing each yin line to yang and vice versa.[5]

The Ten Wings carried this line of exploration further. Among other approaches, they interpreted the texts attached to the hexagrams (the "judgments," *t'uan tz'u* 彖辭) and to their individual lines *(yao tz'u* or *hsiao tz'u* 爻辭) by examining the relations of lines to each other within the hexagram and by their yin-yang associations.

By the first century B.C. such analyses had satisfied most thinkers that the texts were not a motley set randomly attached to the hexagrams. Hidden within the words of each text must be an order identical with that of the corresponding hexagram. The logic of these sixfold binary symbols in some subtle way must determine the words of the judgments and line texts. Scholars became convinced that the complicated and ambiguous ideas in a text could eventually be resolved by careful analysis into simple images and concepts associated in a regular way with the six lines and two trigrams (intermediate three-line symbols) that make up the corresponding hexagram. The hexagrams, in other words, had become decipherable symbols of a manifold reality. The texts had become literary expressions of the truths that the hexagrams express more abstractly and emblematically.

The masters of the first century B.C., especially Meng Hsi 孟喜 (fl. 69) and Ching Fang 京房 (77–37), continued this research. They created an imposing new armamentarium of techniques to carry on established lines of inquiry and to explore the relations between the meaning of each hexagram's texts and the multifarious associations of the eight possible trigrams of which they were built.

Some modern historians have praised Han scholars of the Changes for their contributions to positive science, and others have blamed them for launching science down a dead-end road. Both evaluations grossly overestimate their influence on methods of scientific discovery, and ignore their aims and their actual effect in the history of thought.

The Han mutationists sought to understand the patterns underlying all process: in the external world, in the body, in the recesses of the human heart, in the conscientious action of the individual, and in the ceremonial of the empire. Well-ordered activity in any of these spheres was a manifestation of the one Way. In the fourth century, Mencius and Chuang-tzu did not agree on whether knowledge of these patterns must be ethical. In the Han, no judgment that flouted cosmic patterns could arrive at the Good, and no knowledge that disregarded the Good

---

5. Whether this was the pre-Han order is uncertain. See Nielsen 1995: chap. 3.

could be true. Meng Hsi, Ching Fang, and others like them can hardly be blamed for being men of their time rather than modern scientists.[6]

Their concern as men of their time was the ultimate issue in understanding change: how does all this infinite diversity of natural mutation in Nature, society, and the human psyche arise from the Way, which rests in mystery and does not change at all? What obscure paths do the Tao's "spiritual forces" *(shen* 神*)* travel to keep the cycles turning? The Han experts were reexamining the symbolic notation of the *I* to find regularities. They were not merely trying to narrow down the objective significance, even what they believed to be the objective moral significance, of the symbols. Their goal was a universal nexus of association and correlation, extending the meanings of traditional symbols to create an infinitely rich language they could use to relate everything that people observe, think, feel, contemplate, and imagine. What they valued in this vocabulary was scope, not rigorous definition. They did not see themselves as widening the ambit of the Changes, but as coming to grips with the universality that the Sages had given it. Their demonstration that endless wisdom was stored in the symbols of the Changes confirmed its status in the canon whose transmission was sponsored by, and in turn lent legitimacy to, the dynastic house of Han.

The Book of Changes no doubt originated as a manual of prognostication. Even today it remains a living force in Chinese civilization. As one intellectual in the People's Republic puts it, "Each loosening of political control in China has resulted in a debate involving both Confucius and the *Book of Changes.*"[7]

Over the two and a half millennia in which the Changes has been studied as a key to the future as well as a revelation from the archaic past, forecasting remained only one of its roles. At the same time, suitably interpreted, it served as a philosophical summa and a pillar of orthodoxy. For its most dedicated students, since the time that the Ten Wings were written, the Changes was a model of the Way in all its aspects, to be mastered and contemplated as a guide through the complexity of experience, back to the hidden center in which all tensions and con-

---

6. The best introductions to Han studies of the Changes are Hsü Ch'in-t'ing 徐芹庭, *Liang Han shih-liu chia I chu ch'an wei* 兩漢十六家易注闡微 (Hong Kong, 1975), which collects and discusses the extant texts; Kao Huai-min 高懷民, *Liang Han I hsueh shih* 兩漢易學史 (Taipei, 1970); Ozawa Bunshirô 小澤文四郎, *Kandai Ekigaku no kenkyû* 漢代易學の研究 (Tokyo, 1970); and the first essay in Suzuki, *Kan Eki kenkyû* 漢易研究 (Tokyo, 1963). Ch'ü Wan-li 屈萬里, *Hsien Ch'in Han Wei I li shu p'ing* 先秦漢魏易例述評 (Taipei, 1970), gives a systematic account of techniques used to interpret hexagrams.

7. K'o Yun-lu, cited in Germie Barmé & Linda Jaivin, *New Ghosts, Old Dreams* (New York: Times Books, 1992), p. 376.

tradictions are resolved, and all sound decisions imperceptibly set in motion.

In the Way, which was also Yang Hsiung's Mystery, science and ethics were one. This Han vision was not a reduction of science to subjectivity. Nature and human nature constitute a single order. As Yang argues in his "Evolution" commentary, "Now Heaven and Earth are placed; therefore, the noble and base are ranked. The four seasons proceed [in order]; therefore, the son inherits from the father. The pitchpipes and calendar are set forth; therefore, relations between ruler and subject are orderly."[8] The good society reflects the cosmos, and vice versa, because both are part of the Way. The object of inquiry in this view of Nature differs fundamentally from that of modern physical science. Since historians have almost entirely ignored the methods and aims of the Tao-centered science of early China, few of their generalizations about its character are useful.

Yang Hsiung shared his predecessors' vision of an order that united the cosmos, the sphere of action, and the individual. This vision drew his attention, as it had theirs, to the one scripture in the canon that encompassed every aspect of that order. But Yang was unwilling to join those such as Meng and Ching who were willing to multiply techniques of interpretation until the non-philosophic assertions of the archaic Changes had been explained away. In the orthodoxy he sought, mystery and rational pattern were inseparable and complementary.

A generation before Yang, Chiao Kan 焦贛 had stepped outside the scholastic tradition when he compiled his Forest of the Changes *(I lin* 易林*)*. Chiao rejected

---

8. *T'ai hsuan ching* (under the title *Tien k'an T'ai hsuan* 點勘太玄; Chu tzu chi p'ing 諸子集評 ed. of 1909; reprint, Taipei, 1970, hereafter T), *7:* 7a (p. 1020). This edition contains annotations by Ssu-ma Kuang and others. For its use of earlier recensions see *10:* 10b–11a (pp. 1041–1042). Nylan's complete translation is *The Canon of Supreme Mystery* (SUNY Series in Chinese Philosophy and Culture; Albany, 1993, hereafter N), p. 432. Since Ssu-ma uses Fan Wang's 范望 commentary (ca. 265) sparingly, it is necessary to consult the full version in *T'ai hsuan ching* (Chung-kuo tzu hsueh ming chu chi ch'eng 中國子學名著集成 reprint of the 1524 ed.), cited below as F. Most late annotators add little to the explanations by Han and Six Dynasties scholars preserved in these two editions. Occasionally useful are the *T'ai hsuan pen chih* 太玄本旨 (preface dated 1368) of Yeh Tzu-ch'i 葉子奇, in *Ssu k'u ch'üan shu chen pen* 四庫全書珍本, 3rd collection, and *T'ai hsuan ch'an mi* 太玄闡祕 (pref. end of 1814 or early 1815) of Ch'en Pen-li 陳本禮, in Ch'u hsueh hsuan ts'ung-shu 初學軒叢書, 4th coll. Yeh is mainly concerned with cyclic correlations of the texts, and reads criticisms of Wang Mang into them. Ch'en also attempts to prove Yang did not support Wang's usurpation. Somewhat more than a mere curiosity is the Ch'ien-lung *Yang-tzu t'ai hsuan pieh hsun* 揚子太玄別訓 of Liu Ssu-tsu 劉思祖, which provides a commentary in the form of one or two rhymed tetrasyllabic quatrains for each Head, Appraisal, and Fathoming, and long prose colophons for each of Yang's commentaries.

as "incomplete" the prevalent analysis of hexagrams based on individual moving lines (see below, p. 25) and constituent trigrams. By changing moving lines in the original hexagram to their opposites, inquirers arrived at a new hexagram; but they then interpreted it using the same techniques as they applied to an original hexagram. To better elucidate the dynamic relation between original and derived hexagrams, Chiao provided 4,096 (64^2) rhymed texts. Each text gave meaning to a second-order hexagram. Sixty-four of these could be formed from each original by manipulating all possible combinations of moving lines, from one to six.[9]

Yang went much further. The Canon of Supreme Mystery is a completely new book that in its philosophical coherence and order, and in its overt correspondence of text and emblem, is everything that Yang's predecessors had so laboriously sought to read into the Changes.

Political Background

Yang Hsiung's books took authoritative earlier writings as their starting points, but were not mere imitations. The Mystery made considerable demands on its readers. The clarity of its structure was intentionally balanced by the complexity of language that strives above all for allusiveness. But its notorious difficulty was only one reason that the book came to be neglected while the study of other arcane scriptures flourished. Here is the explanation of Ssu-ma Kuang 司馬光 (1019-1086), the book's most eminent enthusiast, who annotated it and at the end of his life wrote another book in the same tradition:[10]

---

9. Like the *T'ai hsuan ching*, the *I lin* is preserved in the *Cheng-t'ung Tao tsang* 政統道藏. The former is found in vols. 860–862 (hereafter TT; item 1183 in K. M. Schipper, *Concordance du Tao-tsang. Titres des ouvrages*, Paris, 1975, hereafter S), and the latter in TT 1101–1104 (S 1475). Chiao's expansion carried further what Han scholars believed had been the expansion of the eight trigrams to 64 (8^2) hexagrams. On Chiao's motivation see the preface attributed to Pi Chih 費直, a *Chou i* master of the late Western Han. For further discussion of the *I lin* see Suzuki 1963: 431–593, and Kao Huai-min 1970: 126–138. Chiao's authorship has been questioned, but Suzuki confirms it in "Shô shi Ekirin no sakusha ni tsuite 焦氏易林の作者について," in *Tôhô gakkai sôritsu nijûgo shûnen kinen Tôhôgaku ronshû* 東方學會創立二十五週年紀念東方學論集 (Tokyo, 1972), pp. 307–320.

10. From Ssu-ma's prefatory essay to the *T'ai hsuan ching* entitled "On reading the Mystery" *(Tu Hsuan* 讀玄), in TT, prefaces, pp. 1a–3a. Ssu-ma takes the anecdote from HS, *87B*: 5135–5137. Our first ellipsis indicates his omission. His own book in the *Chou i* tradition, greatly influenced by the *T'ai hsuan ching*, is *Ch'ien hsu* 潛虛 (Hidden and empty). The structure of its diagrams is decimal. See *Ssu k'u ch'üan shu tsung mu t'i yao* 四庫全書總目提要 (Kuang-tung shu-chü ed. of 1868), *108*: 7a–8b, and Suzuki 1963: 38–41.

III

THE FIRST NEO-CONFUCIANISM

When I was a youngster I heard of the Mystery, but never succeeded in seeing it. I was able to read only Master Yang's autobiographical preface, which led me to acclaim the Mystery as a splendid work. When Pan Ku 班固 wrote Yang Hsiung's biography, he said,

"Liu Hsin 劉歆 (ca. 53 B.C.–A.D. 23) said [to Yang], 'You have troubled yourself for nothing. These days scholars can hold official position and reap its benefits without being able to understand the Changes. Where does that leave your Mystery? I fear that our successors will use it to cover their saucepots!' Yang laughed but did not reply.... Confucian scholars have sometimes derided Yang on the ground that, in composing a canon although he was not a sage, he was like the lords of Wu and Ch'u in the Spring and Autumn period, who usurped the title of King, a crime that merited executing them and terminating their family lines."

The fact that Pan Ku recorded these anecdotes indicates that although his attitude was more favorable than that of Liu Hsin, he still would not have said that the Mystery is as fine a book as Yang had claimed. I myself thought it strange that Master Yang, rather than contributing to the [study of the] Changes, wrote his own Mystery. The Way expounded in the Changes encompasses all the multiplicity of natural and human phenomena. What could Master Yang have had to add that justified writing a new book? Nor did I know how he meant it to be used. Thus I too was unable to acknowledge that Master Yang was right to compose the Mystery.

When I grew up, in studying the Changes I was greatly troubled by its abstruseness. It occurred to me that since the Mystery was the composition of a wise and compassionate man [rather than the revelation of the Sages], compared to the Changes its meanings should be less deep, and its style more accessible.... I thus wished it were possible to devote myself first to the Mystery, and thus gradually to advance far enough toward the Changes that I could hope, standing on tiptoe, to catch a distant glimpse [of its inner meaning]. From then on I sought [a copy of the Mystery] for years on end. Finally I was able to read it. At first I found it boundless and unfathomable, and could hardly tell where to begin. Then, studying it more closely, I changed my attitude. I set aside social obligations and read the book dozens of times, until finally it seemed that I had at least a limited understanding of its gist. Then I laid it aside and said to myself with a sigh: "What a great and good Confucian this man Yang must have been! Since Confucius' death, who if not Master Yang has comprehended the Way of the sages? Hsun-tzu, I fear, is hardly a model in that respect, much less the others."... Thus I realized that the Mystery *is* a contribution to the [study of the] Changes, and that Yang had not written a separate work in order to compete with it. How superficial was Liu Hsin's and Pan Ku's understanding of him; how profound the wrong they did him!...

If a scholar is able to study the Changes to the exclusion of all else, truly nothing more is needed. But the Changes is heaven, and the Mystery provides what is needed to build a staircase up to it. Are *you* prepared to float up to heaven, leaving the staircase unused?

In this preface Ssu-ma Kuang addresses the major objections levelled against the *T'ai hsuan*.[11] The Mystery, like the Changes, was said to be hopelessly abstruse and of no practical benefit. Liu Hsin's famous jibe seems to have been directed not against Yang Hsiung but against the intellectual limitations of contemporary careerists. To say that the Mystery would not be read in an age that had no interest in the Changes is merely to state the obvious. Pan Ku was using this anecdote—among others—to suggest that Yang's self-esteem could not be threatened by a *bon mot*. But Ssu-ma Kuang redirects it to praise Yang, who was not a cynic like Liu. Liu could not understand that the opportunists' neglect of the Mystery proved its worthiness. Next, Ssu-ma acknowledges the opinion that Yang was presumptuous when he composed a work in canonical form rather than a commentary or a work in some other genre conventional in a post-classical age. Many wrote such imitations, but it was usual to attach the name of some ancient sage rather than to sign one's own. Yang's defender argues that the Mystery was a contribution, rather than a rival, to the study of the Changes.

The identification of intellectual arrogance with political usurpation brings to mind the most explosive issue of all. Yang served the usurper Wang Mang 王莽 (r. 9–23), who, setting aside his fealty to the house of Han, assumed the throne as first emperor of the Hsin 新 dynasty in A.D. 9. Such service, tantamount to treason in the eyes of many historians, not only cast doubt on Yang's character and judgment but profoundly affected the repute of his writings. Possibly Ssu-ma, in claiming that Liu intentionally derided Yang, means to emphasize the difference between Yang and a man notoriously sympathetic to Wang Mang's ambitions.

The evidence that has survived about Yang's involvement with Wang is scant and ambiguous; we will review it in its setting.

Pan Ku avers that the Mystery was completed during the reign of the sickly young Sad Emperor (Ai-ti 哀帝, 7–1 B.C.), when the Ting 丁 and Fu 傅 imperial distaff clans were dominant in the court. Within three months of that emperor's accession, Dowager Empress Fu forced Wang Mang into retirement, awarding him, upon his departure in August, 7, special grants, honors, and an enlarged fief. Powerful opponents quickly neutralized the influence of his supporters, and within two years were speaking of him as "deserving public execution." Wang

---

11. For a typical contemporary attack see Su Hsun 蘇洵 (1009–1066), "*T'ai hsuan* lun 太玄論," in *Chia-yu chi* 嘉祐集 (BSS), 7: 61–72, and on Su's essay, George Hatch in Yves Hervouet (ed.), *A Sung Bibliography* (Hong Kong, 1978), pp. 388–389.

III

THE FIRST NEO-CONFUCIANISM

"shut his gates and preserved himself" on his estate.[12]

During this period, when Yang Hsiung was writing the *T'ai hsuan ching,* Wang's rise to highest power was anything but predictable. In 2 B.C., following the deaths of the dowager empresses Ting and Fu, his aunt the Empress Dowager Wang, on the pretext of an ominous solar eclipse, recalled him to court. After the fortuitous death of the Sad Emperor in 1 B.C., Wang Mang was well placed to influence the selection of the Tranquil Emperor (P'ing-ti 平帝, 1 B.C.–A.D. 6). Without significant rivals, Wang consolidated his position throughout the new reign. Upon the emperor's death, Wang, with his aunt's support, became Regent on behalf of Liu Ying 劉嬰, the new child-emperor (February, 6), Acting Emperor of the Han (July, 6) and finally Emperor of the Hsin (January, 9).

While this tumult was making a court career impossible, Yang Hsiung too was "preserving himself and remaining tranquil." In Pan Ku's supportive biography, Yang's self-preservation was as much a matter of fostering spiritual integrity as of lying low.[13]

Yang Hsiung's feelings toward Wang Mang before and after Wang's rise to power are not clear from the record. As court poet, Yang was expected to write encomiums to his patron, and he did so. For example, he appended to his Model Sayings praise for the Duke who Gives Tranquillity to the Han, the title by which Wang was known from A.D. 1 to 4, when he was considered not a potential usurper but the last pillar of Han stability. As Wang, Liu, and others with whom Yang served were promoted again and again, Yang, according to Pan Ku, "did not change his post during three reigns." He held only nominal office under Wang Mang, which suggests that he indeed held himself aloof from the sycophants who

---

12. HS, *99A:* 5661, trans. Homer H. Dubs in *The History of the Former Han Dynasty* (3 vols., Baltimore, 1938–1955), III, 133–134. For basic information on Yang's relations with Wang, see *87B:* 5133–5137. Attempts to show consistent enthusiasm for Wang in Yang's writing, countered by arguments that many allusions satirize or mock Wang, add up to a tidal wave of scholarship that has not yet abated. The most judicious detailed assessment we have seen is in an unpublished MS by Michael Barnett, "The Han Philosopher Yang Xiong: An Appeal for Unity in an Age of Discord." For Yang's life, see Knechtges, *The Han Shu Biography of Yang Xiong (53 B.C.–A.D. 18)* (Occasional papers, Center for Asian Studies, Arizona State University; Tempe, 1983). See the important references in Knechtges, "Uncovering the Sauce Jar: A Literary Interpretation of Yang Hsiung's 'Chü ch'in mei Hsin,'" in David T. Roy & Tsuen-hsuin Tsien, *Ancient China. Studies in Early Civilization* (Hong Kong, 1978), pp. 229–252, esp. pp. 232–234. Michael Loewe, *Crisis and Conflict in Han China* (London, 1974), esp. p. 301, discusses the political background.

13. HS, *87B:* 5113. Yang uses the term on p. 5120. For what he meant, see p. 14.

were reaping rich benefits from Wang's favor. In 10, falsely implicated in a plot against Wang, he jumped from an upper-story window and nearly died of injuries. Roughly four years later, near the end of his life, he wrote a memorial (later misclassified as a "portent text") praising Wang's Hsin dynasty. A rhymed epigram, Pan tells us, circulated in the capital:

> Wanting purity, stillness
> He threw himself out of a tower;
> Wanting solitude, quiet
> He composed a portent text.

A glance at the quotation that ends the present essay will make the irony clear.[14]

Yang's writings and personal character did not appeal to the careerists of his time, but they won him the esteem of such eminent intellectuals as Huan T'an 桓譚 (43 B.C.–A.D. 28) and Wang Ch'ung 王充 (A.D. 27–97). Pan Ku's History of the Former Han Dynasty accorded him an exceptionally long biography. The bibliographies of the Standard Histories classified the Canon of Supreme Mystery not as a book of divination but as an orthodox writing of the Confucian tradition (ju chia 儒家). Literati often referred to its author as "Master Yang" (Yang-tzu 揚子), a form rarely applied to thinkers after the Chou period. Scholars of the Six Dynasties greatly respected his philosophical preoccupation with fundamental patterns of cyclic change rather than with fluctuation and chance. The Mystery provided inspiration as well as terminology for "studies of the mysteries" (hsuan-hsueh 玄學), a third-century philosophical revival much more eclectic than the orthodoxy of Yang's time.[15]

---

14. HS, *87B*: 5134–5135. Fritz Jäger recognized the allusions to Yang's own writings; "Yang Hiung und Wang Mang," *Sinica-Sonderausgabe*, 1 (1937): 18n19. The memorial, entitled *Chü Ch'in mei Hsin* 劇秦美新, is preserved in *Wen hsuan* 文選, *ch*. 48, and discussed and translated in Knechtges 1978. At least part of it is spurious; see pp. 244–246.

15. For Yang's biography see HS, *ch*. 87A–87B. The term *hsuan-hsueh* suggests both the *T'ai hsuan ching* and the "Three Mysteries" *(san hsuan* 三玄) of the Chou, namely the *Lao-tzu, Chuang-tzu,* and *Chou i*. Yang, following his teacher Chuang Tsun 莊遵 (discussed below), was among the first to draw on the *Lao-tzu* and *Chuang-tzu* selectively along with the Changes. See Yang's *Fa Yen (Hsin pien chu tzu chi ch'eng* 新編諸子集成), 4: 10, 5: 13, 15–16. This unfashionable catholicity is perhaps one reason that Yang was ignored by so many of his contemporaries, a point acknowledged as a sign of worthiness by Huan T'an and other admirers. See Huan's *Hsin lun* 新論, *ch*. 13–15 of *Ch'üan Hou Han wen* 後漢文, in *Ch'üan shang-ku san tai Ch'in Han San-kuo Chin Nan-pei-ch'ao wen* 全上古三代秦漢三國六朝文 (1894 reproduction of 1st ed.), esp. *15*: 8a–8b, trans. Timoteus Pokora in *Hsin-lun (New Treatise) and Other Writings by Huan T'an (43 B.C.–A.D. 28)* (Michigan Papers in Chinese Studies, 20; Ann Arbor, 1975), pp. 172–173, items

III

THE FIRST NEO-CONFUCIANISM

With the revival of a classicist Confucianism from the T'ang on, Yang's views regularly entered discussions of the moral nature of human beings and other central issues. Such pivotal figures as Han Yü 韓愈 (768–824), Wang An-shih 王安石 (1021–1086), and Su Shih 蘇軾 (1037–1101), considered his arguments alongside those of Mencius and Hsun-tzu. In the Northern Sung, as the civil service examination system and the rise of a new official class tied Confucianism more closely to the authority of the state, dynastic legitimacy became a sensitive issue.

Scholars asked afresh which rival regimes in the past had actually held the Mandate of Heaven. They greatly stressed the unwavering loyalty of officials toward the ruling house that had appointed them, an ideal that had not existed among the Western Han elite.[16] Any association with Wang Mang—early or late, long or short—came to imply betrayal of the legitimate Han dynasty. Ideological judgments of this sort lowered the reputations and devalued the writings, not only of Yang Hsiung but of Liu Hsin and his father Hsiang 向 (B.C. 79–8), polymaths whose influence on later scholarship had been enormous. In the passage already quoted, Ssu-ma Kuang testifies to the obscurity into which the Canon of the Supreme Mystery had fallen in the eleventh century.

Chu Hsi 朱熹 (1130–1200) in his influential chronological survey of history recorded Yang's death in a way that any Sung reader would recognize as damning:

163–164 (slightly modified): "People desire what is close to them and admire what is far away. They saw for themselves Yang Tzu-yun's salary, position, and appearance, none of which were impressive, so they thought little of his writings."

In the Six Dynasties Yang's work played an important part in the teaching of the influential Ching-chou 荊州 school, founded by Sung Chung 宋衷 (or 忠, d. 219), famed for his interpretation of the Mystery. This school was the matrix in which *hsuan-hsueh* was formed. See T'ang Yang-t'ung, "Wang Pi's New Interpretation of the *I-ching* and *Lun-yü*," *Harvard Journal of Asiatic Studies*, 10 (1947), 124–161, esp. pp. 129–132, and Yü Ying-shih, "Han Chin chih chi chih hsin tzu-chueh yu hsin ssu-ch'ao 漢晉之際之新自覺與新思潮," *Hsin Ya hsueh-pao* 新亞學報, 4 (1959). 1, 25–144, esp. pp. 86–91. For other evaluations of Yang by later writers, see Knechtges, *The Han Rhapsody. A Study of the Fu of Yang Hsiung* . . . (Cambridge, England, 1976), pp. 109–110, and Alfred Forke, *Geschichte der mittelalterlichen chinesischen Philosophie* (Hamburg, 1934), pp. 78–90, which also discusses the *T'ai hsuan ching*. *T'ai hsuan ch'an mi*, "Wai pien 外編," collects essays in praise of the Mystery by eminent scholars over the centuries.

16. This important point was made by Hsu Fu-kuan 徐復觀 in *Liang Han ssu-hsiang shih* 兩漢思想史 (rev. ed., Taipei, 1976), II, 458–459. See Han Yü, "Tu Hsun 讀荀," in *Han Ch'ang-li chi* 韓昌黎集 (BSS), 3: 72–73; Su Shih, "Yang Hsiung lun 揚雄論," *Su Tung-p'o ying chao chi* 蘇東坡應詔集 (idem), *hsia*, 18: 70–71; and Wang An-shih, "Yang Meng 楊孟," *Wang An-shih ch'üan chi* 王安石全集 (Taipei, 1974), 39: 102–103. On the

III

"Wang Mang's court grandee Yang Hsiung died." In his informal talks with his disciples he goes much further: "Yang Hsiung is the most useless of all, a true rotten pedant. Whenever he gets excited he throws in his lot with the Yellow Lord and Lao-tzu. His judgment is unfailingly inferior and his writing dull in the extreme. He is most laughable." Elsewhere in the same *chüan*, Chu cites Yang's astronomical ideas and praises his personal depth and his ability to reason on such matters as the alternation of yin and yang. His main objection is that Yang uses an uncanonical threefold mode of analysis (based on heaven, earth, and man) rather than the classical dialectic mode. Inertia kept Yang among those worshipped in the state temples to Confucius long after Chu's lifetime, but his contempt blasted Yang's reputation among scholars.[17]

Yang saw his time as a time of chaos. Public life was often catastrophic not only for the individual but for his clan. The danger could be reduced either by currying favor with the clique in power—which could become a fatal liability if the order changed—or by withdrawal—which opportunists would see as failure. Yang chose the latter. He remained aware of the cost; in fact he wrote long and erudite essays to remind his detractors that his obscurity was voluntary.

But withdrawal was not just a matter of rational calculation. An aphorism of his teacher Chuang Tsun 莊遵 (better known as Yen Chün-p'ing 嚴君平, late first century B.C.) comes to mind: "No matter how fast you walk you can't escape your shadow; no matter how loud you speak you can't drown out your echo. But in silence, keeping to the shade, you give no cause for shadow or echo."[18] Yang was no doubt influenced by his teacher, a diviner by occupation, when he chose "mystery" as his metonym for the Way. The obscurity that it implies is a proper attribute for a noble man whose time is out of joint. Yang's essay "An Antidote for Ridicule" *(Chieh ch'ao* 解嘲) makes it clear that he saw renouncing politics not as failure but as fidelity to the Way. In his time "those who say anything unusual are suspected; those who do anything different are punished. . . . Doing what can be done in a time when things can be done results in success; doing what cannot

---

new alignment of ruler and officials see Vol. I, Chap. III, p. 3.

17. *Tzu chih t'ung chien kang-mu* 資治通鑑綱目 (1172, K'ang-hsi palace ed.), *8:* 34a; *Chu-tzu yü lei ta ch'üan* 朱子語類大全 (1973 Kyoto reprint of 1668 Japanese xylograph), *137:* 4b, 15a, 10a, pp. 6782, 6803, 6793. On Yang's complex career as an object of official worship see Thomas A. Wilson, *Genealogy of the Way. The Construction and Uses of the Confucian Tradition in Late Imperial China* (Stanford University Press, 1995).

18. "Tso yu ming 座右銘," in *Ch'üan Han wen, 42:* 13b, in *Ch'üan shang-ku san tai Ch'in Han San-kuo Chin Nan-pei-ch'ao wen.* "Yen" was used to avoid a later Han taboo on "Chuang."

III

THE FIRST NEO-CONFUCIANISM

be done in a time when nothing can be done results in failure." Lacking opportunity to succeed, "silent and alone I keep to my Supreme Mystery."[19]

Yang was forced to make unpalatable choices. What political convictions underlay his allegiances and withdrawals?

The plain sense of his writings, as well as the veiled allusions and subtle ironies that commentators have judged so variously, indicate that Yang was above all a conservative critic of contemporary abuses, and saw himself as loyal to the Han. He was no dissident. He avoided political involvements "in an age that did not favor virtuous action." We believe that he favored the reform of Han rule rather than a new dynasty, even under a vigorous monarch, because he was consistently predisposed toward continuation rather than upheaval. He was concerned above all for the stability of fundamental social institutions and obligations. He saw the mutually beneficial subordination of subject to ruler as analogous to the dynamic relation of the myriad phenomena to the Supreme Mystery, the Way.[20]

The fact remains that Yang publicly praised Wang Mang's *fait accompli*. But that was in Yang's old age, when he had just been restored to office after having been under vehement suspicion of disloyalty. This episode furnishes no warrant for rejecting the Mystery as the work of a toady. The many desperate recantations during the Great Proletarian Cultural Revolution of what had until then been virtues inspire sober reflection on this issue.

Philosophical Background

"Hsuan" carries a range of meaning from "black" to "darkness" to "hidden" to "mystery." Its overtones are stillness, solitude, isolation, nondifferentiation, and inaccessibility by purely rational processes. In Chinese thought the ideas at the philosophical end of this range bear no unpleasant connotations. They express that aspect of experience that can be known only by quiet and deep contemplation, or by illumination. Yang Hsiung uses *hsuan* in his book's title and throughout to mean the profound darkness, silence, ambiguity, and indefiniteness out of

---

19. HS, *87B:* 5119, 5122–5123, trans. Knechtges 1976: 97–101.

20. On Yang's despair about the politics of his time, see the HS biography, esp. *87B:* 3584, 3587. He expresses his views on stability, for instance, in Heads 51, "Constancy," and 53, "Eternal" (N, 314, 323). Knechtges sees even Yang's notorious memorial in praise of the Hsin dynasty as apolitical in its thrust, and more complimentary than critical toward the Han. Its "major theme is not that . . . Hsin [is] especially virtuous, or even that the material achievements of Wang Mang are worthy of praise. All of these points of course are made, but they are secondary to the espousal of a kind of classicism, in which the classics . . . are viewed as the embodiment of all ethical principles" (1978: 251).

III

which creation comes. In cosmogony it is the undifferentiated state out of which yin-yang and eventually the myriad phenomena separate. In Nature as humans experience it, it is the latency out of which individual things are spontaneously born, and out of which events shape themselves. In the sage—that is, the human being as he should be, as the student of the Mystery is striving to be—it is the spiritual inwardness that precedes conscious decision and action and spontaneously accords them with natural process. It is, in other words, the creative aspect of the Way wherever it is manifested. It is described in the famous opening passage of the *Lao-tzu*, which we translate in accordance with the interpretation attributed to Yang's teacher Chuang Tsun:

> The way that can be told is not the common way
> The name that can be named is not the common name
> What has no name is the beginning of heaven and earth
> What has a name is the mother of the myriad creatures
> Those without desires contemplate its secrets
> Those who have desires contemplate its periphery
> These two emerge together, but differ in name
> Being together, they are called Mystery
> Mystery upon mystery
> Gateway to the myriad secrets.

Although it would be unrealistic to expect general agreement on the meaning of this poem, most of those who take it seriously as philosophy discern the mystic Way in two aspects. One is the ineffable fountainhead, outside and prior to nature. The other is the immanent process that forms things and events. Compounding these two mysteries is that of their commonality, the never-broken connection between the change we see, the natural processes that produce orderly change, and the unchanging ground of all process. These lines, like the rest of the *Lao-tzu*, apply equally to the cosmos and the heart and mind of the sage.[21]

It is from Lao-tzu's Mystery that that of Yang derives, although his moral stance differs: "As for Lao-tzu's discussion of the Way and its power, I have drawn

---

21. Chuang Tsun apparently compiled two commentaries to the *Lao-tzu*. The second half of the longer one, *Lao-tzu chih kuei* 老子指歸, is preserved in the *Cheng-t'ung Tao tsang* (TT 376–377, S 693). We cite the reconstituted critical ed. of Chuang's whole *Lao-tzu* text in Shima Kunio's 島邦男 remarkable *Rôshi kôsei* 老子校正 (Tokyo, 1973), p. 55; see also N, 3. It is impossible to reconstruct entirely the text Chuang used. The surviving text differs slightly from the usual *Lao-tzu*, e.g., in the omission of *ch'ang* 常 from lines 5 and 6. Chuang's understanding of the first two lines is idiosyncratic but in keeping with his philosophy of withdrawal. Our translation is to some extent modelled on that of D. C. Lau, based on later commentary traditions: *Chinese Classics. Tao Te Ching* (Hong

III

THE FIRST NEO-CONFUCIANISM

upon it; but from his rejection of Good (jen 仁) and Right (i 義), his elimination of ritual and study, I have taken nothing."[22]

The first lines of the "Evolution" commentary echo the ideas that begin the Lao-tzu:

The Mystery of which we speak in hidden places unfolds the myriad species without revealing a form of its own. It fashions the stuff of Emptiness and Formlessness, giving birth to the regulations. Tied to the gods in Heaven and the spirits on Earth, it fixes the models. It pervades and assimilates past and present, originating the categories. It unfolds and intersperses yin and yang, generating the ch'i 氣. Now severed, now conjoined [through the interaction of yin and yang ch'i, the various aspects of] Heaven-and-Earth are indeed fully provided![23]

Yang's Mystery, like that of Lao-tzu, bridges the gap (in both cosmos and consciousness) between the inexpressible and the concrete. The imagery of this passage is explained in the "Diagram" commentary: "The Way of Heaven is a perfect compass. The Way of Earth is a perfect carpenter's square. The compass in motion describes a complete circle through the sites. The square, unmoving, secures things [in their proper place]. Circling through the sites then makes divine light possible. Securing things then makes congregation by types possible. . . . Now the Mystery is the Way of Heaven, the Way of Earth, and the Way of Man."[24] The Mystery includes not only the yin matrix of creativity but its yang impetus toward form. This idea Yang has added (or at least made explicit), as he has added a typically Han concern with ch'i, the energy or vitality that shapes everything.

For the authors of the Lao-tzu, life is best lived by avoiding structures and obligations that impede access to the Way. In the ideal society individuals interact without demands or sentiment. It was only "when the Great Way declined that Good and Right arose." But Yang, in reply to a question about the archaic Golden Age that "was in good order without models or laws," speaks of the sage's abhor-

---

Kong, 1982), pp. 3, 267.

22. Fa yen, 4: 10. Ma Tsung's 馬總 (d. 823) collectanea I lin yü yao 意林語要 preserves a comment attributed to Yang in the lost Yü Fan 虞翻 (164–233) recension of the Mystery: "Confucius is a sufficient [guide] to human culture; Lord Lao [i.e., Lao-tzu] is a sufficient [guide] to the Mystery" (Ming Chia-ch'ing ed., microfilm of old National Peiping Library rare book collection, no. 248 (7), 3: 8b).

23. T, 7: 5b (p. 1018; N, 429).

24. T, 10: 1b (p. 1032; N, 458); see also the beginning of the "Revelation" commentary, 10: 3b (p. 1034; N, 461).

rence of great chaos.[25]

In emphasizing the immanent and formative aspects of the Mystery, Yang has made a fundamental shift toward Confucian ideals. The Mystery can manifest itself only when society realizes the potentiality of individuals through distinctions in rank and function, reinforced by ritual precepts, sumptuary regulations, and a penal code. Yang not only upholds the need for the Five Constant Relationships *(wu ch'ang* 五常), but stresses those of father to son and ruler to subject, which the *Lao-tzu* condemns. In Yang's thought even the central notion of *wu wei* 無爲 (non-purposive activity which does not interfere with the Way) has come to mean "action suited to one's position in time."[26]

Although the *Lao-tzu* provides Yang with most of his mystical images, the Confucian Five Canons is his inexhaustible font of cosmic and moral wisdom:

> For discussing Heaven, there is no more discerning language *(pien* 辯) than that of the Changes. For discussing events, there is no more discerning language than that of the Documents. For discussing the embodiment of virtue, there is no more discerning language than that of the Rites. For discussing intent, there is no more discerning language than that of the Songs. For discussing inherent patterns *(li* 理), there is no more discerning language than that of the Spring and Autumn Annals. If these [scriptures] are excluded, discerning language is [wasted upon] petty subjects.

Yang similarly assimilates Confucius himself to the Mystery. The greatest of sages makes it possible for his disciples to "daily hear what cannot be heard, and see what cannot be seen."[27] One who has learned through the Master to appreciate the fundamental unity of the Way and the multiplicity of its manifestations is ready to become a full partner in the triad of heaven, earth, and man.

Yang departs from his Confucian models in ways that influenced contemporary trends. First, like most original thinkers of his time, he is openly eclectic, finding support for the canonical teachings in the "Hundred Schools" of Warring States thought, and drawing on the *Lao-tzu* and *Chuang-tzu* for orthodox purposes (see note 15 above). The eclecticism accelerated with the eclipse of Confucian orthodoxy in the Eastern Han and Six Dynasties. Second, he adapts to his philosophic discourses the rhythmic cadences, richly descriptive language, and multivalent meanings peculiar to the Han prose-poem or rhapsody *(fu* 賦). Thus Head 44, "Stove" *(tsao* 竈), uses the image of an empty stove. By analogy with the

---

25. *Lao-tzu*, 18; *Fa yen*, 4: 10.
26. *Fa yen*, 4: 10-11, and *passim* there and in T.
27. *Fa yen*, 7: 19; *11*: 33.

Changes hexagram 50, "Cauldron" *(ting* 鼎), this implies empty, i.e., undeserved, reputation. The stove lacks firewood *(hsin* 薪, a word the extended meaning of which is "official salary" and the synonym of which is *ts'ai* 材, meaning both "lumber" and "talent"). The complex beauty of the Mystery's language, no less than its philosophic power, insured its transmission for ten centuries, when it seems to have been used little if at all for divination.[28] Third, Yang systematically applies contemporary theories regarding the interplay of yin-yang and the Five Phases in his reinterpretation of the Changes, as we will show below.

Arrangement of the Book

The structure of the Canon of Supreme Mystery is best understood by comparing it with that of the Book of Changes. By the first century B.C. the latter consisted of a set of sixty-four texts, each associated with a six-line diagram in which, noting the result of a divination, each line might be solid or broken (considered yang if solid and yin if broken). Under each hexagram and its associated judgment text there are six associated texts. Each corresponds to one line, and is read if the polarity of the line is changing rather than stable. The Ten Wings relate these archaic texts to the moral, cosmological, and epistemological convictions of their authors, who were shaping a new orthodoxy around Confucianism.

In the Canon of Supreme Mystery the corresponding elements were created simultaneously by a single author. There are eighty-one four-line diagrams ("tetragrams"). Yang originated a method of divination with yarrow stalks (see the next section) in which manipulation of thirty-three sticks provides three possibilities for each line rather than the *I ching*'s two. The three were recorded as an unbroken line, correlated with heaven, a line broken once like the yin line of the Changes, representing earth, and a line broken twice, symbolizing man in his triadic relationship with heaven and earth, intermediate between them. Four such lines—that is, four repetitions of the divination procedure—provide eighty-one (3^4) possibilities, a number of the same order of magnitude as the sixty-four (2^6) of the Changes (three lines would correlate with only twenty-seven texts, and five would require 243). Yang's four lines were read from top to bottom rather than from bottom to top as in the Changes. They were associated with a nest of divisions that are at the same time geographic and social:

---

28. It was used for divination in Yang's own time. See, for instance, the anecdote in which Wang Mang received similar oracles from the Changes and the Mystery; F, *8:* 5a (p. 351). Some of those who studied it as philosophy also used it for divination, e.g., the learned southern statesman Lu K'ai 陸凱 (ca. 198–269); see his biography in *San kuo chih* 三國志, "Wu chih" 吳志, *61:* 1400.

III

3 regions *(fang* 方)   27 departments *(pu* 部)
9 provinces *(chou* 州)   81 families *(chia* 家)

The single Supreme Mystery stands for the cosmos as a whole. It occupies the center of the universe and the political realm, as the emperor does, where the three regions of heaven, man, and earth come together. Each of the three regions is divided into three provinces, to correspond to the ideal nine of the Han empire, and each of those into three departments, corresponding to the Han sub-provincial level. The ultimate eighty-one families stand for the multiplicity of individual phenomena in society and Nature.[29]

Each tetragram is associated with a "Head text" *(shou* 首) set out in three parts, a title, an image that refers to yin-yang, and an image related to the "myriad phenomena" or "all things" *(wan wu* 萬物) of the natural order. The title of the tetragram, a single graph, names one aspect of the comprehensive Mystery, such as "Measure" *(tu* 度, Head 52), and "Eternal" *(yung* 永, Head 53), to which humans respond for good or ill. The next line describes in poetic language the evolution of yang or yin *ch'i* during that precise phase in the annual cycle. The remainder of each text describes the effect of that evolution upon the phenomena of Nature. Each Head (by which we mean tetragram and texts together) is associated with a stretch of four and a half days in the cycle of the year. The first forty-one texts, between the winter and summer solstices, speak either first or exclusively of the ascendant yang *ch'i*, while the last forty detail the waxing of the yin *ch'i*. Read in sequence, they provide a remarkable picture of the finely graded steps of cyclic change. Each of the eighty-one Heads is linked to one of the sixty-four hexagrams of the *Changes* (with some duplication, of course) to evoke the old meanings and associations.

For each Head Yang provides nine "Appraisals" *(tsan* 贊) loosely patterned upon the line texts of the Changes. The Appraisals differ from the line texts in ways that increase the flexibility of interpretation when the book is used for divining. In the Changes, as understood in the Han, each line text is tied to the shift in polarity of one line in a hexagram (see p. 25 below). For this fixed correspondence Yang substitutes the point-counterpoint relationship of all the Appraisals to the Head text, whose cosmological theme they link to changing situa-

---

29. The series 1–3–9–27–81 comes from *Li chi* 禮記, which speaks of the one Son of Heaven, the three dukes, the nine ministers, the twenty-seven counsellors, and the eighty-one attendants (5/10, repeated in 44/8, where the emperor is authorized wives and concubines of various classes in the same ratio; citations of this form refer to texts used in the Harvard-Yenching Sinological Index Series). Yang makes this derivation explicit in the "Illumination" commentary, 7: 8b (p. 1021; N, 435).

III

tions. By freeing the Appraisals from the individual lines of his tetragrams (which have their own protocols of interpretation), Yang directs the inquirer's attention to a more capacious relationship, the effect of eternal cosmic patterns upon the changing circumstances that prompted divination. The Appraisals bridge the dominion of fate and the fields of choice and achievement.

On the one hand, the Appraisals, like the Heads, are correlated with the year, with yin and yang, and with the Five Phases. Each Appraisal, as one-ninth of a Head, represents half a day, with alternating Appraisals designated day and night.[30] Two of the Appraisals are not assigned to a Head, but make up the deficiency of $\frac{3}{4}$ day between the $364\frac{1}{2}$ days of Yang's basic structure (81 Heads × $4\frac{1}{2}$ days) and the $365\frac{1}{4}$ days in the solar year. Through their association with night and day, Appraisals come to be considered yin (and in some sense inauspicious) and yang (and therefore auspicious) by turns, with the first Appraisal yang in odd-numbered Heads (which are said to belong to a yang family) and yin in even-numbered Heads. To each Appraisal is also assigned in turn a direction that aligns it with one of the Five Phases.

On the other hand, Yang links the Appraisals to the act of divination. One reads them, unlike the line texts of the Changes, according to what time of day one uses the book. Yang connects the Appraisals to the individual's situation in three additional ways. First, they pertain to successive stages in the situation. The first three Appraisals describe its commencement; Appraisals 4–6, its culmination, usually centered on the fifth; and Appraisals 7–9, its decline. Second, they mark stages in the inquirer's response to the situation, categorized as Reflection (*ssu* 思, the period that precedes action), Felicity (*fu* 福, fruitful activity), and Calamity (*huo* 禍, failure when action is taken too late to succeed), as in this schema:

RESPONSE		SIGNIFICANCE	
Reflection	1 = interior	2 = middle	3 = exterior
Felicity	4 = small	5 = medium	6 = great
Calamity	7 = nascent	8 = median	9 = maximum

---

30. Here we follow Ssu-ma Kuang rather than Fan Wang. Fan oddly reverses the yin/yang values assigned to the alternating Heads and Appraisals in *chüan* 3 and 4 only, identifying yang with even and yin with odd. The reversed yin-yang values are at odds with both the context and Yang's own statements in his "Evolution" commentary. The "Diagram" commentary, *10:* 1b ff. (p. 1032; N, 459–460), sums up the numerological associations of the Appraisals. On the supernumerary appraisals, see note 42 below.

Third, they reflect a symmetric hierarchy of social ranks. The fifth Appraisal is reserved for the ruler, as in Han commentaries to the Changes. The flanking Appraisals, the fourth and sixth, carry implications for his ministerial attendants. The first and ninth Appraisals, those furthest from the son of Heaven, pertain to the "small man."

These correspondences interact to determine which are pertinent to a given situation, and in what way. For example, the relation of yin-yang association to the time of divination determines the prospects for the short, middle, and long term of the situation queried. The criterion is whether the yin-yang characters of the Head and those of each of the three relevant Appraisals are the same (auspicious) or different (inauspicious). This schema shows these time-bound significances of the Appraisals; same is shown as +, and different as – :

TIME OF DIVINATION	FAMILY OF TETRAGRAM	LINES READ	COMMENCEMENT	CULMINATION	DECLINE
Morning	Yang	1, 5, 7	+	+	+
	Yin		–	–	–
Evening	Yang	3, 4, 8	+	–	–
	Yin		–	+	+
Median	Yang	2, 6, 9	–	–	+
	Yin		+	+	–

Consider a divination carried out in the evening, the result of which is an odd-numbered or yang Head. This result corresponds to the third line of the table.[31] Appraisals 3, 4, and 8 would be read because of the time. The first is odd-numbered and thus yang. Its presence in a yang Head makes the outcome auspicious for "commencement," with which the first of the three Appraisals is concerned. The same reasoning makes Appraisals 4 and 8 (yin because even-numbered) inauspicious. Considering them in sequence, the indication for the beginning of the situation is fortunate, and those for its culmination and decline unfortunate.

Good and bad tidings, like all yin-yang orientations, are relational, not abso-

---

31. Yang's "Numbers" commentary, esp. *8:* 1b–2a (pp. 1023–1024; N, 438–445), gives the table's relationships in schematic form. The "Illumination" commentary, *7:* 8a–9b (pp. 1021–1022; N, 434), outlines general principles. For a clear explanation see the *Shuo hsuan* 說玄 (809) of Wang Ya 王涯, appended to F, esp. sec. 4 (pp. 452–453; for the date of this work, p. 444). On the median period, see below, n. 37.

lute. In the value system of the Mystery, a yang affiliation is in principle auspicious. In practice one must evaluate it alongside other factors. A yang entity in conflict with a yin entity may be baleful, and two yin entities in accord may presage good fortune. Since no single factor such as a yin-yang orientation absolutely determines events, Yang made several Appraisals conflict with the relations in the table. What outweighs all else in the outcome of an uncertain situation is action based on individual integrity. "The noble man is inwardly upright, and outwardly compliant, always humbling himself before others. That is why the outcome of his actions is good fortune and not calamity."[32] Yang Hsiung did not mean divination to be an objective science of forecasting. Combining subtle reasoning on cosmic trends with sensitivity toward social and individual circumstances—in other words, making a synthesis of heaven, earth, and man—is a highly skilled art.

In addition to the basic text—the eighty-one Heads and the seven hundred and thirty-one Appraisals—Yang provided ten commentaries analogous to the Ten Wings of the Changes. The "Fathomings" expand upon or explain one or more aspects of each Appraisal. Since the time of the scholiast Fan Wang 范望 (fl. ca. 265), the Fathomings, unlike the other commentaries, have been dispersed in the main text, each following the Appraisal to which it refers (the *Chou i*'s corresponding Commentary on the Images had been similarly dispersed under each hexagram a little earlier).[33] The "Elaboration" commentary discusses only the First tetragram. Yang makes it, like the "Elaborated Teachings" *(Wen yen* 文言) commentary of the Changes, an epitome of the entire book. The remaining commentaries do not interpret individual texts, but assess or illuminate the canon as a whole. On the next page we list all ten in order, with the corresponding *Chou i* Wings.[34]

---

32. Examples of Appraisals that do not accord with the table are Head 1, Appraisal 9, 6/6, 9/9, 20/5, and 23/7; see N, 94, 123, 139, 184, and 196. The exceptions in 1/9 and 9/9 are discussed by Wang Ya in *ibid.*, section 2 (pp. 449–450). The quotation is from the "Illumination" commentary, 7: 9b (p. 1022; N, 437).

33. T'ang Yung-t'ung 1947: 135–138.

34. Ssu-ma Kuang notes correspondences between commentaries of the Changes and the Mystery in an introductory essay to the latter omitted from the 1909 edition; see the TT version, prefatory section, pp. 6b–7a. Yang defines the titles of his commentaries in "Representations," 9: 3a (p. 1030; N, 453–454).

III

COMMENTARY	TRANSLATION	CH.	TEN WINGS COMMENTARY
1. Hsuan ts'e 玄測	Fathomings		3-4. Hsiang chuan 象傳 (Images)
2. Hsuan ch'ung 玄衝	Polar Oppositions	7	9. Hsu kua 序卦 (Sequence)
3. Hsuan ts'o 玄錯	Evolution	7	10. Tsa kua 雜卦 (Interplay of opposites)
4. Hsuan li 玄攡	Evolutions	5-6	Hsi tz'u 繫辭 (appended judgments, "Great Commentary")
5. Hsuan ying 玄瑩	Illumination	5-6	Same
6. Hsuan shu 玄數	Numbers	8	8. Shuo kua 說卦 (discussion of the trigrams)
7. Hsuan wen 玄文	Elaboration	9	7. Wen yen 文言 (elaborated teachings)
8. Hsuan i 玄掜	Representation	9	5-6. Hsi tz'u
9. Hsuan t'u 玄圖	Diagram	10	Same
10. Hsuan kao 玄告	Revelation	10	8. Shuo kua

Method of Divination

The procedure described in the Numbers commentary is a point-by-point modification of that for the Changes as given in its Great Commentary. There are thirty-six yarrow stalks in the *T'ai hsuan* set, of which three are set aside. Yang does not explain clearly why this is so. Later students of the Mystery agree that the three supernumerary sticks correspond to the basic triad of Heaven, Earth, and Man, as the one stick set aside from the *Chou i* set of fifty represents cosmic unity. An additional stick is taken into the left hand, and the remaining thirty-two divided at random into two piles (in the *Chou i* method the one stick, which represents man, is taken up from the right-hand pile after the division). The left and then the right pile are counted off by threes (rather than fours as in the Changes), and the remainder (one, two, or three sticks) added to the stick in the left hand. Using the twenty-seven or thirty sticks that remain, the segregation of one

THE FIRST NEO-CONFUCIANISM

stick, division, and counting off are repeated. Twenty-seven, twenty-four, or twenty-one sticks will remain. Dividing by three, the result will be 9 (equivalent to the twice-broken line of Man), 8 (the divided line of Earth), or 7 (the solid line of Heaven). The first line of the tetragram has now been determined. Three repetitions of the entire procedure complete the tetragram, which directs the user to the appropriate Head text.[35]

In the Changes a "mature" yin or yang line determined by the extreme numbers 6 or 9 (rather than 8 or 7) is considered moving—that is, about to change polarity. A particular divination may yield no such lines or as many as six. One reads the texts corresponding to moving lines and applies them to the question divined about. The diviner may also evaluate a new hexagram, with moving lines changed to their opposites.[36]

Yang's approach, as we have noted in discussing the significance of the Appraisals, was quite different. If the act of divination is carried out in the morning, Appraisals 1, 5, and 7 of the given Head are read and considered; if in the evening, Appraisals 3, 4, and 8; if at "median times" (chung 中), Appraisals 2, 6, and 9.[37] Yang does not specify these periods of time more definitely. We do not know

---

35. Yang outlines the divination procedure in a form reminiscent of the Great Commentary to the Changes in the "Numbers" commentary, 8: 1a–1b (p. 1023). F. van der Blij demonstrated that the results of divination with the Book of Changes are not equiprobable: "Combinatorial Aspects of the Hexagrams in the Chinese Book of Changes," *Scripta mathematica*, XXVIII (1966), 37–49.

36. Such evaluations occur in texts dated as early as 661 B.C. (*Tso chuan* 左傳, Min 1/appendix), although it is unlikely they were recorded then. On early interpretations of the Changes see Hellmut Wilhelm, "I-ching Oracles in the *Tso-chuan* and the *Kuo-yü*," *Journal of the American Oriental Society*, 79 (1959), 275–280; Li Ching-ch'ih, "Tso Kuo chung *I* shih chih yen-chiu 左國中易筮之研究," *Ku shih pien*, III, 171–187, reprinted in Li 1978: 407–421; and Kao Heng, "'Tso chuan' 'Kuo yü' ti 'Chou i' shuo t'ung chieh '左傳' '國語' 的 '周易' 說通解," pp. 70–110 in *Chou i tsa lun* 周易雜論 (Jinan, 1962; reprint, 1969). Shih-chuan Chen, "How to Form a Hexagram and Consult the I-ching," *Journal of the American Oriental Society*, 92 (1972), 237–249, covers similar ground; but cf. Kao, *Chou i ku ching t'ung shuo* 周易古經通說 (Beijing, 1958; reprint, Hong Kong, 1968), pp. 112–130.

37. It is likely that Yang meant one of the two median periods to correspond to what Han time reckoning designated *tung chung* 東中, *jih chung* 日中, and *hsi chung* 夕中 (roughly 9 A.M. to 3 P.M., between *tan* and *hsi)*, and the other to the central three of the five night watches, the length of which varied through the year. This nomenclature was used at least from the time of the Martial Emperor of the Han to the beginning of the Eastern Han. See Ch'en Meng-chia 陳夢家, "Han chien nien-li piao hsu 漢簡年曆表

whether he meant by the median times the afternoon or the periods centered about noon and midnight. Early users of the book apparently were not consistent.

Like any other Chinese divination procedure, this one can succeed only if someone in the correct spiritual state carries it out. The inquirer's mind must be correctly oriented *(chen* 貞 or *cheng* 正*)*. The yarrow stalks will yield no useful result if he lacks integrity *(or* sincerity, *ch'eng* 誠), the quality that unites the individual with the cosmic order.[38] Divination is essentially a communion, which the yarrow stalks can only facilitate.

According to the *T'ai hsuan* there are two prerequisites for communion with the transcendental Way. The first is a genuine will to approach the Mystery. The second is single-minded devotion to living its attributes.

Of the first prerequisite, Yang Hsiung writes in a passage reminiscent of the *Analects,* "Whoever would draw near to the Mystery, the Mystery for its part draws near to him." One emulates the cosmic Way, as a child its parents: "The Sage . . . would match his body with Heaven-and-Earth, aim for the numinosity of the gods, push his transformations to the limit with yin and yang, and participate in the integrity of the four seasons. Contemplating Heaven, he is Heaven; contemplating Earth, he is Earth; contemplating the divinities, he is divine; contemplating time, he is timely."[39]

The Sage achieves identity with the cosmic Way by single-minded concentration *(ching* 精) on virtue—a discipline as much spiritual as intellectual.[40] "If the noble man daily strengthens what is deficient in him [i.e., the good], and eliminates what he possesses in surplus [i.e., the bad], then the Way of the Mystery is nearly approximated indeed!" He refines his innate powers until they are perfectly attuned to those of the creative Mystery: "When one divines with single-minded concentration, the gods prompt the changes [that reveal an answer to the inquiry]. When one deliberates [on this response] with single-minded concentration, one's plans are appropriate. When one establishes what is right with single-minded concentration, no one can overturn it. When one maintains [one's principles] with single-minded concentration, no one can snatch them away." Yang Hsiung's other requirements also emphasize the sacred character of the divination process: "The

---

敘," *K'ao-ku hsueh-pao* 考古學報, 2 (1965), 103–149, esp. pp. 117 ff.

38. On the notion of *ch'eng,* see Wing-tsit Chan, *A Source Book in Chinese Philosophy* (Princeton, 1963), p. 96.

39. T, 7: 7b (p. 1020) and 9: 2b (p. 1029). The first quotation alludes to the *Lun yü,* 13/7/30. For translations of quotations in this paragraph and the next, see N, 29.

40. T, 7: 6a (p. 1019; N, 430), 8b (p. 1021; N, 435; cf. 10: 3b, p. 1034), and 8: 1a (p.

III

THE FIRST NEO-CONFUCIANISM

Way of divination consists in this: If you have not attained single-minded concentration, do not divine. If the issue is not in doubt, do not divine. If [your plan is] improper, do not divine. If you will not act in accordance with the outcome [of divination], it is exactly as if you had not divined."

Interpretation

Drawing upon the elaborate correspondence schemes of Han mutationists, Yang built a coherent and well-wrought system for determining meanings. As he put it in the "Numbers" commentary, "there are four ways to interpret the result of divination: through stars, times, numbers, and phrasing."

Yang does not explain what he means by these terms, merely remarking "if the result of divination is fortunate, [the inquirer] will meet with yang; times, numbers, and phrasing will be in accord. If unfortunate, [the inquirer] will meet with yin; stars, times, numbers, and phrasing will be in discord." If they are to include all the major components from which Yang has structured meaning, "times" must refer to the temporal associations of the Heads and Appraisals, and their relation to the time of divination; "numbers," to numerological significances, especially those of lines within the tetragrams; and "phrasing," to the verbal meanings and implications of the Head and Appraisal texts. "Stars" for a thinker of Yang's time would have meant an elaborate system of correspondences—astronomical, physical, even musical—that he fully incorporated in the Mystery and enumerated in the "Numbers" commentary.[41]

The correlation of Heads and Appraisals with stars (and to some extent with times and numbers) is based on the correspondence of the tetragrams and Head texts to equal divisions of the annual cycle. The beginning of the book corresponds to the Grand Inception *(t'ai ch'u* 太初*)* as defined in the calendar reform of the same name in 104 B.C.[42] It amounts to a new beginning of time, a midnight

---

1023; N, 29).

41. T, *8:* 1a–1b (p. 1023); *T'ai hsuan pen chih, 8:* 4a. See also F, *8:* 1b–2a (pp. 344–345).

42. T, *6:* 12a (p. 1013; N, 421). On the Grand Inception calendar reform see Vol. I, Chap. II, and Christopher Cullen, "Motivations for Scientific Change in Ancient China: Emperor Wu and the Grand Inception Astronomical Reforms of 104 B.C.," *Journal for the History of Astronomy,* 24 (1993), 185–203. Yang does not specify whether the 730th Appraisal, Deficit *(chi* 踦*)*, or the 731st, Surplus *(ying* 嬴*)* corresponds to the one-quarter day that makes up the total of 365¼ days. Fan Wang does not state his opinion on this issue (F, *6:* 24b–25a, pp. 300–301), Ssu-ma Kuang (in T) takes Surplus as the Appraisal with the shortest time span (probably because it comes last), and Suzuki designates Deficit (probably because of its title; 1963: 83).

III

that simultaneously marks the winter solstice (the beginning of the tropical year), the conjunction of sun and moon (first day of the lunar month), and the beginning of the sixty-day cycle. Each of the tetragrams and its Head text represents four and a half days of the year counted off from this epoch (so that the forty-second is the summer solstice). All except one of the seven hundred and thirty-one Appraisals (nine per Head plus one of the two not associated with a head) is associated with half a day in the round of the year.

From this equipartition a great array of correspondences follows. At the winter solstice the sun was by convention located in the first degree of the lunar lodge Drawn Ox. Since there were 365¼ Chinese degrees *(tu* 度) in a circle, the sun moves one degree per day. Each Appraisal applies to an expanse of precisely half a degree, and each Head to four and a half degrees. The twenty-eight lodges are not of equal extent. The second lodge, Serving-maid, comes into play after the eight degrees of Drawn Ox have been assigned, namely at the eighth Appraisal of the second Head, Circuit *(chou* 周). Serving-maid, twelve degrees wide, is succeeded by Tumulus, ten degrees in width, at the fifth Appraisal of the fifth Head, Small *(shao* 少). These stellar correspondences, each with its astrological implications, continue in this way through the round of the sky and the length of the book. Analogously, since the first Head corresponds to the beginning of Drawn Ox at the northmost point of the celestial equator, all the associations of the phase Water and of extreme yin come into play. Another fund of metaphors associated with the annual cycle is the twelve-note gamut of mathematical harmonics, beginning with Yellow Bell *(huang chung* 黃鐘) at the winter solstice. The hours from midnight *(tzu* 子) on are similarly assigned to groups of tetragrams. The inquirer could thus call on a wealth of interconnected entities, each with its symbolic value, organically connected with every Head and Appraisal, and frequently alluded to in their texts. This rich matrix is what Yang means by "stars."

Wang Ya 王涯 (ca. 764-835) explains what Yang Hsiung must have meant by "times": "This refers to whether the time [associated with the result] of divination coincides with the time of year in which [the divination] takes place. For instance, if you divine on the winter solstice and the result corresponds to the tenth lunar month [which contains the solstice] or earlier, the Head is contrary; if it corresponds to a time after the winter solstice, the Head conforms." In other words, a Head whose time association falls after the month in which divination takes place is, all other things being equal, auspicious.[43]

---

43. F, Appendix, sec. 4 (p. 455). Suzuki 1964: 74-83 reconstructs in a detailed table the temporal and stellar correlations of each Head and Appraisal.

III

THE FIRST NEO-CONFUCIANISM

Why is it better to divine early than late? The last paragraph of the "Evolution" commentary defines "what is near the Mystery" as "what advances, but has not yet culminated; what has departed, but has not yet arrived; what has been emptied, and has not yet been filled." The user of the Mystery, if early, has time to adjust his conduct to the Way.

Structure

So many dimensions of meaning can converge on the inquirer's question only if the images and associations of the book are rich enough, and if a well-articulated structure makes them accessible. The Great Commentary and the other Wings brilliantly but speciously read into the original Changes text the fundamental patterns that underlie the realms of heaven, earth, and man. Yang Hsiung, on the other hand, made them the actual structure of his canon. His book applies rigorously and reflects, in its texts and guides to interpretation, the basic seasonal rhythms, the fundamental social relationships, and the functions of yin-yang and the Five Phases that pervade the natural and human worlds. Here we outline how this is accomplished, with special attention to unique features of the Mystery.

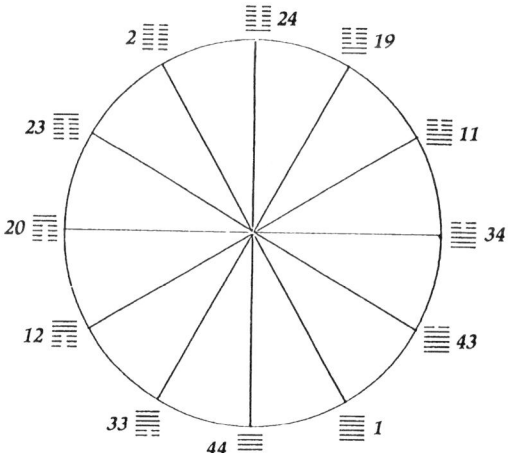

Figure 1. Ching Fang's "waxing and waning" order for twelve hexagrams. The figure is read clockwise.

In interpreting the results of divination with the Changes, the texts associated with the hexagram and moving lines were only one of several sources of meaning. It was also possible to reason from the associations of constituent trigrams and lines, both individually and in relation to the structure of the hexagram that contained them.

Fruitful though such modes of analysis were, Han scholars preoccupied with the question of timeliness were frustrated. Neither the content of the *Chou i* texts nor the order of the hexagrams are explicitly related to temporal sequence. The structure of successive hexagrams does not change in a gradual, regular way that could be interpreted to imply a time cycle. Ching Fang was trying to overcome this difficulty when he proposed his *kua-ch'i* 卦氣 technique for associating hexagrams *(kua)* with the twenty-four solar periods *(ch'i)* that evenly divide the tropical year. In this schema four "standard hexagrams" *(cheng kua* 正卦*)* correspond to the solstices and equinoxes and thus to the four cardinal points on the sun's path. Twelve other hexagrams rule the lunar months (Figure 1). The latter, the "waxing and waning hexagrams" *(hsiao-hsi kua* 消息卦*)*, begin with the pure yin hexagram The Receptive *(k'un* 坤*,* no. 2). It is assigned to the tenth civil month, which contains the winter solstice.

Month by month yang lines increase upward (Return, *fu* 復, no. 24, Approach, *lin* 臨, no. 19, and so on) until the pure yang hexagram The Creative *(ch'ien* 乾*,* no. 1) is formed in the fourth month to govern the summer solstice. Then yin lines multiply from the bottom upward (Coming to Meet, *kou* 姤, no. 44, Retreat, *tun* 遯, no. 33, and so on) until the pattern of The Receptive is restored at the end of the cycle. In pairs of hexagrams separated by six months (e.g., Approach and Retreat), if a given line's polarity is yang, the same line in its counterpart is yin, and vice versa. These twelve, and the remaining forty-eight, correspond to equal intervals of 6.0875 days (one sixtieth of 365¼). The four standard hexagrams, unlike the others, do not correspond to an interval, but rather to the most fundamental markpoints of the sun's annual round. Fixed by astronomical coordinates in space, they move back and forth in time. The sun may pass through one of them on whatever day of the lunar month the solstice or equinox occurs. In a given year each of these points and its standard hexagram may be associated with any one of five ordinary hexagrams whose intervals are distributed through the month. Despite this small inelegance, Ching's system was a workable solution to the problem of relating hexagram structure to time, as its popularity among students of the Changes demonstrates.[44]

Yang Hsiung rendered Ching Fang's approach to symmetry obsolete in the cyclic structure he was creating afresh. Unproblematic symmetries appear both in the content of the texts (where a principle of structure was most conspicuously

---

44. Ch'ü Wan-li 1970: 82–98 gives a detailed account of the *kua-ch'i* technique.

missing in the Changes), and in the construction of the tetragrams. Yang achieves his complex structure, encompassing both text and tetragram, by a combination of gradual cyclic change and artfully distributed opposition.

Cyclic evolution

Heads. All of the Mystery's Heads are assigned to periods of four and a half days. Yang avoids the complication of Ching Fang's *I* schema, in which only sixty hexagrams are involved, by beginning the cycle with Head no. 1 at the winter solstice. That point, not Ching's new moon of the tenth civil month, marked for the Chinese astronomer the beginning of the tropical year. Each text describes the waxing and waning of yin and yang, and their effect on "all things"—the phenomenal world—during each Head's brief dominion. The Head texts, read in sequence, constitute a minute, abstract description of eighty-one phases in the annual cycle, a qualitative graph in the form of a metaphysical prose poem.

This becomes apparent if we simply peruse a series of Heads. Here, for instance, are the first seven:

1. Center   *Chung* 中   N, 84
   Yang *ch'i*, unseen, germinates in the Yellow Palace. Good Faith in every case resides at the center.

2. Full Circle   *Chou* 周   N, 95
   Yang *ch'i* comes full circle. Divine, it returns to the beginning. Things go on to become their kinds.

3. Mired   *Hsien* 礥   N, 102
   Yang *ch'i* stirs slightly. Though stirred, it is mired in yin. "Mired" refers to the difficulty attending the birth of things.

4. Barrier   *Hsien* 閑   N, 107
   Yang *ch'i* is barred by yin. Mired fast, all things are barred.

5. Keeping Small   *Shao* 少   N, 113
   Yang *ch'i*, rippling, spreads through the deep pool. Things like ripplets in its wake can keep themselves very small.

6. Contrariety   *Li* 戾   N, 118
   Yang *ch'i*, newly hatched, is very small. Things, each diverging and separating, find their proper categories.

7. Ascent   *Shang* 上   N, 125
   Yang *ch'i* engenders things in a place below. All things shoot through the earth, climbing to a higher place.

Even without pausing over the details, which Nylan's complete translation explains, it is obvious that Heads 1 through 7 represent step by step the hesitant reawakening of yang energy against the opposition of yin. They alternate images of nascent activity (1, 3, 5?, 7) with reassertions of stasis (2, 4, 6). The phenomena, in the grip of yin, are not perceptibly affected until Head 6 subtly indicates that they have begun to respond to the push of yang. In Head 7 yang has begun to assert itself with unqualified force, which continues to grow until, after the spring equinox, it begins to wane, giving way to the growth of yin.

Looking at the whole sequence of eighty-one Heads, an overall principle of order becomes unmistakable: what Head texts half a year apart say is complementary, although never in a simple-minded way. This rule holds for all Heads, not only the especially significant ones considered above. In most cases the complementarity of language or image is explicit. Here, for instance, is the Head that governs the summer solstice, complementary to no. 1:

41. Response   *Ying* 應   N, 266
Yang *ch'i* culminates on high. Yin faithfully germinates below. High and low mutually respond.

In other instances the complementarity becomes clear when we consider each text in the flow of the series. The opposition is never that of static symbols, but rather of gradual, complex processes that the symbols evoke.

Appraisals. Ssu-ma Kuang's functional definitions of the Heads and Appraisals are parallel in form but significant in their differences:

> It is the Heads that make plain how heaven and earth employ the yin and yang *ch'i* in putting forth and gathering in the myriad creatures, and reveal the laws and rules [underlying seasonal change] to those [who consult the Mystery] . . . It is the Appraisals that reveal how the Sage accords with the order of [change in] Nature as he cultivates his person and governs his state, and reveal good and bad fortune to those [who consult the Canon of Supreme Mystery].

The Heads are a qualitative model of cosmic process, which the Appraisals apply to individual and political action, as well as to thought about one's personal and public future. Kawahara Hideki 川原秀城 makes an equally pertinent contrast between the two types of text: the aspect of permanence is mainly embodied in the Heads, and that of transformation in the Appraisals.[45] From a literary viewpoint, the Heads reflect the ideal Han image of the Changes as the Sage's magisterial view of reality, while the Appraisals echo the actual heterogeneity of the

---

45. T, *1:* 2a (p. 948); Kawahara, "Taigen no kôzôteki haaku 太玄の構造的把握," *Nihon Chûgoku Gakkai hô* 日本中國學會報, *30* (1978), 45–58, esp. pp. 55–57. "Putting forth and gathering in" *(fa lien* 發斂) is an astronomical term for various expansive and contractile effects of the seasonal yang and yin cycles on phenomena. These effects range from the cycle of vegetable growth to the anomalous motion of the sun.

archaic text. The Heads' formulaic, step-by-step snapshots of the state of yin and yang and the phenomena, described in language of relatively uniform poetic texture, provide a backdrop. Against it seven hundred and thirty-one gnomic images—fantastic or prosaic, patent or opaque, sere or allusive—succeed each other in every permutation of linkage. The images and language of the Appraisals often echo those of the Changes and other Confucian classics. Just as the Heads are designed to provide a framework for reflection on cosmic change, the Appraisals in their diversity, which suggests the multiplicity of human experience, constitute a vast repertory of metaphors as fruitful in divination as those of the *Chou i*.

So clear-cut a contrast cannot do justice to so sophisticated a book. In the Head texts, dynamic images are by no means rare. The Appraisals, like the Heads, embody stages of evolution and devolution, and reflect the alternation of yin and yang. Although cosmology is only one of their concerns, they make greater use of cosmic categories dependent on the Five Phases than do the Heads.

The composition of the Appraisals must have been a veritable Chinese puzzle. Not only do we find cycle fitted within cycle, but one clue after another ties the series of Appraisals to every microcosm and system of correspondence important in Han thought. In the first Head alone, the Fathoming of the first Appraisal asserts man's participation, alongside that of heaven and earth, in the Way. There are several reminders of the ages of man. The first Appraisal recalls the womb; the third, the entry of the young adult into an official career; the fifth, culminating accomplishment; the seventh, mature stability; and the ninth, natural death in old age. The symbology of administration appears in the apposition of punishment and virtue, and in allusions to attributes of the ruler (an exemplar in the fifth position) and his vassals (small men because they surround him in the fourth and sixth positions, which are yin). Correlates of Yin-yang and echoes of the Book of Changes abound. Images of darkness, moon, dragon, centering, and fire reflect the significance of each Appraisal in the succession of the Five Phases.

We have just seen that the text of the Mystery reflects directly the cyclic character of natural process (and of political and psychological process to the extent that they are "natural," that is, in accord with the Way). Yang's words describe an evolution, Head by Head and Appraisal by Appraisal. In doing so they reproduce the annual complementarities and symmetries of time and space.

Graphic counterparts of change
The sequence of tetragrams expresses the same evolution, and symbolizes the same oppositions, as the texts do. This conclusion emerges if we examine in order the eighty-one tetragrams. The unbroken line corresponding to heaven is determined when the result of divination is the number 7; the singly broken line correspond-

ing to earth, 8; and the doubly broken line corresponding to man, 9. For reasons that will shortly become clear, when Yang discusses these lines rather than the divination process, he in effect subtracts 6, and writes of them as 1, 2, and 3. For reasons that are Sinologically arbitrary but mathematically of the essence, we will subtract an additional 1 from each digit to yield 0, 1, and 2 respectively.

We transcribe the lines of the first seven tetragrams as 0000, 0001, 0002, 0010, 0011, 0012, and 0020. These are in fact the numbers to the base 3 that correspond to the decimal numbers 0 to 6 (or the ordinal numbers 1 to 7). The remainder of the tetragrams fall into the same sequence. Those that correspond to every tenth Head—0000, 0101, 0202, 1010, 1111, 1212, 2020, 2121, and 2222—correspond to the decimal numbers 0, 10, 20, . . . 80 or the ordinal numbers 1, 11, 21, . . . 81. The sequence of tetragrams is, in other words, simply the regular order of the integers 0 to 80, the first eighty-one numerals, written to the base 3. This is not a surprising conclusion, since the base n is determined by the 3 types of line, and the four lines per tetragram provide a total of n^4 (in this case, 81) possible values.

The tetragrams are thus, arithmetically speaking, a system of notation. They record an unbroken and regular progression, which the sequence of the hexagrams in the Changes does not. The latter, we have seen, fall into pairs the members of which are related by inversion, but the pairs do not follow a discernible order. The first six hexagrams, for instance, may be transcribed analogously in binary notation as 000000, 111111, 011101, 101110, 000101, and 101000, equivalent to decimal 0, 63, 29, 46, 5, and 40.[46] Ching Fang's "waxing and waning" order was a step in Yang's direction, but Ching systematically ordered only a portion of the hexagrams, and not in binary succession (Figure 1). The so-called "prior to the natural order" *(hsien t'ien* 先天) or Fu-hsi sequence rearranges all sixty-four in what can be considered the natural order of binary integers that correspond to decimal 0 to 63. It appeared long after Yang Hsiung's time, possibly as late as ca. 1080 (the date of the oldest extant version).[47] The origin of the Fu-hsi sequence is unknown,

---

46. Although the numbers 0 and 1 are often used by modern scholars wishing to relate the hexagram notation of the Changes to the binary system, the only numbers used by Chinese students of the *Chou i* to record the results of divination were 6, 7, 8, and 9. Use of 0 for notation would have been unthinkable, for it was not considered a yin or yang number; in fact thinkers before Yang Hsiung hesitated to use 1 in that way. See Bernard S. Solomon, "'One is No Number' in China and the West," *Harvard Journal of Asiatic Studies*, XVIII (1954), 253–260.

47. The conventional wisdom asserts that this order first appears in Shao Yung's 邵雍 usually unread *Huang chi ching shih shu* 皇極經世書. The 12-*ch.* version in TT 705–718 (S 1040), generally considered the earliest form of Shao's work, does not include

III

THE FIRST NEO-CONFUCIANISM

but in principle it is merely a translation of the Mystery's much earlier base-3 sequence into what amounts to a binary, six-digit order for the *Chou i* hexagrams (Figures 2, 3). It did not become the basis for an arrangement of the text, nor was it used in divination.

Although the general conception of multiple systems of integers (scales) generated by different bases (radices) did not appear anywhere until the eighteenth century, Yang was aware that his tetragrams constituted a count. As he put it, "Those who study the Changes must peruse the diagram *(kua* 卦) in order to tell [which text to consult], but students of the Mystery determine [the text] by counting *(shu* 數) the lines. The reason that the [tetragram corresponding to] each Head in the Mystery is fourfold is that it is not a diagram but a number *(shu)*."[48] Counting implies a base-3 value for each line in the tetragram. Using the geographical divisions listed above (p. 20), Yang assigns the number 27 ($=3^3$) to the Region, the top line; 9 ($=3^2$) to the Province, the second line; 3 ($=3^1$) to the Department, the third line; and 1 ($=3^0$) to the Family, the bottom line. These are in fact the respective powers of 3 that would be written 1000, 100, 10, and 1 in base-3 notation.

Yang saw the lines of the tetragram, determined by the outcome of divination, as a count. In the "Numbers" commentary he explains how to calculate the number of the Head from the value of each line in the tetragram: "If the Family line [i.e., the bottom line of the tetragram] is 1 [i.e., unbroken], count 1; if 2 [e.g., singly broken], count 2; if 3 [i.e., doubly broken], count 3. If the Department line is 1, do not add anything; if 2, add 3, if 3, add 6. If the Province line is 1, do not add anything; if 2, add 9; if 3, add 18. If the Region line is 1, do not add anything; if 2, add 27; if 3, add 54."[49]

The point of these instructions becomes clear if we apply them to an example.

---

a diagram showing the binary arrangement. One appears in the *Hsing li ta ch'üan* 性理大全 recension (1415), *ch.* 7, with a commentary by Shao's son Po-wen 伯溫 (1057–1134). Anne D. Birdwhistell, *Transition to Neo-Confucianism. Shao Yung on Knowledge and Symbols of Reality* (Stanford University Press, 1989), provides a general view of Shao and his work. For possible predecessors and other details see Kidder Smith, Jr., et al., *Sung Dynasty Uses of the I Ching* (Princeton, 1990), pp. 100–135. For a technical explanation of the "Fu-hsi" order see Martin Gardner, "Mathematical Games. The Combinatorial Basis of the 'I Ching,' the Chinese Book of Divination and Wisdom," *Scientific American,* Jan. 1974, 108–113. The silk MS *I ching* from Ma-wang-tui contains still another order, on which see Liu Dajun (trans. Edward Shaughnessy), "A Preliminary Investigation of the Silk Manuscript *Yijing,*" *Zhouyi Network, 1* (1986), 13–25.

48. HS, *87B:* 5124. The first *shu* is verbal (read in the third tone), the second nominal.

49. T, *8:* 5b (p. 1027; N, 28).

III

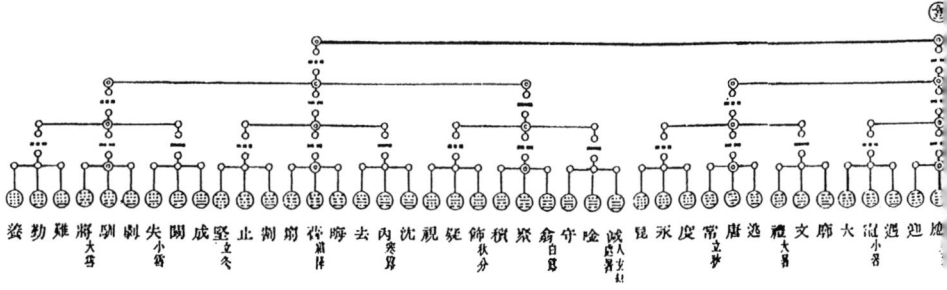

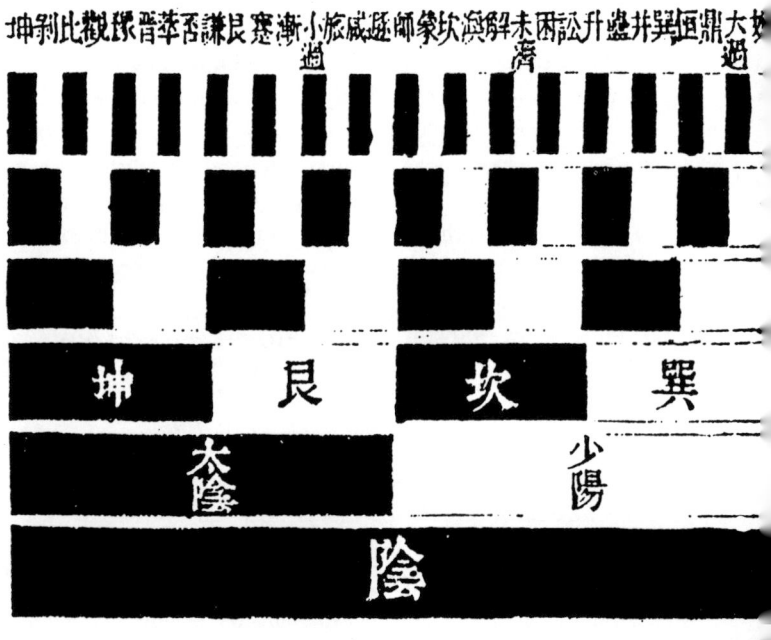

Figure 2 (above). Formation of the eighty-one
T'ai hsuan tetragrams by a series of four trine divisions.
This diagram reads downward. The space above is labeled "Mystery." In each case the four lines (solid, singly broken, or doubly broken) that result from each threefold division combine to give the tetragram corresponding to each Head. Below each tetragram is given the name of the Head, and for each three hexagrams, the corresponding *ch'i* period of the calendar. From Ch'en Pen-li, *T'ai hsuan ch'an mi*, "Hsuan t'u," pp. 9a–13b.

III

THE FIRST NEO-CONFUCIANISM

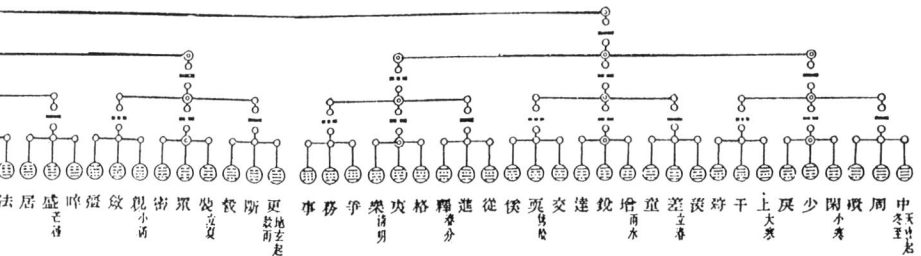

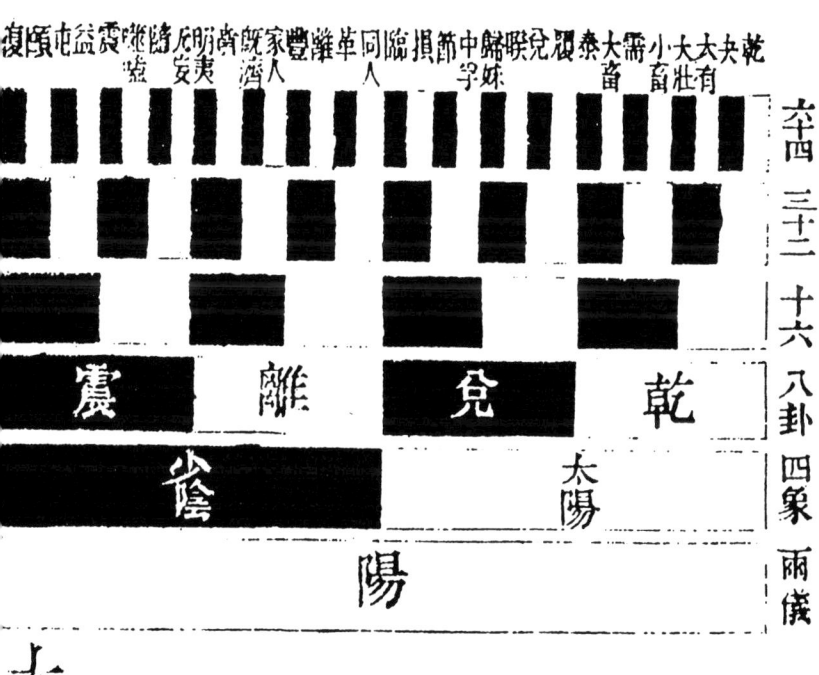

Figure 3 (below). Formation of the so-called Fu-hsi order of the sixty-four *Chou i* hexagrams by a series of six binary divisions, based on Shao Yung (ca. 1050). This diagram reads upward. The white space at the bottom stands for the totality of the "supreme ultimate" *(t'ai chi* 太極). Reading each column of six blocks upward, taking white as 0 and black as 1, gives a binary number between 0 and 63 corresponding to the hexagram named in the uppermost block. From *I t'u ming pien* 易圖明辨 (1700, Shou Shan Ko Ts'ung-shu 守山閣叢書 ed.), 7: 10b-11a.

III

Head 48, "Ritual," would be read as "2 Region, 3 Province, 1 Department, 3 Family," or 2313. Since the Family line's value is 3, we begin with 3. Working upward, we add to this 0, 18, and 27. The result is 48. We see that Yang's code, 2313, corresponds to the base-3 number 1202, equivalent to decimal 47 or ordinal 48. The value of each line in the tetragram is simply 1 higher in each place than the modern base-3 notation. That is why Yang diminishes each value by 1 except that of the Family line. He reads the latter directly because he is interested in the ordinal number of the Head rather than in the modern mathematician's number in the series beginning with 0. It is odd that Yang's explicit manipulation of 3-base numbers two thousand years ago has been ignored while so much speculation has been lavished on the tacit binary ordering of *Chou i* hexagrams, unconnected with calculation or even the divination process, more than a millennium later.

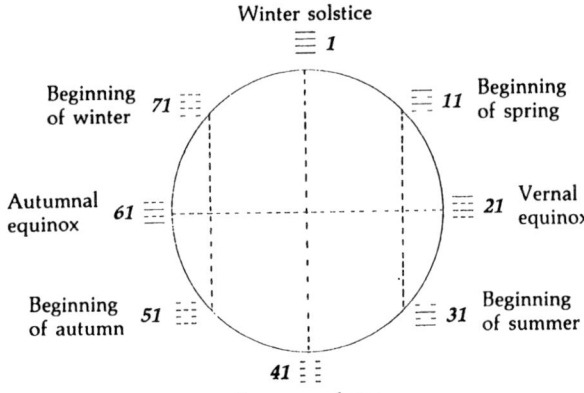

Figure 4. Tetragrams from the *T'ai hsuan ching* corresponding to eight major transitions of the solar year.

The tetragrams also exhibit seasonal symmetries analogous to those we have seen in the corresponding Head texts. In China the solstices and equinoxes are the midpoints, not the beginnings, of the seasons. The eight symbols that correspond to the beginnings of the four seasons and to the solstices and equinoxes are doublets (a two-line pattern repeated, e.g., 2121). This doubled form is significant to any student of the Changes, since it is analogous to the composition of eight especially important hexagrams (including Ching Fang's four standard hexagrams and two of the waxing and waning hexagrams), which are doubled versions of the trigrams that were given their names.

Figure 4 sets out relations of symmetry among the eight tetragrams. We see that the doublets fall into four pairs of tetragrams each of which is an inversion of the other: one pair corresponding to the solstices, one to the equinoxes, one to the beginnings of the yang seasons spring and summer, and one to the beginnings of the yin seasons autumn and winter (the members of each pair are connected by dotted lines in the figure).

Tetragrams half a year apart are related in that the value of each line in the later tetragram is 1 higher than the corresponding line in the earlier symbol. Thus - - - pairs with - -, and - - with — . That is because the ordinal difference between the tetragrams in each pair is 1111 in base 3, corresponding to 40 in base 10 notation. This line-by-line relationship does not hold for all eighty-one tetragrams. It fails in those pairs of which the earlier member contains a doubly broken line, since in base-3 notation 2 + 1 = 10. The consistent pairing of lines in the eight doublets is significant because it is exceptional.

The Mystery asserts its most significant diagrammatic oppositions in the eight tetragrams that hold special meaning for stars, times, and numbers. It embodies point-by-point oppositions through all eighty-one divisions of the year in the texts. Both texts and tetragrams reflect from beginning to end the Head-by-Head continuity of cosmic process. This continuity manifests the Tao's Mystery.

Conclusion

This superficial survey of the *T'ai hsuan ching* can only suggest the scope of Yang Hsiung's vision. The Great Commentary to the *Chou i* had claimed that the archaic scripture, despite the ragtag literary character that commentators did not acknowledge, encompassed every phenomenon in the realms of heaven, earth, and man. The Mystery's systematic internal structure and exquisite gradation of detail made real the Han's ideal conception of the Changes. It described change in the cosmos as manifested in the cyclic order of time. But the Mystery was more than a description; it was a model of natural process.

The Heads set out the alternating growth and attenuation of yin and yang through the round of the seasons, beginning and ending in the darkness and incipience of the winter solstice, a metaphor for the featureless chaos from which the universe ultimately separates. The two stages of differentiation—first into yin and yang, then into the phenomenal world—are analyzed Head by Head into eight-one equally spaced phases. The development from one phase to the next is anything but mechanical. It hesitates and reverses, and is glimpsed from changing viewpoints. The outcome is not a static set of eighty-one classifications, but a synthetic picture of the fine texture of change. Each Head, like earlier concepts used to break down configurations and cycles into their parts—yin-yang, the Five Phases, the eight trigrams—holds a plethora of associations, metaphysical, temporal, spatial, astronomical, musical, and above all, political and moral.

Examining the temporal correlations, we have found them intricately linked in a way that suggests, not only the astronomers' objective time, but the multivalence of time in personal experience. Moments and stretches of time recapitulate each other. Heads half a year apart are related by language and metaphor.

Linkages between the four-line symbols complement the internal resonances of the texts. Special symmetries mark those that correspond to important annual points of transition. The tetragrams echo the concatenation of the Heads. Unlike the hexagrams of the Changes, Yang's symbols are by intent consecutive ordinal numbers. Even when simply scanned as a series of visual images, the flow from one tetragram to the next is unmistakable.

As the Heads and tetragrams set out a scheme of measured change in sky and earth, the seven hundred and thirty-one Appraisals explore the intimate space-time of humans in society moving from reflection to decision to action. The philosophic interest of Yang's Head texts lies in their intricate, nuanced picture of a grand cycle of change, his recognition of complexity within regular order. Conversely, his Appraisals are remarkable because, through highly figurative language and the interplay of cycles within cycles, Yang suggests regular patterns emerging from the inexhaustible variety and ambiguity of moral circumstances.

Each cycle of nine Appraisals related to a single Head can stand alone, whether it sets out the course of a career or a theory of literature. At the same time it is woven into the larger fabric. Allusions echo from one Head to another, artfully recalling images in the Book of Changes. Each set of Appraisals establishes its own rhythm of internal variation. Worked into this congeries of meaning in the sets of nine Appraisals are images related to yin-yang and the five Phases, each precisely placed according to Han numerological usage. Ambiguity of word and image give the texts of the Appraisals meaning not only for the development of the individual but for the fortunes of the state.

Despite the formidable literary and philosophical ingenuity that went into it, the Canon of Supreme Mystery is not art for art's sake. Yang was not inclined toward that European modernist fashion. He publicly renounced the poetic rhapsodies that had brought him literary fame and official rank when he became convinced that the moral content of the genre would not reform the emperor.[50] The *T'ai hsuan ching* became his effort to retrieve a desperate political situation.

To those whose only viewpoint is hindsight, it may seem ridiculous to think of a man nurtured on a philosophy of withdrawal, avoiding court intrigues to write a recondite imitation of the oldest Confucian canon, as politically committed. But the disengagement that Yang was insisting upon was at bottom the psychic distance that in a corrupt time keeps an intellectual's standards intact and his critical gaze steady. Some can keep their inner distance while rubbing elbows with hypocrites and opportunists. Yang found that he could not. Whether his work was a political statement does not depend on that, but on its content.

---

50. Knechtges 1976: 3–4.

III

Despite the superficiality of this essay, we have adduced more than enough evidence to illustrate Yang's aim in writing the Mystery: to instigate and guide the personal striving for integrity that is the only possible basis for a sound polity. This virtue is more than a matter of moral and psychic integration; it involves union with the Way of Nature and its Mystery. As the inquirer aligns his own decisions and actions with the cosmic course, he is able to promote harmony in every external sphere, renovating society on the pattern of the Tao.

Yang looked to the classics as revelations left by the ancient sages to aid the self-cultivation of those who hoped to follow in their footsteps. His own contribution was a guide to the most enigmatic of the orthodox revelations, the *Chou i*. He provided a structure to undergird its amorphousness, and set out in the poetic and philosophic language of the Han what was hidden deepest within it. Thus it was that the greatest poet-philosopher of his generation gave his attempt to inspire those who would reorder civilization the form of a book of fortunes.

The connection between fortune and integrity was for Yang and many other Han Confucians ethical. As his self-professed but posthumous disciple Ying Shao 應劭 (active 165–ca. 204) puts it, "Nothing can disturb the man who returns to integrity and bases himself on what is right, who 'upon self-examination finds himself blameless.' [For such a man] ill fortune changes to good."[51] Integrity and a passion for right determine whether a beginning portends fruition or failure.

That is the conviction that pervades Yang Hsiung's book of divination. It does not offer magical power over nature. It simply aids reflection on the eternal patterns that underlie every aspect of experience and action. Assimilating those patterns, Yang was convinced, could guide the renewal of human creativity and the eventual recovery of order.

> Thus
> Knowing mystery, knowing silence
> Keeping to the middle way of the Tao
> Through purity, stillness
> Roaming the palaces of the gods
> Through solitude, quiet
> Guarding the mansions of virtue
> Times differ, circumstances change
> The Way of man never varies.[52]

---

51. *Feng-su t'ung yi fu t'ung-chien* 風俗通義附通檢 (Centre Franco-chinois d'Etudes Sinologiques, Indexes, 3; Beijing, 1943), 9: 67, allusion to *Lun yü*, 22/12/4.

52. From "An Antidote to Ridicule", HS, *87B:* 5120; cf. Knechtges 1976: 101.

III

Retrospect

Michael Nylan took part in one of my seminars in 1978, when she was a graduate student. I quickly discovered that she shared my enthusiasm for the many dimensions of the Book of Changes, for its several forgotten successors, and in particular for the *T'ai hsuan ching*. This essay, originally for a volume offered to Derk Bodde in 1987, invited attention to a remarkable masterpiece of philosophic literature, comparable in its way to that of Lucretius. The greatest success of this introduction, I would say, was that it convinced Michael to translate the Mystery. Her English version (1993) makes available a book that is superior to the *I ching* for many of the uses to which Europeans have long put the latter, and that deserves to be known as a great work of literature as well as of philosophy. The translation respects Yang Hsiung's intent as well as his language, which few could read as accurately as she has done.

We are pleased that this volume gave us an opportunity to prepare a revised version. The main reason for doing so is that the University of Hong Kong Press gave us no opportunity to read proofs, and did not employ a competent proofreader. The condition of the essay as originally published was an insult to the reader and an embarrassment to us. An additional reason is that, with a complete translation available, this can serve a new purpose as an introduction to it. We have brought the essay up to date, rewriting it, incorporating references to the translation, and removing a couple of sections that are now in effect part of Michael Nylan's book. We have left full references in footnotes.

One reviewer of the original version objected to our use of the term "Neo-Confucianism." We agree with alacrity that in conventional usage it lumps together intellectual themes best kept distinct, and ought to be abandoned. After an additional eight years, however, it is as ubiquitous a cliché as ever. That being the case, we still find it useful to point out that the new Confucianisms that formed from the T'ang were not the first consequential ones, and that studying their neglected predecessor is indispensable for anyone who hopes to understand the evolution of Chinese philosophy.

We are pleased that Derk Bodde remains in fine fettle, so that we can offer this report to him again.

IV

THE MYTH OF THE NATURALISTS

Introduction
Until very recently, historians of Chinese thought have treated the doctrines of yin-yang and the Five Phases *(wu hsing* 五行) as cosmological from the outset. Benjamin Schwartz in *The World of Thought in Ancient China* speculates that the "correlative thought" that these concepts typify "may have existed even in neolithic 'primitive' China," and sees it as a "mode of thought" that corresponds closely to Lévi-Strauss's *"pensée sauvage."* Fung Yu-lan believes that this cosmology originated in a "school of the *Yin-yang,"* and that Tung Chung-shu 董仲舒 first combined it with orthodox socio-political thought ca. 136 B.C.

Joseph Needham, tracing the origins of "the Two Forces in the universe (Yin and Yang) and the Five Elements," goes a good deal further. He traces the latter to Tsou Yen 鄒衍, whom he dates ca. 350-ca. 270 B.C., "the real founder of all Chinese scientific thought. . . . If he was not the sole originator of the Five-Element theory, he systematised and stabilised ideas on the subject which had been floating about, especially in the eastern seaboard States of Chhi [i.e., Ch'i] and Yen, for not more than a century at most before his time." What Fung calls the "school of the *Yin-yang"* Needham rechristens the "School of Naturalists."

Needham's view has provoked, if not a debate, a stark divergence of views. John Knoblock says of *wu hsing* that "its systematization into a comprehensive philosophy appears to have occurred only after 300 in the works of Zou Yan" (i.e., Tsou Yen), but A. C. Graham declares that "one of the few things which may be said with confidence about Tsou Yen is that he belongs to a world right outside the philosophical schools."[1]

A fresh look at the evidence suggests a different approach. First, like a number of colleagues now exploring the history of ideas, I find ancient Chinese categories more useful than modern Western ones in understanding such matters as what yin-yang, *wu hsing,* and *ch'i* 氣 actually meant. Modern rubrics are suggestive in orienting oneself, and essential in explaining meanings to a broad readership, but they often miss the mark if they are allowed to design analyses. Second, I do not believe that the prominence of these concepts in the history of science makes

---

1. Schwartz 1985: 351, Fung 1952-1953: I, 17, Needham et al. 1954- : II, 232, Graham 1986: 12, and Knoblock 1988-1994: I, 215. I am grateful for the comments and suggestions of E. Bruce & A. Taeko Brooks, Derk Bodde, A. C. Graham, G. E. R. Lloyd, and Sarah Queen, and the assistance of Asaf Goldschmidt.

them purely scientific. It seems advisable to reopen the questions first of why and then of how they were used.

I have argued in Chap. I that the cosmological concepts were moral and political from the start, and that in the long sweep of Chinese thought, as they took on new ranges of meaning, they remained moral and political. Their association with cosmology and science came about, not because they were pulled by the demands of science or technology, theoretical or practical, but because they were fitted into various doctrines that (among other aims) legitimated the workings of the unified and centralized Ch'in-Han state as a model of Nature's processes. The Han did not spawn a single tightly integrated orthodoxy. Nevertheless, its rulers and intellectuals learned to yearn for one. This drive ensured that the more or less convergent world view that informed the eclectic philosophies of the time also pervaded the sciences as they gradually emerged.

Third, I find the evidence conventionally used to prove that Tsou Yen founded a school to teach "naturalism" or "scientific thought" unpersuasive. I suggest it is time to ask exactly what the records indicate that his achievement was. Finally, the celebrated Chi-hsia 稷下 Academy, a fixture in practically every discussion of Tsou and other late Warring States philosophers, is remarkably elusive in the early accounts of their careers. Again the evidence calls for skeptical scrutiny.

The Formation of Yin-Yang and Five Phases

Origins. If we ask how Yin-Yang and Five Phases began, we are confronted with a very confused picture. What used to be the conventional datings of the Book of Documents *(Shang shu* 尚書 or *Shu ching* 書經) and the Book of Changes *(Chou i* 周易 or *I ching* 易經), early in the first millennium B.C., made yin-yang and the Five Phases part of an archaic world view. It implied that the Chou was already focussed on the cosmic foundations of the state, and well on its way to the classic formulations. This view can no longer be taken seriously.

As with many other concepts defined before the Han, advances in critical dating have discredited much of the evidence for evolution of yin-yang and the Five Phases without, for the time being, offering a more believable picture. What used to be considered the pertinent early sources tell us little or nothing about thought before the fourth century. The most relevant text in the Book of Documents, although it claims archaic origins, is no earlier than the fourth or early third century. The parts of the Book of Changes conventionally cited for the history of cosmology, the "Ten Wings," are no earlier than the third century. The historical classic that seems to record many early experiments with diverse numerical categories, before such efforts converged on twofold and fivefold sets, the Tso tradition of interpretation of the Spring and Autumn Annals *(Tso chuan* 左傳*)*, has been dated by redoubtable authorities as early as the late fifth century and as late as the early second. In it we find what are almost certainly the earliest cosmological usages of *"yin-yang"* and *"wu-hsing."* E. Bruce Brooks has argued persuasively that the book is better considered a "historical romance" than a chronicle diligently recording events and conversations that took place on the very early dates it assigns to each entry. The similar Discourses of the States *(Kuo yü* 國語*)* has been placed no more exactly within about the same range. Other relevant works traditionally ascribed to a single late Chou author, such as *Chuang-tzu* 莊子, *Kuan-tzu* 管子, and *Mo-tzu* 墨子, now have to be dated part by part.[2]

Within this jumble of sources that we cannot place confidently in historical sequence, we find a great many instances of yin-yang and various fivefold (and other) categories used in quite concrete senses, for instance yin and yang in their archaic senses of "dark" and "light," and the *wu hsing* as authorized objects of ritual offering. We can also see, especially in anecdotes of divination and its interpretation in the Tso Tradition, attempts at abstract twofold, fivefold, sixfold, and other sets of correlations, with little continuity of meaning or content from one set to another. Many of these bear on the seasonal cycle and other aspects of Nature and its use by humans. Whether these are stages in a process that eventually converged in yin-yang and the Five Phases, or whether the author elaborately confected them while inventing an ideal past, remains to be seen. Inventing them all at once implies that a conception of Nature-knowledge as progressive, which datable early sources do not otherwise reflect, guided the author. In our current state of ignorance, the first alternative seems the more likely. The irreducible uncertainty, inconvenient though it is, is preferable to the delusion of certainty.

Original character of *wu hsing*. In 1980 P'ang P'u 庞朴 set to rest the assumption that *wu hsing* since its inception was a naturalistic concept. For more than two thousand years no one knew what to make of Hsun-tzu's 荀子 attack on Mencius and his teacher's teacher, Confucius' grandson and second-generation disciple Tzu-ssu 子思. It is part of his chapter on "Debunking the Twelve Masters":

They vaguely emulate the Former Kings, but do not understand what was

---

2. For the classics the best current datings are summarized in Loewe 1993. See also Shaughnessy 1983, Graham 1978, 1979, Rickett 1985- , Roth 1991, and Nylan 1992. For Brooks's discussions of the *Tso chuan* and *Kuo yü* see 1994: 48–51 and 1993- : Note 70 (1994.4.1), Queries 23 (1993.12.28), 29 (1994.7.15), and 52 (1995.7.24). He dates them 315/305 and 304/296 respectively. Yates 1994: 98ff adduces new evidence for early stages in the evolution of yin-yang and the Five Phases. Knoblock stated in 1982–1983: 29 the general principle that what have been treated as pre-Han classics are compilations.

abiding in their traditions *(t'ung* 統*)*. Still[3] their capacities were considerable, their ambitions large, and what they had learned by seeing and hearing was varied and broad. They draw on antiquity to fabricate *(tsao* 造*)* a doctrine that they call *wu hsing*. This is peculiar, contradictory, and lacks proper categories; obscure, arcane, and with no apparent argument; closed, vague, and without adequate explanations. This done, they polished their verbiage and presented it reverentially, announcing 'These are truly the words of the former gentleman.'

Tzu-ssu sang this tune and Mencius harmonized. Vulgar "Confucians" stupidly made a great fuss over [this doctrine], unaware of where its fallacies lie. They were initiated into and transmitted it, believing that because of it Confucius and Tzu-kung 子弓 would be venerated by later generations. That is indeed reason to hold Tzu-ssu and Mencius guilty.

The "former gentleman" is Confucius. Tzu-kung was a disciple who, Hsun-tzu believed, transmitted the authentic Confucian teachings. Mencius is supposed to have studied with one or more of Tzu-ssu's disciples.[4]

Tzu-ssu's writings are long lost, and there is nothing in Mencius that resembles the Five Phases, much less the term *wu hsing*. P'ang's clue is a set of categories that Mencius does use, namely the Four Beginnings *(ssu tuan* 四端*)*: "Feelings *(hsin* 心*)* of commiseration are the beginning of concern for others *(jen* 仁*)*; feelings of shame and revulsion are the beginning of righteousness *(i* 義*)*; feelings of modesty and yielding are the beginning of propriety *(li* 禮*)*; and feelings of approval and disapproval are the beginning of judgment *(chih* 智*)*. Human beings have these Four Beginnings just as they have four limbs."

This grounding of the Way in individual consciousness has nothing to do with Hsun-tzu's own *wu hsing*. The term occurs only once in his work, where he sees the Way embodied in a rural drinking ceremony: "clarifying distinctions of rank, distinguishing lavish and simple [according to rank], harmonious enjoyment without undisciplined behavior, distinction of seniority [in drinking] with no one left out, and serene feasting without disorder: these *wu hsing* are all that is needed to

---

3. I emend provisionally following Lu Wen-ch'ao 盧文弨 (Knoblock 1988–1994: I, 303n44). The text Englished as "still" might be translated without emendation as "they give an impression of languidity, but . . . "

4. *Hsun-tzu,* 6.10–14. Knoblock's interpretation (1988–1994: I, 224; see also 214–215) makes contemporary "vulgar Confucians" the villains because they have done an injustice to Tzu-ssu and Mencius. It rejects the notion that Hsun-tzu would have blamed others for championing five virtues that he himself accepted. The passage is not concerned with the individual virtues, however, but with their use as an ensemble. The form of the chapter, which attacks prominent philosophers by name, a pair at a time, renders this explanation improbable. Knoblock does not cite P'ang's research.

rectify the individual and make the state peaceful and secure." Graham is correct in understanding this usage of *wu hsing* as "five courses of action." In this and many other arguments we see Hsun-tzu's lack of sympathy for Mencius' "psychological" approach and its supporting conceptions.

Another clue is *wu ch'ang* 五常 or "five constants," the virtues associated with Mencius' Four Beginnings plus good faith *(hsin* 信*)*. Tung Chung-shu defined this term in a memorial submitted to the Martial Emperor ca. 136 B.C..

As P'ang showed, two tractates buried in 168 B.C. and excavated at Ma-wang-tui, Honan, in 1973 solve this problem. The first (sufficient for our purpose) is, like the *Mo-tzu,* in the form of a canon and its explanation. It speaks of not only five *hsing* but also four: "The *hsing* of virtue are five, which taken together we call virtue. The four *hsing* taken together we call good *(shan* 善*)*." The set of four, which it also calls *ssu shan,* or Four Good Behaviors, are precisely Mencius' Four Beginnings. The set of five, which it also calls *wu te,* or Five Virtues, adds sagacity *(sheng* 聖*)*, rather than the good faith of the *wu ch'ang,* to concern for others, righteousness, propriety, and judgment. One can conclude without reviewing P'ang's further arguments that at the end of the Chou dynasty, in certain contexts, *wu hsing* was a set of moral categories that could plausibly be linked to Mencius. Whether a tradition attributed use of the term to him is not pertinent to this inquiry. The word would be most accurately translated "Five Activities."[5]

The Transition. The crucial period for the evolution of what became Han cosmological naturalism is from the middle of the fourth to the end of the first century B.C. In that time three similarly crucial wider transitions took place.

The first put individual self-cultivation at the center of the philosophic quest, and focussed its disciplines increasingly on nurturing the "all-encompassing *ch'i*" of the cosmos within oneself. This emphasis first appeared, as in the *Meng-tzu* 孟子 and—with different terminology—the *Chuang-tzu,* without an explicit cosmology, but one soon became part of the picture.

The second change, equally gradual, adapted this new emphasis to the urgent task of inventing a new political order as the old one collapsed once and for all. From at least 235, in the Springs and Autumns of Master Lü *(Lü shih ch'un-ch'iu* 呂氏春秋*)*, we find proposals that the emperor, using esoteric techniques, person-

---

5. *Meng-tzu* 2A/6; cf. Fung 1952–1953: I, 121–122. *Hsun-tzu* 20.48–49; cf. Knoblock 1988–1994: III, 86. For Tung's memorial see *Han shu, 56:* 2505, which gives 誼 for *i.* Graham 1986: 76 gives other examples of *wu hsing* as a social category from *Lü shih ch'un-ch'iu* and *Huai-nan-tzu* 淮南子. Brooks argues that Hsun-tzu is objecting to tendencies toward "a mechanistic universe" in Tzu-ssu and Mencius, a point on which I cannot follow him (1993– : Query 50.R1, 1995.7.10).

ally cultivate special powers that would attune his realm to the Way of the cosmos, ensuring that it would endure. His bureaucracy as well was designed according to cosmic symmetries described in highly abstract language. From early in the Han, the government protected and supported the teaching, within the court, of a diverse assortment of canons. These represented the varied traditions on which the elite intellectuals who staffed the civil service drew.

Beginning 135 B.C. with the new hegemony of Confucian teachings under imperial patronage, largely due to Tung Chung-shu, and ending a little over a century later with the collapse of the Western Han, several attempts at a single state ideology included this doctrine of cosmological monarchy and other equally adaptable currents of thought. Most of these components of what Michael Nylan and I have called "the first Neo-Confucianism" (Chap. III) would have seemed grotesque to Confucius and his immediate successors.

The third change eventually integrated yin-yang and Five Phases, redefining them as aspects of *ch'i*, to form the mature synthesis that I describe below. Exactly when this happened is debatable. The oldest book in which it can be found is Copious Dew on the Spring and Autumn Annals *(Ch'un-ch'iu fan lu* 春秋繁露*)*, before 130, but there is reason to suspect the authenticity of the few chapters that incorporate the Five Phases.[6] The first surviving source for this mature cosmology, maintaining the political and moral dimensions that shaped it, seems to be the medical Inner Canon of the Yellow Lord *(Huang-ti nei ching* 黃帝內經, probably first compiled in the first century B.C.).[7]

Mature phase. By the end of the first century, yin and yang were not forces, and *wu hsing* were not elements. They were, rather, sets of qualifiers used to describe either two or five aspects of *ch'i*. They shared certain important characteristics: they were dynamic (accounting for change), relational (an attribute is defined by its relation to others within the system of two or five), and aspectual (the choice between yin-yang and Five-phases analysis depends on what aspect of the phenomena one wants to discuss).[8]

---

6. Sarah Queen raises serious doubts about the authenticity of the few *p'ien* of *Ch'un-ch'iu fan lu* in which the term appears. This is not an iconoclastic interpretation. Chinese scholars for centuries have been raising doubts about the ascription. Queen (1991) has focussed the question more sharply.

7. It is possible that the *T'ai hsuan ching* of ca. 4 B.C., which also integrates these concepts, was earlier. See Loewe 1993: 199–201 and Chap. III.

8. For detail on these characteristics of *ch'i* see Sivin 1987: 43–80. Although Needham continued to use "Five Elements," conventional in the 1950's, for *wu hsing*, he noted that it is incorrect and misleading (II, 244).

IV

THE MYTH OF THE NATURALISTS

As the sciences evolved from roughly the first century on, this approach to thinking about natural and social phenomena became usual, although never standardized to the point that it could be called a paradigm. It survived in traditional Chinese medicine to the 1960's, when the cosmology of *ch'i* began to be "modernized" on a large scale, haphazardly using dialectical materialism and scientific positivism of a very old-fashioned sort.

The various scientific traditions—astrology, medicine, alchemy, siting, and so on—were built on foundations of yin-yang and Five Phases, each interpreting and integrating them differently. But to think of these simply as specialized technical concepts is to ignore their ubiquity in everyday thought at every level of society, and to overlook their social and moral meaning even when explaining physical phenomena that from a modern viewpoint have nothing to do with human norms. The cost of this neglect is an impoverished understanding of Chinese science. Despite its formidable abstract and concrete accomplishments it never, from first to last, strove or pretended to be value-free.

Tsou Yen's setting. Angus Graham has provocatively dissented from the prevalent estimates of Tsou Yen as putative founder of the Yin-yang School of Philosophy. He notes that "cosmological speculation, which is at the beginnings of Greek philosophy, entered the main current of Chinese thought only at the very end of the classical period. It is possible to spend a long time studying the philosophers from Confucius (551–479 B.C.) to Han Fei 韓非 (died 233 B.C.) without ever having to come to terms with it." He claims, in fact, that "the Yin and Yang are fully established in the philosophical literature as the two fundamental principles by about 300 B.C., but without yet being fitted into correlative schemes. As for the Five Phases, both the Mohist Canons and the military classic *Sun-tzu* declare flatly that 'The Five Phases have no regular conquests' . . . and Han Fei mentions them only among methods of divination which he is deriding."

Graham draws from these important data a couple of conclusions that are not at all intuitively obvious. He begins sagaciously by observing that "one should not think of Chinese correlative thinking as the application of the metaphysical theories about Yin and Yang and the Five Phases," and arguing that what preceded systematic thought using the latter was a welter of miscellaneous correlations. They are frequent in what he considers very early arguments in the feudal courts by diviners, physicians, and others in such sources as the Tso Tradition and Discourses of the States. He goes on to assert that these intellectual explorations had nothing to do with philosophy. They point not to a phase in the evolution of cosmology and metaphysics, but to a separate trend that remains distinct from them.

At the center of this argument stands Tsou Yen. Graham denies outright that he was a philosopher, assigning him without further ado to the non-philosophical courtier group. He is then willing when he looks back over the entire history of Chinese philosophy to "risk the generalisation that it was fully committed to correlative thinking only during the Han, and that in its great periods (Neo-Taoism, Neo-Confucianism) systems of correspondances [sic] are always marginal to it rather than central."[9] This argument, as a whole and in detail, hangs on a narrow modern definition of "philosophy." It also disregards an enormous Han and post-Han literature on correspondences (e.g., the Inner Canon) that few specialists in the history of philosophy feel a professional obligation to read.

Tsou Yen does not require such an implausible dismissal in order for Graham to make his point, namely denying that Tsou was a naturalist before his time. The issue, I believe, is not whether Tsou would be eligible for appointment in a modern Department of Philosophy. Was he concerned with issues that were focal for other Chinese thinkers whom specialists consider eligible for the modern label of philosopher, or did he want only to perform technical services like those that diviners and physicians provided to their patrons? The evidence, which I will shortly review, points to the first alternative.

One may ask, of course, whether imposing the rubric of philosophy on Chinese thought does anything but distort the historical picture. Some distortion is, to be sure, inevitable. Nothing could be more distorted than the old prejudice that the only real philosophy was European. The scope of "philosophy" is roughly similar to, although narrower than, that of the Chinese bibliographic categories of "Confucian teachings" (ju 儒) and "Masters" (chu-tzu 諸子). These categories were distinct in classical thought, and the latter included many masters of studies utterly unrelated to philosophy. But within the two one can find a range of concerns quite comparable to, although different from, those of any other literate culture.[10]

What is the point of putting an end to "great" philosophy just as science is emerging, and identifying as "great" only those eras in which cosmology and other initiatives were nearly submerged in a tide of fundamentalist humanism? If this restrictive definition of philosophy is to be taken as more than personal taste, it requires an explicit justification that Graham did not provide.

Tsou Yen
One can hardly avoid choosing a position on the spectrum between Needham's avowal that Tsou was the real founder of all Chinese scientific thought and

---

9. Graham 1986: 1; 1989: 9, 70, 12–13, 15. "Cosmological speculation" seems to refer to use of yin-yang and similar concepts.

10. It is interesting that Ch'ien Mu, in the title of his important study of pre-Han

THE MYTH OF THE NATURALISTS

Graham's unwillingness to concede that he was a philosopher. The difficulty lies in the fact that Tsou's writings were lost very early, leaving only a few fragments. Scant personal information about him has survived. He was a native of Ch'i, and probably lived ca. 347–ca. 276. He entered the patronage network, still on a small scale, in his native state, probably after Mencius departed in 313. He probably spent his entire career as a client in a succession of courts. There is no reason to believe that he ever held an official position.[11]

Almost all the inferences about Tsou's position in the history of Chinese thought, aside from a few enigmatic fragments, derive from a brief, ambiguous biographical notice in Ssu-ma Ch'ien's Memoirs of the Grand Astrologer. It tells us that

> He was later [as a recipient of support from the King of Ch'i] than Mencius. He saw that those who possessed states were becoming increasingly dissolute, unable to value virtue. This corruption he compared to a real gentleman's [resolve to] perfect [virtue] in himself and apply it to the common people.[12] He deeply contemplated the waxing and waning of yin and yang, and set down his writings on ends and beginnings and great sagehood *(chung shih ta sheng chih p'ien* 終始大聖之篇), dealing with extraordinary transformations, in more than a hundred thousand words.[13] What he said was grandiose and unorthodox. He would first confirm some small matter, and then push and enlarge [his assertion] until it extended to the boundless.
>
> First he established a sequence, from the present back to the Yellow Lord, of what scholars had collectively recorded. Summing up broadly the rises and declines over the ages, he set down in conformity [with his sequence] the por-

---

philosophers, used *chu-tzu* in the modern sense, to include both early categories.

11. The best reconstruction of Tsou's moves to various courts is Brooks 1993– : Query 23 (1993.12.28), pp. 19–20. See the discussion of patronage below, p. 20.

12. In interpreting *ta ya* 大雅 to mean a paradigmatic Confucian gentleman I follow Wing-Tsit Chan (1963: 246–248, citing p. 247). The word is usually taken in a different sense to refer to part of the Book of Songs, so that the characteristics would be a quotation or paraphrase from one of them. No attempt to find such a source is convincing.

13. *Shih chi*, 74: 2344. The Former Han history *(30:* 1733) lists a Book of Master Tsou, *Tsou-tzu* 鄒子, and a Master Tsou's Ends and Beginnings, *Tsou-tzu chung shih* 鄒子終始. Neither was in the imperial library in the Later Han, but Ju Ch'un apparently read them in the third century A.D. (see below, p. 11). Kaltenmark 1949 has argued that *Chung shih ta sheng* means "the succession of the sages," which may be true. My translation does not assume that either pair or their combination is a title. What I have translated as "extraordinary transformations" may also be a title. The process of compiling and recompiling short writings into a book of stable content was gradual. In many cases it was not completed by the end of the Western Han period.

tents and institutions [of each]. He pushed this back to remote times, tracing its origins to the time before sky and earth were born, to the formlessness that cannot be traced to an origin of its own.

First he set out the eminent mountains, great watercourses, extensive valleys, birds and beasts, what grows in the water and on the earth, and what among the species is considered precious, of the Central States [Chung-kuo 中國, the area of the early polities]. On the basis [of this array] he pushed [his account] further, beyond the seas to places that men are unable to set eyes on.

He cited the cyclic shifts *(or* cycles and shifts) of the Five Virtues *(or* Powers, *wu te* 五德) from the time that sky and earth separated, and the rule appropriate to each, so that everything tallied [with everything else] in this manner. He held that what scholars *(ju)* call the Central States is only one part in eighty-one of All under Heaven. He called the Central States 'the divine province of the red district.' Within this 'divine province of the red district' are its own nine provinces, the nine that Yü laid out. But these are not [Tsou's] nine provinces. Outside the Central States there are nine like 'the divine province of the red district.' It is these that he means by the nine provinces. That much settled, [each large 'province'] has a minor sea *(p'i hai* 裨海) encircling it, so that the people, beasts and birds are isolated from [those in the other provinces]. The part that corresponds to one region *(ch'ü* 區) is one province, and there are nine of those. And then there is a vast ocean circling the outside, where sky and earth meet.

Ssu-ma added his own estimate:

Tsou's arts were of this sort. As for the gist of [his doctrine], it rests on the exercise of benevolence, righteousness, self-discipline, frugality, and [hierarchic alignment of] monarch and vassal, superior and inferior, and the six kin relationships. Only the starting point [of his philosophy] was extravagant. Monarchs and great officials, when they first saw his method, were struck by it, gave it much attention, and were converted to it. Afterward they were unable to apply it.[14]

The addendum, when it says "saw" rather than "heard about," reminds us that the notables it mentions encountered Tsou's doctrines in writing.

Fragments. Is this a trustworthy account of Tsou's thought? That issue will no doubt remain open for some time to come. In order to evaluate its content to the extent that this study demands, let us consider seven surviving fragments of Tsou's writings.

The first two are from a lost history of the Han, written shortly after its end by Ju Ch'un 如淳 (fl. 221-265). They are preserved in a commentary on a passage

---

14. In addition to Chan, cf. Graham 1986: 15 and Needham et al. 1954– : II, 232.

of Records of the Grand Astrologer that we will later scrutinize (p. 15).[15]

1. "Now among his books there is an Ends and Beginnings of the Five Virtues *(or* there are 'Ends and Beginnings' and 'The Five Virtues,' *Wu te chung shih* 五德終始). Each virtue acts according to what it conquers. Ch'in said that Chou corresponded to *(or* acts by, *wei* 爲) the virtue of fire. Water is what extinguishes fire, so [Ch'in] claimed the virtue of water."

2. "Now among his books there is a Cycles of Rule *(Chu yun* 主運). The Five Phases *(wu hsing)*, in sequence, cyclically take charge. [The dynastic houses] follow the [sequence of] directional orientations in appointing [court] vestments."

These two passages are referring to what became hallmarks of Five Phases doctrine, sequential orders of characteristic activity that "rule" each phase of a process (in this instance, what came to be called the Mutual Conquest order) and a web of correspondences that included directions (elements of, among other things, court ritual) and colors of costumes. These statements taken together suggest a large and systematic framework of correlations—geographic and historic, with not a whiff of cosmology—rather than the loose *ad hoc* associations in the Tso Tradition's tales of diviners. "Ends and Beginnings" implies that these schemata were, among other things, cyclic.[16] The citations also indicate that at least two of Tsou's books survived into Ju Ch'un's lifetime.

The next two citations are presented as actual fragments of Tsou's writing in a famous anthology of ca. 530.[17]

3. "Tsou Yen said 'The four corners are unquiet.'"

4. "The *Tsou-tzu [or* Tsou-tzu] says 'The Five Virtues are followed by what they cannot conquer. Yü 虞, [the dynasty ruled by Shun, was associated with the virtue] Earth; Hsia, Wood; Yin [i.e., Shang], Metal; Chou, Fire.'"

It is impossible as usual to be sure whether "says" *(yueh* 曰) signifies direct quotation or paraphrase, but at least these are one or the other. The first sentence in no. 4 is confusing, and historians have generally paraphrased rather than translated

---

15. *Shih chi* 史記, 28: 1368n3, 11. Ma Kuo-han 馬國翰 gathered Tsou's fragments in his great collection of quotations from lost books, *Yü han shan fang chi i shu* 玉函山房輯佚書, LXX, 2828-2829). Ma accepted ten quotations as authentic, but the reasoning behind his identifications of three is impossible to credit. The implausibility of the three can be sampled in Fung, who reproduces with approval Ma's argument for one (1952: I, 162). See translations of nine quotations in Needham et al. 1954- : II, 236-238.

16. *P'ien* 48 of *Ch'un-chiu fan lu*, entitled "Yin-yang chung shih 陰陽終始," is about yin-yang cycles.

17. *Wen hsuan* 文選 (Choice of literature), 6: 3b, 59: 18a.

IV

it. It is, however, perfectly consistent with the canonical mutual conquest order in the second sentence. Since Metal conquers Wood (e.g., a hatchet can chop down a tree), Metal is "what Wood cannot conquer."

The next fragment comes from Cheng Hsuan's 鄭玄 Han commentary on the Rites of Chou:

5. "In spring make the fire with elm and willow wood; in summer, with jujube and apricot wood; at midsummer, with mulberry and cudrania wood; in autumn, with two kinds of oak; and in winter, with pagoda tree wood and sandalwood."[18]

In the biography of Yen An 嚴安 (fl. ca. 134 B.C.) in the Han History, a memorial by Yen refers to Tsou or his book:

6. "I hear that the *Tsou-tzu* (or Tsou-tzu) said: 'Administrative regulations or education, refinement *(wen* 文*)* or inner simplicity *(chih* 直*)* [according to the need of the time] is what lets one speak of redeeming [the excesses of the previous age]. When the time is right, use [government of an appropriate kind]. When the time is past, abandon it. When there is change, change it. Thus those who insist on consistency and do not change never experience perfection in the art of ruling.'"[19]

The last quotation is in an annotation to a place name in the Water Canon:[20]

7. "Tsou Yen said 'My reason for going up the city wall at Min was to gaze at the capital of Sung.'"

Now let us take stock of these fragments. Nothing in the first two excerpts suggests that they are direct quotations from Tsou Yen's writings. Since Ju Ch'un apparently read the two books, his characterizations are useful, but they tell us nothing about Tsou's own terminology. No. 1, on the other hand, is consistent with the use of what later became known as the mutual conquest sequence of the Five Phases in no. 4. The formulation in no. 4, however, as "following what cannot conquer," is not a fixture in the Han. Its early and abortive career perhaps signals an early stage in the definition of the Five Phases sequences. As we will see below, historical evidence connects it with Tsou Yen (p. 17).

What in any case can we conclude from this evidence?

The putative founder of the "school of the *Yin-yang*," oddly enough, does not mention yin-yang. Ssu-ma Ch'ien says that he pondered the notion, but this is as

---

18. *Chou li* 周禮, 7: 24b, citing Cheng Chung's 鄭眾 (d. A.D. 83) quotation from the *Tsou-tzu*. *Che* 柘 is Cudrania tricuspidata (Carr.) Bur., a small East Asian fruit tree. On interpretations of this passage see Ch'ien Mu 1935: 442.

19. *Han shu, 64B*: 2809. The first sentence partly follows Queen 1991: 324.

20. *Shui ching chu* 水經注, *8:* 25b.

THE MYTH OF THE NATURALISTS

likely to be a Han way of putting it as it is to be a term from Tsou's writing.

The term *wu hsing* does not appear in the Records. *Wu te* is used instead. I have translated it "Five Virtues" but *te* is close to Latin *virtus,* and could equally well be Englished as "powers." I chose the first alternative to keep the moral implications of the term, often slighted by historians, in plain view.

In these data there is no evidence of either inductive study aimed at generalization about Nature, or deductive reasoning about natural principles. Claims to the contrary either turn the confirmation of small matters into a grand Baconian enterprise, for instance, "he invariably examined small objects and extended this to larger and larger ones until infinity," or indulge in wishful thinking, e.g., an observation on no. 5 that "these admonitions probably have regard to the varying nature of the woods, how far hygroscopic, etc."[21] The admonitions more likely have to do with procedures in fire-kindling ritual, which is after all the topic.

Ssu-ma Ch'ien's account begins and ends with goals of ethical and political reform. This link to conventional morality, although no doubt exaggerated, is hardly to be dismissed as puffery. The book he was summarizing still existed in his time. Ssu-ma, although consciously didactic, is not an orthodox pedant. These are not goals that he routinely reads into eccentric doctrines, particularly those that he admits are "unorthodox." The quotation in the sixth fragment is consistent with his account. If we adopt the least laborious assumption, namely that ethical and political reform appears because it is Tsou Yen's goal, in that respect he is in the main stream of philosophy at the end of the Chou dynasty.[22] This stream flowed from, among other springs, the writings of Confucius and his successors. The fragments of Tsou's writings are entirely consistent with this understanding.

Now what sort of nascent science is this grand extrapolation? Not the same sort, certainly, as Democritus' leap of abstraction that carried him from the world that we see and taste to a cosmos that contains nothing but atoms bombinating in the void. Into that world, we remember, only the Epicureans among his successors in natural philosophy were willing to follow him. Aristotle found neither atoms nor a void thinkable.

---

21. Chan 1963: 247 (cf. p. 9 above). Needham et al. 1954- : II, 236.

22. Graham does not discredit them, but rather ignores them. *Pu ching* 不經, which I translate "unorthodox," may simply assert, as Brooks suggests, that Tsou's writings differ in form from the aphoristic writings usually called *ching* in the fourth century. He concludes from this and other evidence that Tsou's innovation was "the consecutive essay," produced on a large scale (1993–: Query 23, p. 19). I question his reading of *pu ching,* but not his larger point. Ssu-ma, writing ca. 100, is not necessarily using *ching* in the fourth-century sense. My translation attempts to convey the ambiguity.

IV

14

Tsou's quest begins not with observations of Nature but with a catalogue of China's contents, its human institutions and the natural resources that support them, with the tidy relation of society to the space that it occupies and that surrounds it. He then extrapolates to yield a sequence of shifts in political order that holds for the whole age of the earth. It encompasses parts of the world that no one had ever seen or, he assures us, could hope to see. But there is nothing abstract about it, only more people, more institutions, more resources, until they are all swallowed up in the formlessness—not at all abstract—that gave birth to them.

As for whether Tsou gathered this information on his travels, Ssu-ma has told us that his account of history gave equal weight to the unobservable, and was drawn from "what scholars had collectively recorded." Plato would call it "right belief," and moderns would call it hearsay. This is a remarkably untrammeled vision of time continuous from the mythic origin of the world, and of space inhabited by peoples isolated from one another but knowable by inference from the society of China. The universe of discourse throughout is human society, not the world of Nature.

In this expansive vision there is no cosmos. The stars in their courses, the rhythms of the seasons, the changing courses of rivers and the rising and subsidence of mountains, the *ch'i* that the earth and its lands and waters float on—the stock in trade of later natural speculation—play no part. We are hardly in a position to insist that these were not mentioned at all in books as long as Tsou's. Nor are we free to indulge in pipe dreams about what must have been in them.

Tsou's Five Virtues, according to the fragments, are more sophisticated than similar categories in the Tso Tradition and Discourses of the States, and appear in this respect (pending a clearer chronology) to be transitional between them and the later Five Phases. They are ordered by the mutual conquest sequence, one of the many sequences that in the mature Five Phases doctrine came to explain change. But one can hardly claim that Tsou's mutual conquest sequence is aspectual (p. 6) as the later mutual conquest series is, since it is the only sequence in the evidence. His cyclical shifts are part of a space-time manifold of human experience that is ultimately ethical and political.

We would do well to consider an ethical meaning for *wu te*. Since such a meaning was already well established for *wu hsing* and *wu ch'ang*, we can see Tsou as extrapolating from the Five Virtues to create, not scientific thought, but a philosophy of history, principally of dynastic change. His patrons who schemed to succeed the Chou king were bound to find it useful. The laconic mention of vestments, and the fragment on woods used in kindling fires, indicate that he went further and adapted this view of political change to new rituals that would reassert in the periodic ceremonies of each new dynasty the images of its legitimacy. In

IV

THE MYTH OF THE NATURALISTS

that sense he is a direct predecessor not of the astronomers and physicians but of the theorists of the new state—Lü Pu-wei 呂不韋 (d. 235), Tung Chung-shu (ca. 179-ca. 104), Liu An, King of Huai-nan 淮南王劉安 (179-122), and others—who went beyond his schema of political time and space to incorporate in their doctrines a real cosmology. It is from this cosmology that we can directly trace the separation of the sciences from general philosophy in the Han.

What did the political application of Tsou's doctrine turn out to be? The answer lies in a chapter of Records of the Grand Astrologer that traces the evolution of the imperial sacrifices to the great cosmic deities.

> From the time of Kings Wei and Hsuan of Ch'i (378-314), the followers of Tsou-tzu discussed and wrote on ends and beginnings and the cycles of the Five Virtues *(or* wrote Ends and Beginnings, the Five Virtues, and Cycles of Rule). When [the king of] Ch'in took the title of emperor, men of Ch'i sent memorials about them [to the court, *tsou* 奏]. The First Emperor (221-213) selected and used them.
>
> But Sung Wu-chi 宋毋忌, Cheng Po-ch'iao 正*(or* 征) 伯僑, Ch'ung Shang 充尚, and Hsien-men Kao 羨門高 were men of Yen. They practiced [techniques of?] the Way of the Immortals, rebirth from one's cadaver, dissolution and transformation, and establishing relations to the gods.
>
> Tsou Yen had been well known to the feudal lords [before the Ch'in conquest] for his yin-yang and his cycles of rule. The technicians *(fang-shih* 方士) of the Yen and Ch'i coasts passed down his methods without being able to comprehend them. From then on, those capable of fantastic prodigies, skilled in flattery, and willing to do anything to win people over, flourished [in the court] in numbers too great to count.[23]

When read with normal care, this passage is not as enlightening as it has often been taken to be. It asserts no link whatever between Tsou's scholarly disciples in Ch'i in the first paragraph and the legendary immortals, somewhere in the mists of Yen's past, in the second. The verb *tsou* apparently refers to the submission of Tsou Yen's writings, to which, we will see just below, the emperor had access.

Both paragraphs are there to explain why it was technicians of Ch'i and Yen who sought imperial patronage. These "experts" traded on Tsou's reputation to gain entrée to the court. They had access to his writings but did not understand them, which implies that they were not initiated into Tsou's lineage of teaching. That in turn suggests that by the third century formal transmission had broken

---

23. *Shih chi, 28:* 1368-1369. Cf. Needham et al. 1954- : II, 240-241, and Fung 1952: I, 169. This passage is rather corrupt. For a translation of the entire treatise see Watson 1961: II, 13-69. No commentator has made sense of *tsui hou* 最後; I do not translate.

down, and the books were available to the uninitiated.[24] These exoterics from the northern coastal states became the emperor's occultists.

The confidence games of the men of Ch'i and Yen are too well known to need reviewing here. One cannot conclude from these elaborate distractions, however, that Tsou's actual teachings played no part when the First Emperor consolidated his conquest in 221 B.C. His very first move was to signal the changed reality by changing names: new words for "decree" and "statute," "First Emperor" instead of "king." The next step was to prove that the mandate had passed from Chou to Ch'in, and that a new and lasting political order had begun:

> The First Emperor inferred from the cyclic sequence of Ends and Beginnings and the Five Virtues that Chou had held the virtue of Fire. Ch'in would replace the virtue of Chou by what it could not conquer. The present moment would be the inception of the virtue of Water. The beginning of the year would be altered, so that felicitations in audience would always begin at the new moon of the tenth month[, the civil new year]. In robes, pennants, and banners, black would be the preferred color. Of the numbers, six [the numerological correlate of Water] would be the standard. Tallies for contracts and official caps [in the style of the King of Ch'u, *fa-kuan* 法冠] would measure six inches, and carriages would be six feet long. Six feet would be one pace. Six horses would make up a carriage team. The name of the Yellow River would be changed to Water of Virtue, for it was regarded as where the virtue of Water begins (?). [It was appropriate to be] strict and decisive, letting all matters be judged by the law strictly, without 'benevolence, favor, harmonization, or righteousness.' The result would be accord with the regular character *(shu* 數*)* of the Five Virtues. Henceforth [the state] would be zealous in exercise of the law. For a long time there would be no amnesties.

These ritual matters took priority. Only afterward did the emperor get round to settling the borders, collecting weapons, standardizing the empire's diverse weights and measures, and so on.[25] Why was court ritual (of which legal policy was a part) so urgent, and what did the changes in it amount to?

The First Emperor, until he succeeded, was just a member of one of the many lines of local potentates that had been competing for power against each other, their number decreasing over the centuries. His state, Ch'in, was bold enough finally to end the spiritual supremacy of the Chou house. By wiping it out in 256, Ch'in left no ritual center that could orient the realm with respect to Heaven. No previous warlord, no matter how powerful, had dared seize that spiritual author-

---

24. On the role of written documents in such lineages see Sivin 1995.
25. *Shih chi,* 6: 236–239; cf. Chavannes 1895–1969: I, 128–130.

IV

THE MYTH OF THE NATURALISTS

ity. When the First Emperor triumphed over his last rivals a generation later, he was ready to create a new ritual center for the first time in nearly a millennium. To claim the legitimacy of the Chou meant claiming the mandate that (according to legend, if not history) it had asserted when it humbled the Shang. How could one prove that the mandate had passed, that Ch'in was not a mere usurper? That was the pressing problem that Tsou Yen's doctrine solved.

The terminology of Ends and Beginnings, Five Virtues, a virtue being replaced by what it cannot conquer, and so on in the passage just quoted is unmistakably Tsou Yen's.[26] The First Emperor was legislating a web of correlations that signalled the proper and legitimate replacement of Fire by Water. The colors of official regalia, the standards pegged to six, the emphasis on unyielding enforcement of strict legal standards as an expression of Water's associations, were joined to convey this message. One cannot say whether these highly articulated frameworks of symbol were taken literally from Tsou's lost writings. In Ssu-ma Ch'ien's account, Ends and Beginnings and The Five Virtues may in fact refer to the book or books. But there is no need to pursue this thread further.

We can conclude that, first, Tsou's philosophy of history was explicitly designed not as a tool of physical research but, as he insists, as a means to political reform. Second, it was indeed put to political use, by a conqueror who frankly despised government based on the moral principles that defined Tsou's reform. Third, the First Emperor used Tsou's writings because he learned from them that ritual legitimates change. This was hardly a new idea, but the documents suggest that Tsou was offering an immensely more coherent and elaborate ideology than anything hinted at in the Tso Tradition. It was a great deal more adaptable to political realities than the teachings of Confucius and his direct successors.

Now perhaps we can come to grips with Graham's reluctance to grant Tsou the title of philosopher. He is probably correct that Tsou's direct predecessors in the use of correlative schemata were court diviners, and that what they were doing was not philosophic inquiry. He is quite correct that Han thinkers, always eager to find intellectual ancestors, are curiously silent about Tsou. Why this is so deserves some reflection.

Ssu-ma T'an's 司馬談 (d. 110 B.C.) "On the Essentials of the Teachings of the Six Schools," included in Memoirs of the Grand Astrologer, became the definitive account of philosophic categories and their progenitors, despite its unabashed

---

26. A corrupt passage of the Treatise on Astrology *(26:* 1259) moves directly from Tsou Yen's musings on the Five Virtues to the First Emperor's ritual alignment of his state with Water. Despite misreadings, it does not say Tsou made astrological predictions.

IV

partiality.[27] It summarizes the strengths and weaknesses of five of the "schools," in order to demonstrate that the sixth, that of the Tao, combined the strengths of the others without their weaknesses. Here is the essay's account of *yin-yang chia* 陰陽家 (Fung's "School of the *Yin-yang*," Needham's "School of Naturalists"):

> According to yin-yang,[28] there are teachings and ordinances for the four seasons, the eight directions, the twelve measures, and the twenty-four nodes. Those who conform to these flourish, and those who defy them, if they do not die, fail *(?wang* 亡*)*. This is not necessarily true. Therefore I say that they inhibit people and make them fearful. [But] generation in spring, growth in summer, harvest in autumn, and latency in winter is the great norm of the Way of Heaven. Those who do not follow it are unable to provide guidelines for the empire. Therefore I say that one cannot ignore the paramount conformity [of the yin-yang schema] to the four seasons.

This document does not mention any of Tsou's characteristic teachings. Its account is based not upon what Ssu-ma knew about Tsou's writings, but upon the "Twelve Records" in the Springs and Autumns of Master Lü that give elaborate sets of correspondences to govern the royal rituals for each month.[29]

The disregard of Tsou in the official writing of the Han had nothing to do with the merits of his writings as philosophy. His fame, and his influence on the beginnings of an imperial ideology before the Han, arose not because he founded a school, whatever that may mean, or because he was a naturalist. They have to do, rather, with his scheme of historic change that proved irresistible to certain feudal lords, and ultimately to the First Emperor, as a new basis for legitimacy that was urgently needed in the generation between the abolition of the Chou and the first unification. The association of Tsou's schema with the First Emperor made it an undesirable ancestor in the Han. Philosophers of that time also did not acknowledge the great influence of Lü Pu-wei due to his association with Ch'in. Once the linkage of dynastic character with ritual was incorporated in the Springs and Autumns of Master Lü and in early Han teachings, there was no further need for Tsou as a progenitor. In ancient China progenitors needed disciples, and in this

---

27. *Shih chi,* 130: 7–14, trans. Watson 1958, 43–48.

28. Although the translators add the conventional but misleading "school," the corresponding Chinese word *chia* 家 does not accompany either occurrence of *yin-yang*.

29. *Shih chi,* 130: 3290. The commentary (n. 1) suggests that the eight positions are those that correspond to the trigrams of the Book of Changes, and the twelve measures are the twelve divisions of the Jupiter Cycle. These appear to be guesses. The twenty-four nodes are the equal divisions of the tropical year, which indeed appear in *Lü shih ch'un-ch'iu*. The "twelve records" are the "Monthly Ordinances" *(yueh ling* 月令*)* of the *Li chi* 禮記, but that version is later than Lü's. See Loewe 1993: 295.

THE MYTH OF THE NATURALISTS

sense too his succession seems to have been barren. Surely the courtiers who "passed down his methods without being able to comprehend them" could not form the basis of a lineage that philosophers (more fastidious about discipleship than rulers were) would take seriously. They preferred to claim more prestigious intellectual ancestors such as Confucius and the Yellow Lord.

The Imaginary Academy

Intimately related to the myth of a School of Naturalists is that of the "Chi-hsia Academy." An up-to-date historian of philosophy typically describes it as a think tank before its time: "The Academy was a meeting place for intellectuals and can in some respects be likened to some of the larger government funded research centres today." Another calls it "a government sponsored academy, where wise men of every stripe received salaries and gathered disciples without any political responsibilities whatever . . . Hsun Tzu taught at this academy where he was revered as senior teacher." He remarks on "the many scholastic sects gathered at Chi-hsia during the third century B.C., each of them busily engaged in programs of self-cultivation and philosophical speculation, vigorously disputing among one another to earn preeminence among the academic schools, attract students from the community and beyond, and retain or increase the stipends granted them by the rulers of Ch'i" until the Ch'in conquest. A translator speaks of "the intellectual center of ancient China, the Jixia [i.e., Chi-hsia 稷下] Academy," where the young Hsun-tzu "was obliged to master the doctrines of the many schools, the forms of argumentation, and the techniques of rhetoric at which the Jixia scholars excelled" and asserts that "in Xunzi's [i.e., Hsun-tzu's] day . . . it was a department of government in which distinguished scholars held official rank as grand officers."[30] No one has noted what in ancient China was bound to be a discrepancy between official position and freedom from political responsibility.

The confidence of these assertions contrasts with the ambiguity of the sources. No document contains any word that corresponds to "academy." When it appears, it is there because historians have added it. No early writings describe the putative programs of this academy. Scholars infer them with the indispensable help of creative analogy. I will argue that many of these analogies to modern institutions fit the contemporary evidence badly, and have led to much mischief.

Patronage. The word "patronage" can mean a great many things. In looking

---

30. Makeham 1994: 170; Eno 1990: 48, 137; Knoblock 1988–1994: I, 4, 261n10. Some Chinese scholars writing in Western languages have been more circumspect. Fung Yu-lan 1937–1953: I, 132–133, writes only of Chi-hsia as a place, and Chan Wing-tsit 1963:116 writes simply of "Ch'i, where scholars congregated at the time."

IV

at Chinese history the most obvious meaning is support depending on personal favor, sometimes combined with honorary position that does not entail official responsibility. This quintessentially private relation with the mighty implied that if official status was occasionally offered, it could as quickly be withdrawn. That was its implication in Europe in the sixteenth and seventeenth centuries, when official court appointments were a very important source of support for intellectuals and artists. As Mario Biagioli has shown, the risk of losing favor made possible the hope of sudden fame and fortune. For some to rise, others had to fall.[31]

In China the counterparts of Renaissance clients were the "guests" (k'o or pin-k'o 賓客) at courts of local rulers from a little before 400 B.C. until late in the second century. By that point this institution had been largely replaced by official appointment in the Han imperial government, save for an occasional client of an emperor who was so inclined.

Patronage is a large topic. Failure to study it systematically has led to a great deal of guessing about its character. Work in progress (see Chap. I) suggests that several broad characteristics consistently appear in the sources:

1. Much of the prestige of patronage came from collecting clients wholesale. Benefactors wanted it known that they could afford such expensive pursuits, and that their reputation for largesse and taste could lure desirable dependents from their rivals. T'ien Wen, lord of Meng-ch'ang 孟嘗君田文 and prime minister of three kingdoms, cited the typical quota that serious patrons strove for: "I feed three thousand guests." He did not amass this formidable collection by searching out philosophers individually, but by "attracting the guests of the feudal lords as well as refugees accused of some crime."[32]

2. Most of the "guests," as we know from a wide range of anecdotes, were clever adventurers, mercenary military trainers, experts on useful arts, occultists, assassins, people qualified to carry out miscellaneous commissions, and advisors on rulership, strategy, and dirty tricks. Very few of these were inquirers after wisdom. Ch'ien Mu 錢穆, whose detailed study of Chi-hsia we will consider below, found there only fourteen philosophers definitely associated with the patronage system that at its peak accommodated "several thousand" clients at a time.

For the survival of small states, wisdom is desirable, but no more so than

---

31. By far the best analysis of the workings of patronage in Italy in this period is Biagioli 1993. For elsewhere in Europe see Moran 1991.

32. Shih chi, 75: 2360, 2353. For a war of conspicuous consumption between two munificent patrons, won not by wisdom honored but by jeweled shoes, see 78: 2395. Some local lords collected books on a grand scale, a matter that deserves study.

THE MYTH OF THE NATURALISTS

knowledge of political realities, mastery of the halberd, or skill at deception. The Lord of Meng-ch'ang, for instance, especially honored two minor members of his entourage whose special skills saved his life. One was a burglar who disguised himself as a dog. The other could crow exactly like a rooster.[33]

Historians, early and modern, devoutly convinced that thinkers are more consequential than assassins and tricksters, have greatly exaggerated the importance of philosophers in the courts. At the same time, the early ones preserved ample data to allow a more adequate perspective. Not only were intellectuals a tiny minority in the various Warring States courts, the evidence does not indicate that they set the style of the patronage system.

3. Only a minority of clients became salaried officials. They subsisted on the largesse of the ruler, and had no status in the court beyond what favor gave them.

4. Philosophers rarely carried on oral debates with living rivals. The main purpose of their lineages by 300 was to preserve and transmit authoritative teachings. Unlike the Greek situation, polemics were not good for public relations. Livelihood depended on access to the patron. This was not normally an interactive discourse. The patron drew his own judgment about merits. He was not constrained to approve or use others' ideas, or reveal what he intended to do with them.[34]

With those generalities posted, let us look at the record.

Ch'i as a center of patronage. It is difficult to trace the beginnings of patronage in the Ch'i capital. Ca. 400, at least a generation before the T'ien clan usurped the marquisate and its rulers began calling themselves king (357), we find their ancestor the high minister T'ien Ch'ang 田常 treating "guests" with exceptional hospitality and reaping private benefit: "He selected women at least five feet tall from throughout Ch'i for his seraglio; they were counted in the hundreds. He arranged for his guests *(pin-k'o)* and retainers *(she-jen* 舍人) not to be stopped when they went in and out of the seraglio. By the time he died, he had more than seventy sons."[35] One may assume that he had not chosen his guests for their skill at

---

33. Ibid., p. 2355.

34. Although Graham organized his recent history of late Chou philosophy under the title "Disputers of the Tao," the word "debate" appears in it only once. I have discussed this and other germane points in Sivin 1992.

35. *Shih chi, 46:* 1885. *She-jen* (Hucker 1985: item 5136, "Houseman") were not menials, but "squire-like dependents." Seven *ch'ih* 七尺 is too round a figure to convert precisely, but would average between four and a half and five feet, a little above average for the time. The archaistic *Huang-ti nei ching* gives 8 *ch'ih* as a round height for a noble male, and 7.5 *ch'ih* for a male commoner; women would presumably be a little shorter. See Yamada 1991: 40, 42.

philosophic disputation. Patronage eventually became a family hobby:

> King Hsuan [of Ch'i, r. 319–301] was fond of gentleman-retainers who were literary scholars or itinerant advisors *(wen-hsueh yu-shui chih shih* 文學游說之士). Among his seventy-six were Tsou Yen, Ch'un-yü K'un 淳于髡, T'ien Pien 田駢, Chieh Yü 接予, Shen Tao 慎到, Huan Yuan 環淵, and their like. He granted them all mansions and made them Senior Grand Masters. They did not govern, but took part in court deliberations *(i lun* 議論). As a result of this, 'the scholarly gentlemen of Chi-hsia' once again flourished, amounting eventually to several hundred and then several thousand."[36]

This passage lists some illustrious philosophical names, but any attentive reader will have noted that they are identified for literary or political, not philosophic, skill. Those are not, of course, mutually exclusive categories, but they tend to be treated by historians of philosophy as though they were. The documents indicate that "guests" were kept as advisors for the new ruling clan's plots against and with Ch'i's neighbors. It asserts that this was not the first time Ch'i clients had a scholarly (not necessarily a philosophical) reputation. But no source claims they were there, as Knoblock puts it, "to deliberate and propound learned theories." None, in fact, contains any word corresponding to "theories."

The passage also fails to mention any organization. Unlike modern historians, late Chou and early Han sources speak only of guests, never of institutions. Tempting though it may be to envisage a Chinese Library of Alexandria, the only buildings that sources based on contemporary documents mention are residences.

The biography of Tsou Yen provides significant details. It asserts that approximately the same group of denizens

> all wrote books speaking of matters that impose order on disorder, so as to bring themselves to the attention of the rulers of the time. How can one do justice to them all? . . . They greatly pleased the King of Ch'i. He gave them, from Ch'un-yü K'un down, the title of Adjunct Grand Master *(lieh ta-fu* 列大夫), and established mansions for them at the best locations on the thoroughfares, with high gates and great halls, to show his respect and favor. He reviewed the guest lists of all the feudal lords, saying that Ch'i could attract worthy gentlemen-retainers from the whole realm.[37]

This description clearly emphasizes the prestige of writing, not skill in oral

---

36. *Shih chi, 46:* 1895. "Gentleman-retainer" is my translation of *shih* 士, not an official title here. Fung's "scholar," Hucker's "Serviceman," etc., are too narrow.

37. *Shih chi, 74:* 2346–2347. On the title see Hucker 1985: item 3700. Hucker defines it as "a title of honorary nobility awarded to deserving subjects," but is aware of its use only in the Han. *Lieh* generally marks titles that differ from regular ones (item 3697).

IV

THE MYTH OF THE NATURALISTS

argument, as the key to support. That obviously was not the case for all clients (for instance, burglars or mercenaries), but this document is concerned only with the few authors in the Ch'i entourage.

"Senior Grand Master" in the first quotation is ambiguous. It was as often an honorary title as a working one.[38] The "adjunct" version in this quotation and the next one is clearly honorary, and thus settles the question. Its holders, in other words, were not regular officials, but personal dependents of the monarch. Such appointments, as this passage tells us, attracted clients desirable to other patrons.

In the biographical account of Hsun-tzu, his relationship to Chi-hsia repays examination:

> Hsun Ch'ing 荀卿 was a man of Chao. At the age of fifty he first arrived in Ch'i as an itinerant scholar. . . . When T'ien Pien and his colleagues all died, in the time of King Hsiang (r. 284–265), Hsun-tzu was the senior of the masters. Because Ch'i was still filling the vacancies in the ranks of the Adjunct Grand Masters, Hsun served three times as Libationer *(chi-chiu* 祭酒*)* there.

A modern chronology of his life imaginatively phrases his status during this decade as "active as Head of the restored Jixia Academy," but in his biography there is not a word about an academy or his headship of anything. Again we see that the vacancies are among the honorary Adjunct Grand Masters, not in the civil service.[39] This is a fair sampling of the early evidence. All of it is consistent with the four generalizations listed above (p. 20). It does not mention an academy. Nothing in it supports the notion that there was an organized institution that could be so called. It does not say, and offers no basis on which to conclude,

---

38. Hucker 1985: item 5939.

39. *Shih chi, 74:* 2348. Knoblock has argued, on the basis of late sources, that "fifty" is an inversion of "fifteen"; see 1982–1983: 33, and *passim* for the chronology. That would make the point not Hsun-tzu's advanced age, but (by Chinese standards) the youth at which he was so honored. In the second sentence I have maintained an ambiguity, since the time phrase may refer to the death or Hsun-tzu's age. See Brooks 1993– : Query 50, R5, for a forceful argument that Hsun-tzu was in Ch'i 257–255. *Chi-chiu* is ambiguous. It may be an honorary title, or the title of a high educational official. Hucker 1985: item 542 recommends "Libationer" for the first sense and "Chancellor" for the second. When Graham asserts (1989: 237) that Hsun-tzu "was the most eminent of the teachers in Ch'i, and three times performed the wine sacrifice as head of the Academy," he is overtranslating the first phrase *(tsui wei lao shih* 最爲老師*)*, and translating the second twice, using both senses! Since no source mentions an academy, there is no warrant for taking *chi-chiu* here as the title of its principal, nor does other early evidence support that interpretation. Liu Hsiang's 劉向 preface to the extant recension of *Hsun-tzu* which he edited at the end of the first century B.C., paraphrases this passage. Knoblock 1988–1994: III, 271–274.

that Tsou Yen, Hsun-tzu, and their like were regular civil servants in Ch'i.

It remains to ask, then, how this myth, a fixture in Occidental histories of Chinese philosophy, arose. Like most products of the imagination, it was not purely imaginary.

Evidence for an academy. The best histories of Chinese thought written in the West over the past half century have drawn for biographic data on the massive erudition of Ch'ien Mu. Ch'ien found in obscure and fragmentary sources, all notably later than Memoirs of the Grand Astrologer, evidence pertinent to Chi-hsia. He concluded that the "scholarly gentlemen-retainers" of Chi-hsia had disciples, that they received stipends large enough to support not only themselves but their pupils, and that one or more buildings were provided for teaching.[40]

In his factual conclusions, I believe, Ch'ien was quite correct. His first two points do not require comment, since patronage was expected to support a large household, and disciples were considered part of one's household (as were servants, both in numbers proportional to one's prestige). Teaching, if not very remunerative, was an acceptable hedge against the uncertainties of royal favor. These and other sources do not say that patrons encouraged their clients to teach.

Ch'ien did not attach to the activities he documented any word that might be translated "academy," nor did he speculate about an official institution. He described the buildings with prudent vagueness as "the organizational aspect of Chi-hsia." But it is not difficult to understand how Occidental Sinologists familiar with the schools, academies, and universities of Europe before modern times should be tempted to impose such notions on the frequently enigmatic shards of quotation that he recovered.

We must ask what Ch'ien's data on offices and other physical facilities actually reveal about organization. I translate, in chronological order, five citations that he considered pertinent.[41] Numbers 1 and 4 are also concerned with the origin of the name Chi-hsia (literally, "below Chi").

1. After 26 B.C.: "There was a Chi gate in the Ch'i capital; it was the gate in the western city wall. Outside it were school buildings *(hsueh-t'ang* 學堂*)*, which were the places established for studies by King Hsuan of Ch'i [see p. 22]. That was

---

40. Ch'ien 1935: 232–235 includes a table of seventeen philosophers generally associated with Chi-hsia, fourteen of them reliably.

41. The five quotations are from *Pieh lu* 別錄, *Wu ching i i* 五經異義, *Chung lun* 中論, 2. 18: 27a, a lost commentary by Yü Hsi 虞喜 (fl. 307–338), and *Ch'i ti chi* 齊地記. Nos. 1, 4, and 5 are quoted in *So yin* 索隱 commentary to *Shih chi*, 46: 1895, n. 6. I cannot find the passage in the only available ed. of *Wu ching i i.*

why [people] spoke of studies at Chi-hsia."⁴²

2. Ca. A.D. 100: "In the Warring States period, Ch'i established an office for Erudites." Ch'ien and others have suggested that this refers to Chi-hsia but that is only a guess. The pre-Han meaning of *po-shih* 博士 is poorly documented. In the Han it was used not only for officials responsible for teaching texts to disciples, but for ritual specialists subordinate to the Chamberlain for Ceremonials.⁴³

3. Ca. A.D. 200, from a discussion of patrons who sought clients but who did not make use of them: "In olden times Lord Huan of Ch'i (r. 363–357?) established the office at Chi-hsia and instituted the title of Grand Master in order to attract worthy people, to whom he [and his successors?] showed respect and favor. Meng-tzu and people like him sojourned in Ch'i. . . ."⁴⁴

4. Ca. A.D. 325: "There was a Chi Hill [near] in Ch'i *(or* near the Ch'i capital). Below it was established a guest house to accommodate itinerant gentleman-retainers." This passage indeed names a physical facility, but has nothing to do with instruction. The relation of the guest house to the mansions noted elsewhere is unknown. This may be late lore based on the name of the hill.

5. Ca. A.D. 400: "Next to the western gate of the Ch'i capital city wall, on both sides of the Hsi 系 stream, were lecture halls, the ruins of which used to be found there."⁴⁵

Just as the early sources mentioned clients without formal organization or utilitarian buildings, these late texts allude to buildings without details about occupants or formal organization. The earliest is from roughly two centuries after the last of the Ch'i kings; the latest, more than six hundred years after. There are notable inconsistencies. For instance, no. 3, not a historical work, claims that the organizational forms of patronage originated much earlier than any older source does. There is no detail to indicate what the "office" mentioned in nos. 2 or 3 did. No. 2 is almost certainly irrelevant to Chi-hsia. No. 3, on the simplest hypothesis, refers to an early bureau assigned to attract and cater to "worthy people." The alternative interpretation, that it was a bureau staffed with clients on regular

---

42. *Hsueh* 學, which I translate "study," is ambiguous. It may also mean "school," but in that sense it generally refers to an institution rather than a building.

43. Hucker 1985: item 4746.

44. Ch'ien's quotation reads not *kuan* but *kung* 宮, "mansions."

45. Knoblock translates "ruins of which still survive" (1988–1994: I, 261n9). *Wang wang* 往往 is ambiguous, but I do not believe "still" is among its possible meanings. Knoblock's translation of *chiang t'ang* 講堂 as "debating chamber" uses a Buddhist sense absent from China until after A.D. 420. The sense "lecture hall" is pre-Han.

appointments, is not backed by evidence, and does not tally with the earlier assertion that they "did not govern."

We are left with the obviously interesting assertions in nos. 1 and 5 about buildings used for teaching. They call for circumspection, because they are about ruins and stories told about them.

Three hypotheses come readily to mind. One, the conventional one, is that they prove a formal state institution existed, whether for philosophical research or instruction or both. The second is that they are late, useless lore. The third explanation takes seriously the consistency of nos. 1 and 5, the only two of the five fragments germane to the issue. It posits that they do not imply an institution, simply an amenity available to the few clients who had too many disciples to teach in their mansions. In this view the "lecture halls" were not the heart of Chihsia, but one more fringe benefit. Unlike the residences, impressive on the main thoroughfares, these halls were apparently outside the city wall. These interpretations differ fundamentally.

It is not easy to decide between these three possibilities. Reflecting on early education makes the third, all other things being equal, more probable.

Warring States education. The Han state sponsored instruction because of specific circumstances. The Ch'in government had made it a matter of urgent policy to remove books from circulation among the public. Due to this effort and the devastation of war in the fourth and third centuries, the transmission of texts, the bases of ritual and intellectual lineages, was in many cases decisively interrupted. Most people came to think of teaching as an uncertain enterprise. Early in the Han the government began to protect the founding texts of important traditions, and thus represented itself as essential to the security of scholarship. The appointment of Erudites, each to teach a single text, aimed to guarantee transmission. This system also provided the government with experts in ritual, governance, and other matters of administrative importance. In the second century this sponsorship became in principle exclusive rather than inclusive, as the empire established and enforced an orthodoxy.

These exigencies, which led to what has been called an imperial university, can hardly be read back into the local domains of the fourth century and their rivalry for "guests." As H. S. Galt in his *History of Chinese Educational Institutions* put it, prior to the Han, and in large measure afterward as well, education remained "an undifferentiated part of the social process."[46] No anecdote in the early sources suggests that patrons were concerned about, or even curious about, education. On the

---

46. Galt 1951: I, 98.

IV

THE MYTH OF THE NATURALISTS 27

other hand, the most successful were deeply concerned about contenting their clients, so that their rivals could not lure them away. The school buildings, which a source too late to be convincing calls lecture halls, were more likely provided as an amenity than integrated into something analogous to the Han "National University." They were hardly a think tank before its time.

Summing up. "Academy" has many meanings—first and foremost, of course, the name of the garden in which Plato taught. The dictionary gives three senses, depending on purpose, that may be pertinent: an educational institution, generally intermediate between school and university; a place of training, such as a riding academy; or a society or institution such as the Royal Academy founded to promote or cultivate the arts and sciences (paraphrasing *OED*, senses 3–6).

Neither education nor training were explicit purposes of patronage at Chi-hsia. The Ch'i rulers supported very few practitioners of the arts and sciences. The cultivation of such subjects was negligible compared to those that served the kings' immediate purposes. Practical advice on policy matters, miscellaneous feats of interstate intrigue, and the prestige that came from a large, prominent, mixed cast of favorites, do not define an academy. A few late patrons elsewhere reconciled this means to prestige with academic interests, notably Lü Pu-wei just before the Ch'in conquest, and the King of Huai-nan in decisively changed circumstances a century later. By proving that the two purposes were reconcilable, they underline the lack of effort to reconcile them at Chi-hsia.

There is another way to assess the mythical quality of this ancient "academy." Rather than ask whether the word applies, we might ask how solid are the specific claims conventionally made in connection with it, for instance those cited at the beginning of this section.

To call Chi-hsia "the intellectual center of ancient China" is defensible only in an extremely narrow sense. In the more than a century of its existence, we can be confident that at least fourteen intellectuals taken seriously by historians of philosophy were among the thousands, rarely chosen for their intellectual qualities, who partook of the patronage offered in Ch'i. One must grant, however, that at no other center, at least before that of Lü Pu-wei, can we be sure that there were more than fourteen.

Was there at Chi-hsia "a department of government in which distinguished scholars held official rank as grand officers"? No. This way of putting it simply ignores the centrality of patronage. Clients occasionally received official appointments, but it was the fickleness of royal sponsorship that defined Chi-hsia. Hsun-tzu's civil service post is a confusion of two types of Grand Masters.

Were the disciples of sojourners "obliged to master the doctrines of the many schools, the forms of argumentation, and the techniques of rhetoric at which the

Jixia scholars excelled"? This implies a collective curriculum cutting across the boundaries of scholarly lineages. Such a claim is so opposed to what we know about early education that detailed proof is hardly optional. There was nothing so intellectually promiscuous, by Chinese standards, before the nineteenth century. A disciple ideally had one master, and in practice usually had only one at a time. We can read the writings of many individuals with a broad command of classical traditions, but such breadth was not, so far as scholarship has documented, what schooling aimed to provide.[47]

To speak of "many scholastic sects," each with a program of self-cultivation and textual interpretation distinct from that of the others, is quite a different matter, and does better justice to the record. Among other advantages, it avoids the notorious vagueness of the cliché "school."

A Chi-hsia academy, in any normal sense of the last word, is simply a myth. Sinologues have wishfully imposed this medley of Chinese imperial and modern Western institutions on a chaotic, creative set of circumstances that prevailed shortly before the empire came into being. We can find in the particularities of the Ch'i capital and other centers of patronage some new facets of the sociology of knowledge in the last phase of the Warring States. Studies that set myths aside and look closely at the actual relationships of patronage, the interaction of scholars in circumstances that polarized their intellectual lives around the tastes and whims of a royal benefactor, and the richly documented insecurity that was inseparable from the opportunities for recognition, are likely to be more productive.

Conclusion

Modern scholars have combined Ssu-ma T'an's ahistorical "yin-yang school" with a scientific reading of Tsou Yen, whom Ssu-ma did not find relevant to this "school," to invent the seminal Naturalists. To house Tsou and others they have, as the buzzword puts it, socially constructed a Chi-hsia Academy out of a tiny subset of the Ch'i kings' decorative clients. These feats are typical of the natural tendency toward anachronism and Westernization that have added many other confusions to the history of Chinese thought. The same distraction is responsible for efforts to explain the evolution of Chinese thought and institutions in terms of isms, among the most quintessentially modern constructs. I take up that topic in Chaps. VI and VII.

It is unrealistic to expect scholars to slough off, once and for all, the assumptions that they learned, growing up, as common sense. The effort to do so, how-

---

47. Pre-Han education is a badly neglected topic; see Sivin 1995.

ever, is well spent for those of us who follow our curiosity to times and places fundamentally unlike our own. To the extent that we can set aside such parochialisms, we have an opportunity to see fascinating patterns that have little in common with those prevalent in the modern world, or at least in those non-Chinese parts of it that tend to breed students of China.

Finally, I suggest that the alternatives propounded here are considerably more interesting as a basis for historical inquiry than the myths themselves.

We can now substitute for a "School of Naturalists" an evolution of concepts built on resonant categories. Beginning with a not yet dated phase of experimentation with a great range of possibilities, by the first century B.C., a coherent theory of yin-yang and the Five Phases as aspects of *ch'i* emerged. Like its Han predecessors, it was simultaneously political, moral and naturalistic. Tsou Yen did not interrupt this gradual metamorphosis with a full-blown theory before its time. He did organize an assortment of marginal moral and political notions into a historical dimension that firmly tied them to great issues of the time, especially the foundations of political morality and the rise and fall of states.

What is interesting about Chi-hsia, once the myth is subtracted, is the large role of Ch'i in forming institutions of patronage. The kings supported large numbers of intelligent and able people who, in a largely aristocratic society, would otherwise have found no support for their talents. Whether they were mostly philosophers, or even intellectuals, is beside the point. The rulers found room for a handful of remarkable philosophers among their crowd of weapons consultants, experts in treachery, and so on. Their example, although not the first, made the practice significant and launched later large-scale patrons. That is momentous enough without exaggerating the academic importance of Chi-hsia. We can see in its realities ample reason to probe how patronage worked, what it accomplished, and why. Perhaps we will even learn, more generally, how much more we can understand when we give up the modern habit of shutting thought and society in mutually exclusive compartments.

References

Early Chinese Sources

Sources not listed here are cited by the texts in standard concordances, except that the Standard Histories are cited in the Chung Hwa Book Co. 1974 series.

Chou li 周禮 (Rites of Chou). Anonymous, fourth century or earlier, extensively revised by Liu Hsin 劉歆 at the end of the first. Commentary cited from Han Fen Lou reprint of *Shih-san ching chu su* 十三經注疏.

Ch'un-chiu fan lu 春秋繁露 (Abundant Dew on the Spring and Autumn Annals). Tung Chung-shu 董仲舒, parts written 156/130 B.C. *Ch'un-chiu fan lu i cheng* 春秋繁露義證.

Chung lun 中論 (Discourses that hit the mark). Hsu Kan 徐幹, ca. A.D. 200. Surviving portions in *Han Wei ts'ung-shu* 漢魏叢書.

Huang-ti nei ching 黃帝內經 (Inner Canon of the Yellow Lord). Anonymous, probably first century B.C. In Jen Ying-ch'iu 1986.

Kuo yü 國語 (Discourses of the States). Anonymous, late fifth to second century B.C. In Wolfgang Bauer, *Kuo yü yin-te* 國語引得 (2 vols., Taipei: Chinese Materials and Research Aids Service Center, 1973).

Lü shih ch'un-ch'iu 呂氏春秋 (Springs and autumns of Master Lü). Compiled under patronage of Lü Pu-wei 呂不韋, ca. 239 B.C. In Ch'en Ch'i-yu 1984. Cited by *chüan*, section, *p'ien*, and page.

Pieh lu 別錄 (Separate records). Liu Hsiang 劉向, after A.D. 26. Lost; cited passage from *T'ai-p'ing huan yü chi* 太平環宇記 (983; Chin-ling Shu-chü ed.), *18*: 6a.

Shui ching chu 水經注 (The canon of waterways, annotated). Li Tao-yuan 酈道元, A.D. 527. In *Ho chiao shui ching chu* 合校水經注.

Wen hsuan 文選 (Choice of literature). Compiled by Hsiao T'ung 蕭統, 526/531. Hui Wen T'ang Shu-chü ed.

Wu ching i i 五經異義 (Alternative interpretations of passages in the Five Classics). Hsu Shen 許慎, ca. A.D. 100. Lost; fragments in *Wu ching i i shu cheng* 疏證 (*Huang Ch'ing ching chieh* 皇清經解, vols. 313–315).

Yü han shan fang chi i shu 玉函山房輯佚書 (Reconstituted books from the Jade Box Mountain Studio). Ma Kuo-han 馬國翰, 1853. Wen Hai, 1970, ed.

Modern Sources

Biagioli, Mario. 1993. *Galileo, Courtier. The Practice of Science in the Culture of Absolutism*. Science and its Conceptual Foundations. University of Chicago Press.

Brooks, E. Bruce, et al. 1993–. Notes and Queries of the Warring States Working Group, University of Massachusetts, Amherst. Privately circulated.

Brooks, E. Bruce. 1994. Review Article: The Present State and Future Prospects of Pre-Han Text Studies. *Sino-Platonic Papers*, 46: 1–74. Review of Loewe 1993, mostly devoted to an outline of Brooks's own studies of the coming together of early texts.

Chan, Wing-tsit. 1963. *A Source Book in Chinese Philosophy*. Princeton University Press.

Chavannes, Édouard. 1895–1969. *Les mémoires historiques de Se-Ma Tsien*. 6 vols. Paris: Ernest Leroux. Partial translation of *Shih chi*.

Ch'ien Mu 錢穆. 1935. *Hsien Ch'in chu-tzu hsi nien* 先秦諸子繫年 (Chronological studies of pre-Han philosophers). Shanghai: Commercial Press. Rev. ed., Hong Kong University Press, 1956.

Eno, Robert. 1990. *The Confucian Creation of Heaven: Philosophy and the Defense of Ritual Mastery*. State University of New York Series in Chinese Philosophy and Culture. Albany: State University of New York Press.

Fung Yu-lan. 1952–1953. *A History of Chinese Philosophy*. Trans. Derk Bodde. 2 vols. Princeton University Press.

Galt, Howard S. 1951. *A History of Chinese Educational Institutions*. London: Arthur Probsthain. Only vol. I, to A.D. 960, published.

Graham, A. C. 1978. *Later Mohist Logic, Ethics, and Science*. Hong Kong: Chinese University Press. A study of the text, its philosophy and language, and an edited text and translation.

Graham, A. C. 1979 (publ. 1980). How Much of *Chuang-tzu* did Chuang-tzu Write? *Journal of the American Academy of Religion*, 47. 3: 459–501. From a 1976 conference. Reprinted in Graham 1990: 283–321.

Graham, A. C. 1986. *Yin-Yang and the Nature of Correlative Thinking*. Occasional Paper and Monograph Series, vol. 6. Singapore: Institute of East Asian Philosophies.

Graham, A. C. 1989. *Disputers of the Tao. Philosophical Argument in Ancient China*. La Salle, IL: Open Court.

Graham, A. C. 1990. *Studies in Chinese Philosophy and Philosophical Literature*. State University of New York Series in Chinese Philosophy and Culture. Albany: State University of New York Press.

Hsu Chung-shu 徐中舒, editor. 1989 (publ. 1990). *Chia-ku-wen tzu-tien* 甲骨文字典 (Dictionary of oracle script). Chengtu: Ssu-ch'uan Tz'u-shu Ch'u-pan-she.

Hucker, Charles O. 1985. *A Dictionary of Official Titles in Imperial China*. Stanford University Press.

Jen Ying-ch'iu 任應秋, editor. 1986. *Huang-ti nei ching chang-chü so-yin* 黃帝內經章句索引 (Phrase index to the Inner Canon of the Yellow Lord). Beijing: Jen-min Wei-sheng Ch'u-pan-she.

Kaltenmark, Max. 1949. Les Tch'an-wei. *Han-hiue*, 2. 4: 363–373.

Knoblock, John. 1982–1983. The Chronology of Xunzi's works. *Early China*, 8: 29–52.

Knoblock, John. 1988–1994. *Xunzi. A Translation and Study of the Complete Works*. 3 vols. Stanford University Press. Excellent study and good translation of *Hsun-tzu*.

Loewe, Michael, editor. 1993 (publ. 1994). *Early Chinese Texts. A Bibliographical Guide*. Early China Special Monograph Series, 2. Berkeley: Society for the Study of Early China and Institute of East Asian Studies, University of California.

Makeham, John. 1994. *Name and Actuality in Early Chinese Thought*. State University of New York Series in Chinese Philosophy and Culture. Albany: State University of New York Press.

Moran, Bruce, editor. 1991. *Patronage and Institutions. Science, Technology, and Medicine at the European Court, 1500–1750*. Rochester: Boydell Press. Mostly papers from 1989 Hamburg Congress, largely descriptive.

Needham, Joseph, et al. 1954– . *Science and Civilisation in China*. Cambridge: At the University Press.

Nylan, Michael. 1992. *The Shifting Center. The Original "Great Plan" and Later Readings*. Monumenta Serica Monograph Series, vol. 24. Nettetal: Steyler Verlag. Primarily a study of shifting meanings of the text.

P'ang P'u 庞朴. 1980. *Po shu wu-hsing p'ien yen-chiu* 帛书五行篇研究. N. p.: Ch'i-lu Shu She.

Queen, Sarah. 1991. From Chronicle to Canon: The Hermeneutics of the *Spring and Autumn Annals* according to Tung Chung-shu. Ph.D. diss., History and East Asian Languages, Harvard University. Revised version in press.

Rickett, W. Allyn. 1985– . *Guanzi. Political, Economic, and Philosophical Essays from Early China*. 2 vols., 1 to date. Princeton University Press.

Roth, Harold David. 1992. *The Textual History of the Huai-nan Tzu*. Monograph Series, 46. Ann Arbor: Association for Asian Studies.

Shaughnessy, Edward L. 1983. The Composition of the *Zhouyi*. Ph.D. diss., Asian Languages, Stanford University. On the early strata of the Book of Changes.

Schwartz, Benjamin I. 1985. *The World of Thought in Ancient China*. Cambridge, MA: Belknap Press.

Sivin, Nathan. 1987. *Traditional Medicine in Contemporary China. A Partial Translation of Revised Outline of Chinese Medicine (1972), with an Introductory Study on Change in Present-day and Early Medicine*. Science, Medicine and Technology in East Asia, 2. Ann Arbor: Center for Chinese Studies, University of Michigan.

Sivin, Nathan. 1992. Ruminations on the Tao and its Disputers. *Philosophy East and West*, 42. 1: 21–29. Remarks on Graham 1989.

Sivin, Nathan. 1995. Text and Experience in Classical Chinese Medicine. In *Knowledge and the Scholarly Medical Traditions*, ed. Don G. Bates, pp. 177–204. Cambridge University Press.

Watson, Burton. 1958. *Ssu-ma Ch'ien. Grand Historian of China*. New York: Columbia University Press.

Watson, Burton. 1961. *Records of the Grand Historian of China. Translated from the Shi chi of Ssu-ma Ch'ien*. 2 vols. New York: Columbia University Press.

Yates, Robin D.S. 1994. The Yin-Yang Texts from Yinqueshan: An Introduction and Partial Reconstruction, with Notes on their Significance in Relation to Huang-Lao Daoism. *Early China*, 19: 75–144.

V

On the Limits of Empirical Knowledge in Chinese and Western Science*

My theme is the limits of scientific inquiry, that is, ancient Chinese concerns about whether nature can be comprehended fully by rational, empirical investigation.[1]

We find the limitations of observational knowledge taken up regularly in writings on astronomy, the most exact of the ancient sciences, but not in astronomy alone. Because this theme mainly appears in technical discussions rather than in writings of a general kind, we can avoid the dangerous assumption that the opinions of philosophers determined what scientists thought in China. What does emerge from the writings of fifteen hundred years is an abiding interest in the idea that the scale of the cosmos is too large, and the texture of nature is too fine, too subtle, too closely intermeshed (*wei, miao* and so forth) for phenomena to be fully predictable. This proposition denies that the physical world can be fully penetrated by study or fully described in words or numbers. The cognitive strategy behind it evolved, and its history can be traced.

This is not the indeterminacy of contemporary theoretical physics, a point to which I will return in the conclusion, but a range of qualitative convictions drawn from mundane experience. A philosopher might divide this gamut into ontological and epistemological indeterminacy. Epistemological indeterminacy denies that it is possible to comprehend the order and regularity of the universe through study. Ontological indeterminacy asserts

---

* An earlier version of this article has been published in *Time, Science, and Society in China and the West* (*The Study of Time V*), J.T. Fraser et al., eds. (Amherst: University of Massachusetts Press, 1986.) It appears herein courtesy of the Publisher and the Editors.

that, beyond a certain point, the universe lacks the order and regularity that empirical study strives to find. The Chinese did not make such a distinction, which has analytic uses that will become clear when in my conclusion I summarize the evolution of thought about the limits of knowledge.

Before looking at this idea historically, let me introduce two short but typical statements about astronomy's inherent limitation as a science. Here is an early assertion of this idea, which the polymath Cai Yong wrote in A.D. 175:

> The astronomical regularities are demanding in their subtlety, and we are far removed from the Sages [who founded this art]. Success and failure take their turns, and no technique can be correct forever. ... The motions of the sun, moon, and planets vary in speed and in divergence from the mean; they cannot be treated as uniform. When the technical experts trace them through computation, they can do no more than accord with [the observations of] their own time. Thus there come to be [differences between] the techniques of various periods.[2]

Cai's lack of confidence should not be dismissed as a simple reflection of the crude techniques available in the second century, as we will see when we return to that period. First, let us take a passing look at a much later time, when Western astronomy was widely known. Perhaps the last such statement on the part of a scholar well qualified in astronomy comes from Dai Zhen (1724–1777), the leading philologist and in many respects the most influential intellectual of his time, in his essay on solar theory:

> In all prediction of celestial phenomena, as time passes there are bound to be errors that are due neither to inaccuracies in positional data nor to the need for periodic revision of computational methods. The sphere of the sky is so enormous that number and measure cannot get to the end of it, just as when we measure something an inch or a grain at a time, there is bound to be discrepancy by the time we have counted up to a foot or an ounce. Because this is so, we define units of time and observe phenomena so as to make the most of our techniques. Our best course is to continue using a technique so long as its inaccuracies remain imperceptible, and to correct it once they have been noticed. This is a matter of indeterminacy, as error accumulates over a long period.[3]

V

Chinese and Western Science

Now let us look at some of the early philosophical ideas that may have formed the backdrop to statements about cosmic indeterminacy. Then we can consider the historical development of such statements themselves, in astronomy and in other departments of knowledge. Finally, we can ask what light this theme casts on the character and history of prediction as a goal of Chinese science.

It is important to look at each statement, not as a great idea that must be taken at face value, but as a reflection of the viewpoint of someone with certain interests in certain historical circumstances. Since this is only a sketch of work in progress, I will consider only a few sources, and summarily indicate their circumstances.

In the pre-Han classics, it is remarkable how seldom words such as *wei*, *miao* and *xuan*, which later imply subtlety and indeterminacy, refer to the possibility of knowledge. One pertinent treatise is the Great Commentary to the Book of Changes, the major source of orthodox cosmology from the Han on. There the word *wei* refers to the gentleman's sensitivity to the ethical implications of a situation as soon as they begin to evolve, long before they become obvious. Its statements are clearly not about factual or theoretical knowledge of the natural world.

In the *Laozi* we find several other pertinent ideas. The Way itself in its constant and unchanging aspect, we are told, is shadowy and indistinct, and cannot be described. *Wei* and similar words are never clearly applied to the empirical world or to theoretical knowledge, but *wei*, *miao* and *xuan* appear together in one line that describes the exemplary gentleman:

> Of old those adept in the Way
> Their mastery recondite, subtle, and mysterious,
> Were too profound to be known ...

One might guess from familiarity with the *Laozi* as a whole that the Sage becomes indeterminate as he models himself on the indeterminate Dao that he contemplates, but the text does not go quite that far.[4]

To sum up, by 300 B.C. certain aspects of the Dao were described as indeterminate, but these aspects are not identified with the phenomenal world, which can be described. Words implying indeterminacy rather than ineffability are used to describe the character of the ideal person rather than that of the cosmos. Not surprisingly, the key words above, which later appear in astronomical discussions, do not occur in any germane sense in the *Zhuangzi*. That book consistently rejects the humanistic orientation that we have found in the *Laozi* and the Great Commentary, and finds the logical description of experience useless.[5]

The indeterminacy of the cosmos finally appears in less ambiguous form in the *Chunqiu fan lu* (135 B.C.), Dong Zhongshu's attempt to construct for the Han state a new intellectual orthodoxy that used the cosmic order to undergird the political order:

> The Ancients had a saying that if you do not know the future you can see it in the past. Now, in the study of the Spring and Autumn Annals, statements about the past are used to clarify the future. But because its words embody the subtlety of the natural order [*tian zhi we*], they are hard to comprehend.[6]

Dong is using the subtlety of nature, its resistance to being understood, as a metaphor for the arcane language of the orthodox classic.

Indeterminacy in Astronomy

Now let us pass on to the earliest statements about the limits of astronomical prediction. Most such assertions appear in the Standard Histories that chronicle the affairs of each dynasty. Computing the ephemeris and interpreting ominous phenomena were matters of concern to the state, which attempted to center this activity in its Astronomical Bureau and Imperial Observatory. Once astronomy was thus tied to the imperial charisma, a succession of computational systems for predicting the positions and chief phenomena of the sun, moon and planets was officially adopted; there were nearly fifty such systems between the beginning of the first century B.C. and the middle of the seventeenth century A.D. Improvement in technique did not lead to revision of the system in use, but to its complete replacement by a new one. This, at least, was the principle; in practice we find occasional traces of piecemeal revision, and several replacements that were no improvement at all. New systems were sometimes ordered up to signal a new dynasty or to announce a "new deal." On such occasions innovation was too much to expect.

V

Chinese and Western Science

Nearly all the judgments about astronomical systems in the Standard Histories were set down when a new system was presented for adoption, or when an old system was regularly failing to give accurate predictions. Today a sensible person who plans to buy an automobile, and who wants to find out about the limitations of a certain design, would not ask a salesman who sells that model, but would consult someone who has been driving one for some time. In astronomy, as well, familiarity breeds frankness. We usually find doubts about the extension of knowledge expressed, not with respect to a system newly presented for adoption, but when the shortcomings of an established system have become apparent; and this is all the more true when, because a competitor is in the offing, the tenure of the established system seems limited.

Probably the most serious period of crisis and reassessment in early mathematical astronomy began shortly before the end of the first century A.D., when the Grand Inception system (*Taichu li*), adopted in 104 B.C. and greatly developed as the Triple Concordance system (*Santong li*) a century later, was about to be replaced.[7] In A.D. 92, Jia Kui presented the throne with the first major document of this crisis, his "Discussion of Calendrical Astronomy", (*Lun li*). He writes of what we would call the imprecision of constants. Even the constants instituted by the legendary Sages who founded astronomy in the Golden Age,

> unable to endure unchanged [lit., "run through"] for thousands and myriads of years, must be altered and replaced. We [in later times] first determine angular measures and numerical quantities from the observations made over long intervals, and select those that accord with the positions of the sun, moon and planets. ... [Our] methods will thus differ from one period to another. The Grand Inception system [of two centuries earlier] cannot give accurate predictions for the present day; nor can the new system provide correct computations back to the beginning of the Han period. The computational methods of a single school can only be applicable within an interval of three hundred years. ... When the Han first attained power, it would have been appropriate to adopt the Grand Inception system [because time had come for renewal]; but there was no such reform until 104 B.C., 102 years later. Thus, early in the dynasty, there were lunar conjunctions the day before the last day of the month [i.e., two days before mean lunation], but by the

time of Emperors Cheng and Ai (32 B.C. to A.D. 1), the second day of the month was being taken as the day of lunation [i.e., the civil month was routinely set back one day], so that most conjunctions would occur on the last day of the month [which was allowable in the early Han]. This is clear proof [that calendar reform is periodically necessary].[8]

Why cannot even the Sages discover constants precise enough to be used forever? That Jia explained earlier in his report: "The Celestial Way being irregular, lacking uniformity, there are bound to be remainders. These remainders will have their own disparities, which cannot be made uniform."[9] Imprecision is not a characteristic of the constants, that is, but of the universe.

By the end of the second century, as my earlier quotation from Cai Yong indicates, the implications of indeterminacy had become much broader than in Jia Kui's time, and were affecting prediction in ways that did not depend only on the precision of constants.

Although the sun moves along the ecliptic, and the moon is never more than six degrees from it, Han astronomers measured their positions along the equator. The ability to convert mathematically from positions on the ecliptic and the lunar orbit to equatorial right ascension and vice versa was beyond the simple linear techniques then in use. This made major improvement in eclipse theory — the central problem in traditional astronomy — seem hopeless. A report of lunar-eclipse prediction of the late second century outlines this difficulty at some length, and then draws an eloquent conclusion:

In view of this [limited feasibility of mathematical solution], there is no point in rejecting any method that does not conflict with observation, nor in adopting any method whose utility has not been demonstrated in practice. The Celestial Way is so subtle, precise measurement so difficult, computational methods so varying in approach and chronological schemas so lacking in unanimity, that we can never be sure that a technique is correct until it has been confirmed in practice, nor that it is inadequate until discrepancies have shown up. Once a method is known to be inadequate, we change it; once it is known to be correct, we adopt it: this is called "sincerely holding to the mean."[10]

V

Chinese and Western Science 171

The anonymous author is expressing resignation in the face of the crisis I have referred to. Imprecise constants could always be revised, but it was now clear — puzzlingly clear — that Han assumptions about the character of the celestial phenomena were beginning to break down. It was beginning to be apparent that certain phenomena, especially eclipses, could not be described by simple cyclic or linear methods. Finally, when this difficulty could not be resolved over several centuries, astronomers stopped trying. Cosmological hypotheses no longer ordered their computational techniques. They bought the power to predict in the simplest possible way, at the cost of the power to explain. This is a cost that greatly limits the power to predict in the long run, a lesson that did not become apparent until the seventeenth century, when the best astronomers of the time enthusiastically recognized the explanatory power of the geometric models introduced by the Jesuits.[11]

Not everyone had reason to accept the idea of astronomical indeterminacy. We might expect people defending new astronomical systems rather than criticizing old ones to argue against it; indeed, examples are not hard to find. There is the spirited and rather exasperated rejoinder of Zu Chongzhi, one of China's greatest mathematical astronomers, against the attack on his new system by Dai Faxing.

In or near 463, Dai developed an extensive argument along the lines of those I have quoted. Zu's defense, pragmatic rather than theoretical, is too long to cite completely:

> The writings of the Xia, Shang and earlier dynasties have been lost; but the historical chronicles of the Spring and Autumn period and the Han period record eclipses and lunations with care for detail; they constitute clear evidence. Testing my astronomical system by their use, I find the data entirely in accord [with my computations]. There is truly nothing speculative in [my system]. It takes precision as far as possible, so that over a span of a thousand years there is no discrepancy; whatever it be, far away though it be, it can be known. Now, I have studied all the ancient methods, and there I find much that is inexact. Computations are off by as much as three days, and the beginnings of *qi* periods by as much as seven hours. I know of no [ancient system] that can accurately predict the phenomena of the present time.[12]

Zu's claim that there would be no discrepancy in predictions over a thousand years was excessive. It did not accord with informed opinion, could not be proven in practice, and he did not make it persuasive in principle. For reasons as much political as technical, he did not carry the day. Despite its excellence, his system was not officially adopted for fifty years.

In 729, Ixing discussed the technique for predicting lunations in his new Great Expansion system (*Dayan li*). He politely suggested that, even if the course of the cosmos were inherently too irregular to be fully comprehended — and he did not minimize its irregularity — that would be irrelevant to the work of prediction:

> If the anomalies in the celestial positions [of the moon] actually fluctuated with time, providing rebukes [to the ruler] that the regularity of astronomical constants cannot encompass, and substituting for regularity a mutability [that derives] from the inaccessible [fine structure of the cosmos], this would be a matter beyond even [the ability of] Sages to assess. It can hardly lie within the scope of mathematical astronomy.

This is a more meaningful statement than its brevity and skeptical tone make it look. In the first place, it reminds us that Ixing was anticipating exactly the sort of argument that Zu Chongzhi had had to fight off. The idea that astronomical knowledge was inherently limited was now being used even against the best new systems rather than just to explain the failure of old ones. Second, for Ixing the conceptual crisis I have mentioned was long over, and a disinterest in cosmology was the norm among astronomers. They rarely took up questions of the actual spatial relations or physical realities underlying the phenomena. It is curious, considering Ixing's lack of interest in these questions, that he remains the last great astronomer to give cosmology an important role in his computational system. His cosmology was not physical, however, but drew in a curiously antiquarian way on the numerology of the Great Commentary to the Book of Changes.[13] Still, his curiosity in such matters was much narrower than that of his Han predecessors. The astronomical systems that followed Ixing's were narrower still from the viewpoint of cosmology.

Decreasing interest in the metaphysical and physical patterns that underlie such phenomena is not necessarily associated with more "scientific" trends in astronomy, either in China or in the

West, because in scientific work (as distinguished from certain ideal schemes of philosophers and historians of science) analysis of data and thought about their ultimate significance interact. It was the demand for a coherent and intelligible cosmic order that motivated Copernicus, Galileo and Kepler to innovate in directions that became decisive for modern science.

New Issues in the Song

By the Northern Song period, discussions of the sort I have summarized were either too rare or too familiar to record. Many of the difficulties that had originally suggested inherent limitations to knowledge were no longer difficulties; for instance, the time of lunar eclipses could be predicted with some confidence. The issue for the working astronomer, as I have said, had become not knowability but technical progress. Whether some day his science might reach those ultimate limits, or would always fall short, was not an urgent problem.

In the Song period, the idea of indeterminacy suggested ultimate questions of a new kind. These questions came from astronomers better prepared than their predecessors to explore methodological and epistemological aspects of their science, and from philosophers whose main interest was those aspects. My examples will be Shen Kuo (or Shen Gua, 1031-1095), who counted professional astronomy as one of his enormous range of accomplishments, and Zhu Xi (1130-1200) and Cai Yuanding (1135-1198), to represent the philosophers.

In his Brush Talks from Dream Brook (*Mengqi bitan*), Shen summarizes the lost preface to his Oblatory Epoch system (*Fengyuan li*), an innovative document:

> Those who discourse on numbers [by which he means all regularities that make prediction possible], it seems, [can only] deal with their crude after-traces [*ji*]. There is a very subtle [*wei*] aspect to numbers that those who rely on mathematical astronomy are unable to know; [what they can know of] this aspect is, all the more, only after-traces. As for the ability [of the sagely mind as exemplified in the Book of Changes] "when stimulated to encompass every situation in the realm," after-traces can play no role in that [wisdom]. That is why "the spirituality that makes foreknowledge [possible]" cannot readily be

> sought through after-traces, especially when one has access only to the crudest ones. As for the very subtle traces I have mentioned, those who in our time discuss the celestial bodies depend on mathematical astronomy to know them, but astronomy is no more than the product of speculation [*yi*].

Shen proceeds to develop an epistemological point that comes up several times in his writing, namely that in order to know, we break the continuity of nature into blocks of time that we treat as though each were uniform. As he puts it,

> The uninitiated say that, mathematical knowledge of the heavenly bodies being difficult to be sure of, only correlations between the Five Phases and time periods are reliable, but this is also untrue. The uninitiated who discuss the cyclic alternations [*xiaozhang*] of the Five Phases consider only the year. Thus [they know that], after the winter solstice the sun's motion is in the phase of Expansion [i.e., the equation of center is negative] and thus yin, and at the equinoxes corresponds to the mean. They do not realize that in the course of a month there is also an alternation. Before opposition, the moon's motion is in the phase of Expansion and thus yang; after opposition, it is in the phase of Contraction and thus yin; and at the quadratures, it corresponds to the mean.

As for the associations of Spring with Wood, Summer with Fire, Autumn with Metal and Winter with Water, these are also true of the month, and not only of the month but of the day. The "Basic Questions" of the Inner Canon of the Yellow Lord [*Huangdi nei jing su wen*] says "when the disorder is in the hepatic system, the onset [of an attack] is between 3 and 7 A.M., and the most serious time is between 3 and 7 P.M. When it is in the cardiac system, the onset is between 9 A.M. and 1 P.M., and the most serious time is between 9 P.M. and 1 A.M." Thus a single day has four seasons of its own. How do we know that there are not four seasons in each hour, or in each mark,[14] each minute, each instant? And how do we know that there are not a greater four seasons in each decade, century, Era cycle, Coincidence cycle and Epoch cycle? As for the association of Spring with Wood, within a period of ninety days there must be one [completed] cycle of alternation within another. It is impossible that the last hour of the 30th of the third month should belong to Wood, and the first hour of the next day abruptly belong to Fire. Matters of this sort are not to be settled by the methods abroad in the world.

In this second part of his short essay, Shen is writing about techniques of foreknowledge that depend, not upon observations of celestial events, but on associating in rotation the Five Phases with periods of time (for instance, the year, month, day and hour of birth), in order to yield interpretations of the latter. This simple, repetitive approach may have begun with astrology, but has been completely abstracted from what happens in the sky. Such methods cannot be reliable, Shen argues, because they imply regular and abrupt transitions from one block of time with its corresponding phase to the next. But such "quantum" transitions belie the continuous variation in motion of the celestial bodies from which the validity of the methods ultimately derives. This underlying continuity of variation, ignore it though we may, pervades time at every level from the fleetest instant to the long cycles of calendrical reckoning (the Epoch Cycle of the Han was 4,560 years).

Anyone familiar with the philosopher of physics Alfred North Whitehead (1861-1947) will find this line of reasoning familiar. Shen, like Whitehead nine centuries later, was saying that a central problem for science is the gap that seems to separate our unconnected experiences from the unitary causal world that lies veiled in back of them.[15]

Scientific mensuration is necessarily an act of abstraction. Near the beginning of his proposal of 1074 for a new armillary sphere, Shen argues this point with great clarity — for the first time in history, I believe:

> Degrees [on the equator and ecliptic] are invisible; what is visible are stars. [The paths] followed by the sun, moon and planets are occupied by stars. Twenty-eight stars are located [exactly] on a degree division; they are called "mansions" [*she*]. It is mansions that make it possible to measure degrees, and degrees that make it possible to create numerical regularities. Degrees are things that exist in the sky. When we make an armillary sphere [to measure intervals between real bodies], the degrees exist in the instrument. Once the degrees are in the instrument, then the sun, moon and planets can be isolated [*tuan*] in the instrument, and the sky no longer is involved. It is because the sky is no longer involved that what is in the sky is not difficult to know.[16]

Shen implies that one can know either about the organismic universe as a whole ("the sky") or about particular phenomena in it. Observational, empirical science can yield only knowledge of the second kind (a point about which Whitehead would disagree). In doing so it rules out perceptions of the first kind. They can be reached only by other kinds of knowledge — intuition, illumination and so on — in which Shen is equally interested.

Cai Yuanding, another polymath, did away with one of the basic confusions of the Han astronomers. Cai wrote at least one book on mathematical astronomy. This detailed study of Ixing's astronomical system is lost, but certain important arguments are preserved in the conversations of Cai's mentor and friend, Zhu Xi, with Zhu's disciples. It is clear from Zhu's paraphrases that Cai believed inaccuracies of prediction do not imply indeterminacy. Beneath an irregularity may lie a more complicated regularity waiting to be discovered:

> When an astronomical system is first being designed, the discrepant measures of the celestial rotations are combined and included in the computations. So, many years later, there will be discrepancies of so many fractions of a degree, and after so many additional years, of so many degrees. If, from these discrepant quantities, correct quantities are computed, and this process is repeated to the limit [i.e., until the magnitude of the correction becomes negligible, *jintou*], the astronomical system can be made essentially correct and free of discrepancies.
> People today, never having reached a comprehensive and correct understanding [*da tong zheng*], simply claim that the discrepancies are inherent in the celestial rotations. They make systems of computation seeking accord with the celestial phenomena, but their ephemerides become increasingly discrepant. The point is this: they do not understand that, if the sky is able to manifest a certain discrepancy, it is precisely because the celestial rotation must be of that kind.

A discrepancy does not, as Cai's contemporaries think, reflect an anomaly in nature, but rather a more complicated regularity than originally assumed. Ad hoc technical adjustments simply obscure the discrepancies. In doing so they also obscure the underlying complex pattern that will keep generating discrepancies until it is understood.

This is an important perception about method. When we

remember the gradual discovery in Europe of the various inequalities that complicate the moon's motion, we are reminded that the failure of Hipparchus' (fl. ca. 130–150 B.C.) first inequality to give perfect predictions suggested to Ptolemy (ca. A.D. 100 – ca. 165) the evection, the second inequality. The discrepancies for which the evection could not account suggested to Tycho Brahe (1546–1601) the third and fourth inequalities.

Similar processes of discovery can be traced in the history of Chinese astronomy, but Cai was the first (at least the first reflected in the surviving record) to make the point explicit. His own attitude toward the determinate character of the phenomena was decidedly nuanced, even though he did not accept his contemporaries' reasoning about what implies indeterminacy. Zhu quotes him elsewhere to the effect that "there is no constancy in the celestial rotations; the sun, moon, and planets are accumulations of *qi*; they are all moving things [*dong wu*]. Their angular motions may be faster or slower, beyond the mean or short of it; they are not naturally uniform."

We have already seen that Cai considers these motions predictable. As Zhu remarks, "Cai was not saying that there is nothing determinate in the rotation of the sky, but that the angular motions of the luminaries are as they are."[17]

The last quotation from Cai is best understood in the light of similar beliefs held by such Occidental luminaries as Plato and Ptolemy, for whom the planets are divine and self-propelled. This view was an alternative to the idea that the planets are passively driven in their rounds by some common source of motion that determines the speed of each. A philosopher who finds no evidence for mechanical linkages powering the celestial luminaries is likely to find the idea that each planet is the source of its own motion more plausible. Its velocity is thus internally determined and arbitrary with respect to those of other planets. If constant, it is arbitrarily constant. The 'erratic' retrogradations of the planets are thus accounted for; they could not be explained by those who considered the planets passive.

In Greece and Hellenistic Egypt, a source that determined its own motion would be divine; in China, it was an animal-like "moving thing." Neither implies that its motion must necessarily be irregular or unknowable. The astronomer simply attempts to impose order upon whatever irregularities his observations reveal. What matters about Cai's attitude is that he faced the issue of indeterminacy instead of making assumptions that render it all the more problematic. He could thus imply that, even when taken seriously, it need not impede astronomy.

Zhu Xi, like Cai Yuanding, did not believe that there were inherent limits to the astronomer's power to predict. His attitude emerges in several discussions of a chapter from the Mencius (Mengzi, 4B:26), which, in explicating the innate moral nature of man (xing), refers to the work of the astronomer. As Zhu explicates the relevant passage, it would mean: "Consider the sky so high, and its markpoints so distant; if we seek the traces of actual events [gu], without leaving our seats we can bring before us the solstices of a thousand years."

There is no basis for reading into Mencius' casual statement a pronouncement on the limits of empirical knowledge in astronomy. But Zhu Xi, in one of several conversations about the chapter, relates this passage to that question:

> Mathematical astronomers computing backward from the present day are able to proceed without error even to the moment when the physical cosmos was formed. This is possible only because they follow traces of actual events [i ran zhi ji]. There are sometimes irregularities in the true motions of the sky and of the sun, moon and planets, but as time passes these recur spontaneously to the norm.

Here Zhu Xi understands gu as traces of what exists or has existed; in other conversations, referring to other occurrences in the same chapter, he explains the word more subtly as "why something is so" (suoyi ran) and "what something does" (suo wei). He is using Mencius' undefined gu, relating it to Shen Kuo's undefined "traces," to refer to phenomenological patterns. In a society never touched by Plato's opposition of phenomenon and reality, this is a more original step than it might appear.[18]

In the discussion of astronomical prediction that provided the long paraphrase from Cai Yuanding quoted above, Zhu Xi begins more or less at the point where Cai stopped, but moves off in a significant new direction:

> Someone asked why calendrical systems are repeatedly inaccurate. "How can it be that in ancient and modern times no one has studied this matter thoroughly?" Zhu Xi replied: "It is precisely because no one has studied this

matter thoroughly enough to rule out further change that there are repeated discrepancies. If it were studied with enough precision to yield a definitive method of computation, there would be no further discrepancies. ... The astronomical techniques of the Ancients were imprecise [shukuo, lit., loose], but there were few discrepancies. The more precise [mi, lit., tight] the systems of today are, the more discrepancies appear!"

At this point, he measured off one side of his desk with his hands, saying, "For instance, if we divide this breadth into four sections, each is limited in width by its borders with the others. If a discrepancy [between the widths] appears, it will be restricted to one of the sections. Large though [the discrepancy] may be, even so extreme that it involved a second or third section, it would still be restricted to the four sections. So it would be easily computed, and any discrepancy could easily be seen. The astronomical systems of today [in effect] divide these four sections into eight, and the eight into sixteen. As the limits [of the sections] become more precise, the frequency of discrepancies becomes greater. Why is this? Because, as the limits become more precise, they are increasingly overstepped. The discrepancy may be identical, but the precision of ancient and modern systems differs."[19]

Zhu Xi is saying, if I understand him correctly, that increases of precision have led to greater expectations of accuracy, and that it is against these expectations that recent systems were failing; early systems satisfied lower standards. But that is not my point. This is the first clear explanation in Chinese, I believe, of the difference between accuracy and precision. This is not a small contribution to the methodology of the exact sciences. This was certainly not Zhu's aim, but it is hardly a by-product.

These concerns with method and with theory of knowledge, although ignored by modern historians, were carried on by the leading scholars of the Qing period. In a recent book, John Henderson traces the growing importance in Ming and Qing philosophy of arguments from mathematical astronomy. Henderson shows that these concepts of quantitative origin largely replaced earlier conceptions such as yin-yang and the Five Phases. For example, prominent humanists between the mid-seventeenth and late eighteenth century became aware, through Western astronomical writings which they studied eagerly, of such secular changes as the slow decrease in the obliquity of the ecliptic. They came to believe, as did many Europeans in the later part of the same period, that these were not entirely predictable, and their magnitudes could only be known through observation:

> A number of Ch'ing [i.e., Qing] scholars of varying scholastic affiliations thus identified several of the astronomical anomalies and deviations known to them as basically indeterminate, frequently drawing the conclusion that the patterns of the cosmos in general shifted in an irregular and even capricious fashion. They even regarded anomalies not so much as departures from a predictable order as themselves constitutive of the fundamental order, or disorder, of the cosmos.[20]

Limits of Inquiry in the Qualitative Sciences

It is not surprising that the issue of indeterminacy should have arisen mainly in astronomy, the one science that was both quantitative and concerned with prediction. The idea that empirical knowledge and understanding may be inherently limited also turns up in areas of inquiry that are not computational. Sometimes it is brought up by polymaths who are aware of the issue within astronomy. Shen Kuo, for instance, discusses a case in which lightning, striking a house, left its wooden structure unharmed but melted metal objects inside it.

> People insist that fire will burn things of vegetable origin before it melts things made from metals or minerals, but in this instance the latter all fused while not one of the former was destroyed by fire. This is not a matter that human capacities can fathom. A Buddhist treatise says "Water makes the Naga fire blaze up, but puts out the human fire." How true that is! People only know about matters in the realm of mankind. Outside that realm what limit can matters have? We may aspire, by our insignificant worldly wisdom and common sense, to get to the bottom of ultimate truths, but that is hardly possible.[21]

And I have already quoted one of Shen's references to medical theory.

We also find Fang Yizhi, in his Little Notes on the Principles of Things (completed 1643/1650), using what he had learned from Jesuit missionaries about optics to argue that the tendency of light rays to diverge and of shadows and images to converge renders certain optical phenomena unexplainable. Fang describes an experiment in which a piece of paper with four or five small holes yields multiple images of the sun. But, as the paper is moved upward, away from the surface on which they are projected, the multiple images blend — in a way that puzzles him — to form a single image of the sun. "Sound and light," he argues, "are always more subtle than the 'number' of things." By "number" Fang means amenability to exact quantitative description, as "by acute angles and straight lines," i.e., geometric constructions.[22]

In medicine, the idea that one can hope to understand only so much about the vital processes of the human body is natural enough. Ever since the sixth century, medicine has been strongly influenced by Buddhist ethics. Since its beginnings, physicians and medical scholars have drawn on numerology, yin-yang, and Five Phases cosmology, and even on astronomy, in order to investigate the links between the internal order of the body and the order of nature that surrounds it.[23]

An obvious example is Zhang Jiebin's statement, in his Collected Treatises *Jingyue quanshu* (preface dated 1593), about what is needed to comprehend vital processes: "Anyone who does not possess transcendent wisdom is not prepared to master their subtleties [*weimiao*]; anyone who does not possess clarity of moral judgment is not prepared to make fine distinctions concerning what is correct." Empirical observation, in other words, must be supplemented by self-cultivation.

In some such statements, Buddhist influence is plain to see. Typical is Yin Zhiyi's preface to his father, Yin Zhongchun's, little handbook of diagnosis and therapy, entitled Mental Dharmas of Eruptive Disorders (probably shortly before 1621):

> "Medicine" [*yi*] means "meaning" [*yi*]. [The inner meanings of medicine, the patterns of vital processes] may be apprehended by the mind, but cannot be transmitted in words. Because these inherent patterns attain such arcane subtlety [*weiao*], even though the mind may achieve great constancy [in contemplating them], in [therapeutic] doctrine there can be no fixed rules. The interaction of hot and moist as governed by yin and yang, the relations of mutual production and overcoming among the Five Phases, change from one moment to the next. ...

Yin begins with a familiar punning definition of medicine, and moves immediately to the Chan Buddhist notion of wordless teaching. Yin's word for therapeutic doctrine or method [*fa*] is the same as the term for dharma in the title of his father's book (in which "mental dharmas" means at one level "doctrines or truths to be grasped by the mind").

Such instances from therapeutic manuals could readily be multiplied. In astronomy, as we have seen, the limit of observation was a live and evolving issue. In medicine, however, what we find is reiteration of a familiar theme — a formula, more or less — that is seldom examined critically. My preliminary conclusion, pending deeper study and reflection, is that indeterminacy in medical writings is less significant for the history of medical thought than for epistemology in general.[24]

Conclusion

The sources on which I have drawn indicate that, in Han astronomy, themes of both epistemological and ontological limits on knowledge appear, but that, as experience led to confidence, the limit came to be seen consistently as one of imprecision rather than of inherent disorder. This situation continued until the two postulates were again combined in the attack on yin-yang and the Five Phases, as the basis of Confucian orthodoxy that John Henderson has documented.[25]

I have suggested that ideas of astronomical indeterminacy first arose to explain what would now be considered the failure of crude predictive techniques, and that this idea gradually became established to account for what historians would now explain as the failure of crude assumptions about the character of the celestial motions. As these assumptions were given up, and the crisis subsided, it seems that ideas of indeterminacy were for a while used more as a weapon to beat back innovation than as a means to reexamine past failures. These ideas began playing a productive role once more from the Song on, when they were used for diverse purposes, among them to direct critical attention to issues of what we would now call epistemology and method. Some who used them, including Cai Yuanding and Zhu Xi, did not accept the idea that empirical knowledge was necessarily limited.

Why should a notion that looks so obscurantist, so opposed to the idea of progress, have played such an enduring collection of roles in the history of science? To take first things first, since the idea of progress entered Western scientific thought in the eighteenth century, and that of China much later, how progressive a given early idea seems to moderns is beside the point. The point is rather what it meant and how it was used in its time.

I suggest that the idea of indeterminacy was the one proposition that consistently challenged astronomers to come to grips with the distinction between two issues: First, what is involved in predicting future observational data from past observational data? And second, what is involved in making intelligible the nature from which we draw observational data?

This is the difference between astronomy as a collection of data and techniques, and astronomy as a science. Despite the crisis in astronomy that began in the Later Han, the urge to make astronomy a science again never entirely subsided. It became a strong motivation from the eleventh century on, as impulses from philosophy stimulated astronomers, and vice versa. I think that, ultimately, it will be possible to show that discussions of the limitations of inquiry in the Ming and early Qing encouraged both the concern for causes and explanations in the seventeenth century and the prompt and positive response of leading Chinese astronomers to Western astronomy.[26]

The issue I have outlined does not seem to have been important in Western science after Heraclitus and Parmenides. The concept of quantum indeterminacy in modern physics is concerned with a quite different, purely mathematical, issue. It states that there is a small, constant limit on the combined precision with which we can simultaneously measure two parameters, for instance position and momentum, of a particle event. The better we know one, the worse we know the other, up to that limit. In translating the equations into ordinary language, some popularizers have read into this very abstract scenario portentous implications about the subjectivity of the observer, and have even concluded that, at the limit, theoretical physics collapses (or rises, as the case may be) into mysticism. Sympathetic though one may be toward attempts to combat a mechanistic arrogance for which there is also no warrant in the equations, these reinterpretations are not a triumphant revival of an old theme, but rather bad philosophy of science. They are irrelevant to the present topic.

For Plato, observation of phenomena alone cannot lead to

knowledge of reality. It can yield only a third-hand reflection of the abstract Ideas that real knowledge is about. The study of mathematics helps us toward them, but direct apprehension of the ideas is a contemplative, not an empirical, process. Aristotle believed that the reality of things was within them, and could be deduced directly from them, but "the advances made by the arts and sciences in each civilization were the fulfillment of the potentialities of their natural form beyond which they could not go."[27] The Skeptics denied that one could know with certainty; but their discussions of this point served to suspend judgment on all matters. This was not a doctrine that could greatly influence natural science.

The Stoics were the school closest in intellectual temper to Chinese cosmology, and they were influential in science, especially medicine. The empiricists among them opposed Skepticism and, thus, indeterminacy. They were much concerned with the possibilities of knowledge, and considered all truth built up from what the senses deliver, judged by what Stoics called "right reason."

In the European Middle Ages, the analogous issue — a weak analogue — was the relationship of faith and reason. In the midthirteenth century, we find St. Thomas Aquinas quoting with approval the words that St. Hilary of Poiters had set down nine hundred years earlier:

> For he who devoutly follows in pursuit of the infinite, though he never come up with it, will always advance by setting forth. Yet pry not into that secret, and meddle not in the mystery of the birth of the infinite, nor presume to grasp that which is the summit of understanding: but understand that there are things thou canst not grasp.[28]

The faith in unlimited knowledge, in untrammeled understanding, is not a characteristically Western faith; it is a modern faith. It is as welcome today in Beijing as in New York, perhaps more so as skepticism gains ground in the overdeveloped nations.

In seeking valid Western analogies to the role of indeterminacy in Chinese intellectual history, I would take an entirely different direction. I would prefer to ask whether we can find ideas that appear irrelevant or "unscientific" from a vulgar positivist point of view, but that nevertheless played enduring roles in encouraging discussions of scientific issues. It is not hard, in fact, to

Chinese and Western Science   185

think of examples. One is Zeno's paradoxes. It is well known that every important discussion of the continuity of points on a line, from the Greeks to the end of the nineteenth century (Georg Cantor, 1845–1918), focused on those paradoxes.[29]

One more important contrast between East and West remains to be drawn. Whether one begins with Parmenides or Plato, classical European philosophers who wished to find a way past mere speculation about Nature insisted on asking, and arguing about, what they saw as the most fundamental question: How can knowledge be certain? What we know with varying degrees of likelihood, what is merely probable, has no place in science. This axiom dominated Western science until the advent of statistical thermodynamics made it meaningless. The modern vision of the world that we experience as merely the summation of innumerable random atomic, nuclear and subnuclear phenomena, left this drive toward certainty a quaint relic of a dead faith.

It had no analogue in Chinese thought. Empirical knowledge is neither certain nor probable, merely given. The pattern one discerns may or may not be objectively there, but that is no more than to say that one may be empirically right or wrong. For certainty, one looks to illumination, introspection and other alternatives to purely cognitive processes. Certainty is, in the last analysis, a spiritual and moral stance.

In conclusion, let me return to the beginning of this essay. J.T. Fraser once wrote that limits of inquiry divide the world into those phenomena that are predictable and those that are not. The dividing line between these phenomena constitutes a statement of belief, one of those irrationalities on which rationality must always rest. That demarcation usually amounts to a claim about the nature of time.[30]

As we consider the many Chinese statements that we have reviewed through history, the issue is indeed in one sense the domain of prediction. Even more fundamentally, it is what the activity of prediction rules out. What it rules out is an uninterrupted response, at once intuitive and rational, to the concreteness and endless variety of phenomena in nature. That response is what theory in the traditional qualitative sciences, such as medicine, alchemy and siting, seems always to have striven for.

Theory is necessarily abstract, based on rigorously defined concepts. Chinese scientists looked for a balance of concept and phenomenon, for accounts of nature that did justice to its richness. That, not Occam's razor, nor the necessity of geometric demonstration, was their aesthetic criterion.

No one familiar with the sources would argue that their stance is irrational. I would hesitate to say that it is less rational than that of the European positivists of half a century ago, whose starry-eyed faith led them to draw and defend an ultimately indefensible line between positive knowledge and the outcomes of all other mental activity. The Chinese thinkers I have cited were very much concerned with the theoretical ordering of phenomena, and with prediction, in astronomy and medicine. What we see them saying, more and more explicitly, is that prediction is a reductive act and thus an inherently limiting one.

Notes

1. *The Study of Time V* (Amherst: The University of Massachusetts Press, 1986), pp. 151-169 For Chinese characters, see the 1986 version.
2. *Hou Han shu*, "Lüli zhi," 2:1492. Astronomical and astrological treatises of the Standard Histories are cited from the series *Lidai tianwen luli deng zhi huibian*, 10 vols. (Beijing, 1975-1976).
3. "Ying ri tui ce ji," pp. 113-118, in *Dai Zhen ji* (Shanghai: Shanghai Guji Chubanshe, 1980); esp. p. 115. Let me reiterate that my use of the philosophical term "indeterminacy" here and below merely translates Chinese assertions that there are inherent limitations to observational knowledge.
4. *Zhou yi*, "Xici da zhuan," B:4-5; Wilhelm, *The I Ching*, pp. 367, 370, where *wei* is translated as "that which is hidden" and "first imperceptible beginning." *Laozi*, 21, 1 and 15; cf. the translation of line 15 with that of D.C. Lau, *Chinese Classics. Tao Te Ching* (Hong Kong: The Chinese University Press, 1982), p. 21.
5. The famous anecdote about Ding the Cook, in Chapter 3 of the *Zhuangzi*, is about a distantly related conviction, namely, that manual skill is not a matter of technical rules but rather of being "in touch through the daemonic in me," as A.C. Graham's translation puts it. A number of similar "knack passages" occur in chapters of the *Zhuangzi* outside the original corpus. In the story of Bian the Wheelwright in Chapter 13, for instance, the intuitive skill that comes from long practice is adduced to argue against learning from books, not just from orthodox classics, but from all books that purport to transmit human experience. For exceptionally perceptive translations, see Graham, *Chuang-tzu: The Seven Inner Chapters and Other Writings from the Book Chuang-tzu* (London: George Allen and Unwin, 1981), pp. 63-64, 135-141.
6. *Chunqiu fan lu* (Si bu bei yao ed.), 3: 10a.

7. The discussion of Han astronomy that follows is documented in Sivin, *Cosmos and Computation in Early Chinese Mathematical Astronomy* (Leiden: E.J. Brill, 1969). See also the insights of Yabuuti Kiyosi (Yabuuchi Kiyoshi), in "The Calendar Reforms in the Han Dynasties and Ideas in their Background," *Archives internationales d'histoire des sciences*, 2(1974), 51–65, and the monograph by Yabuuchi and Nōda Chūryō, *Kansho ritsurekishi no kenkyū* (Kyoto: Zenkoku Shobō, 1947).
8. *Hou Han shu*, "Lüli zhi," 2: 1482.
9. Ibid.
10. Ibid., p. 1496. For a fuller translation, see Sivin, *Cosmos and Computation*, pp. 61–62. The last line contains an allusion to the Confucian *Analects*, 20:1.
11. Sivin, "Copernicus in China," *Studia Copernicana* (Warsaw), 6(1973), 63–122, esp. pp. 72–73.
12. *Sung shu*, 13:1768–1769 *et passim*. The beginning date of each *qi* period, of which there were twenty-four in a tropical year, was a basic element of the Chinese ephemeris.
13. *Xin Tang shu*, 27A:2177. Ixing's computational system was in fact attacked in 733, four years after it was adopted and six years after the astronomer died. Ixing was not charged with technical inadequacy – perhaps because he had anticipated criticisms on that count – but with plagiarism from Indian sources. The best account is in Christopher Cullen, "An Eighth Century Chinese Table of Tangents," *Chinese Science*, 5(1982), 1–33, esp. 30–32. On the numerological cosmology of the Great Expansion system, see the section of Ixing's treatise on the "Rationale of the Basis of the Ephemeris," in *Xin Tang shu*, 27A:2169–2173, trans. in Ang Tian Se, "I-hsing (A.D. 683–727): His Life and Scientific Work," unpublished Ph. D. dissertation, University of Malaya, 1979, pp. 419–445.
14. A mark (*ke*) is 0.01 day, approximately fifteen minutes.
15. *Mengqi bitan*, Item 123 (*Mengqi bitan jiaozheng*; rev.ed., Beijing, 1960; I:292). Of the two quotations in the first part, one quotes the Great Commentary, A. 9, and the other alludes to it (see Wilhelm, trans., I:339). The *Huang ti nei ching su wen* citation is to 7(22): 125 in the Shanghai, 1954, edition. On the Era and other long cycles, see Sivin, *Cosmos and Computation*, pp. 12–21. For an extremely satisfactory discussion of Whitehead, see *Dictionary of Scientific Biography*, s.v.
16. The *Hun yi yi* is preserved in *Song shi*, "Tianwen zhi," 48: 800–808; I cite the critical text in *Mengqi bitan*, Item 127, note 6 (I: 297), and accept the emendation of *zhou* to *hua*. Shen is using *tuan* in a technical sense that draws on several of its early meanings: to shape into a ball with the hands, to gather, to tie in a bundle, exclusive. For further reflections on the complementarity of scientific and other modes of knowledge in Shen's thought, see Sivin, "Shen Kua," *Dictionary of Scientific Biography*, s.v.
17. *Zhuzi yu lei* (*Zhuzi yu lei da quan*, Kyoto, 1668, reprint by Zhongwen Chubanshe, Kyoto, 1973), 86: 12a; 2:14b–15a. There is an interesting

discussion of these and other passages in Zhu Xi's writings in Yamada Keiji, *Shushi no shizengaku* (Zhu Xi's studies of nature; Tokyo: Iwanami Shoten, 1978), pp. 279–301. Cai Yuanding's astronomical monograph was entitled *Dayan xiang shuo.* See his biography by Rulan Chao Pian, pp. 1037–1039, in Herbert Franke, ed., *Sung Biographies,* 3 vols. (Wiesbaden: Munchener ostasiatische Studien, 16, 1976).

18. These conversations are recorded in *Zhuzi yu lei,* 57: 14a–17a. James Legge, who in his 1861 translation of the *Mencius* often relied on Zhu Xi rather than on more philologically rigorous later commentators, translates *gu* as phenomena (Hong Kong: The Chinese Classics, 1861), II: 206–207. For an especially penetrating discussion of the passage from *Mencius,* see Patrick E. Moran, "Key Psychological and Cosmological Terms in Chinese Philosophy: Their History from the Beginning to Chu Hsi (1130–1200)," unpublished Ph.D. dissertation, University of Pennsylvania, 1983, pp. 68–71.

19. *Zhuzi yu lei,* 86: 11b–12a. "Definitive method of computation" is a tentative translation for *ding shu,* which may mean nothing more elaborate than "definitive constants," or conceivably (but in my opinion less likely), "measures corresponding to true rather than mean motions." I translate *cha* as discrepancies rather than anomalies, since the example concerns error in measurement rather than inequality of motion. A pertinent essay is Hashimoto Keizō, "Seido no shisō to dentō Chūgoku no temmongaku" (Ideas of precision and traditional Chinese astronomy), *Kansai Daigaku Shakaigakubu Kiyō,* 1979, 11. 1:93–114.

20. *The Development and Decline of Chinese Cosmology* (New York: Columbia University Press, 1984), p. 249. I am grateful to Prof. Henderson for his comments on an early draft of this essay. His doctoral dissertation, from which the book is extensively revised, was in part responsible for inspiring me to take up this topic, and provided useful references. I do not agree with some of Henderson's interpretations, but his work takes up heretofore neglected questions and demonstrates the importance of astronomical writing in research in Chinese intellectual history. id=hen

21. *Mengqi bitan,* Item 347. Mark Elvin has remarked on the implications of this passage for "the probable limitations of human understanding" in *The Pattern of the Chinese Past* (Stanford: Stanford University Press, 1973), p. 233, note.

22. *Wuli xiao zhi* (1st ed. of 1664), I: 34a–b. The whole passage is translated (with some misunderstandings) by Willard J. Peterson in "Fang I-chih: Western Learning and the 'Investigation of Things,'" pp. 369–409, in Wm. Theodore De Bary et al., *The Unfolding of Neo-Confucianism* (New York: Columbia University Press, 1975), esp. p. 391.

23. See, for instance, Sivin, *Traditional Medicine in Contemporary China* (Science, Medicine, and Technology in East Asia, 2; Ann

Chinese and Western Science 189

Arbor: Center for Chinese Studies, University of Michigan, 1987 [published 1988]), pp. 43-94.

24. Zhang, *Jingyue quanshu* (photolithographic reprint of 1624 ed., Taipei, 1972), 3: 75b; mistranslated in Paul U. Unschuld, *Medical Ethics in Imperial China: A Study of Historical Anthropology* (Berkeley: University of California Press, 1979), p. 82; Yin, *Zhenzi xin fa* or *Shazhen xin fa*, printed with Yin Dachun's *Yizang shumu* (Shanghai, 1955), pp. 107-108. The title of the latter work, meaning "Bibliography of the Medical Triptaka," also draws on Buddhist imagery.

25. For a number of additional quotations, see Henderson, ibid. (note 19), esp. pp. 246-253.

26. For useful data in this connection, see Henderson, and Benjamin Elman, *From Philosophy to Philology. Intellectual and Social Aspects of Change in Late Imperial China* (Harvard East Asian Monographs, 110; Cambridge: Harvard University Press, 1984), esp. pp. 29-32.

27. A.C. Crombie, "Some Attitudes to Scientific Progress: Ancient, Medieval and Early Modern," *History of Science*, 13(1975), 213-230. See also on this topic, which is best distinguished from the one discussed in the present essay, E.R. Dodds, *The Ancient Concept of Progress and Other Essays* (Oxford: Oxford University Press, 1973).

28. St. Hilary, *De trinitatise* 2: 10, 11, cited in *The Summa Contra Gentiles of Saint Thomas Aquinas Literally Translated by the English Dominican Fathers from the Latest Leonine Edition* (London: Oates, 1924), 1.8 (I, 16).

29. No adequately detailed monograph on the role of Zeno's paradoxes has been published, but see G.E.L. Owen, "Zeno and the Mathematicians," *Proceedings of the Aristotelian Society*, n.s., 58(1957-1958), 199-222.

30. Personal communication, 14 October 1982.

RETROSPECT

This essay emerged from studying a series of intriguing assertions that I encountered in various astronomical texts. The first time I read through the literature I did not notice them. Over a number of years, my increasing curiosity about how astronomy looked to its practitioners gradually alerted me to the frequent remarks on indeterminacy. In gauging the scope of this idea, I found that it was not a rare theme among philosophers who thought that Nature was important. Indeterminacy was not a common notion in medicine because of the emphasis on practice. Doctors did not need, and seldom wanted, to remind each other about the irreducible uncertainty of diagnosis and therapy. Still, medical examples were easy enough to find.

Like most authors, I write to find out what I think about a given topic, and what it is possible to say that makes sense. Writing is of course an excellent guide to further inquiry, since it makes the soft spots and gaps clear. I undertook this essay for one of the grand general meetings of the International Society for the Study of Time, held at the Castello di Gargonza in Tuscany in 1983. My initial hypothesis was that the idea of several complementary kinds of knowing is closely connected to the diversity of time concepts in China. Inquiry, reflection, and writing bore it out.

The essay was published in 1986, as indicated in n. 1. When the editors of the 1989 issue of the yearbook *Philosophy and Religion,* devoted to the limits of rationality, asked me to contribute a revised version, I left out most of the discussion of time concepts, which was not relevant to the new theme. It is available in the 1986 version for those who want to consult it.

Correction

In Pinyin romanization, "Ixing" should be "Yixing," and "*Sung shu*" should be "*Song shu.*" Note that the Index below uses Wade-Giles romanization. Cross-references are provided.

VI

On the Word 'Taoist' as a Source of Perplexity. With Special Reference to the Relations of Science and Religion in Traditional China

I have been asked to contribute to this special volume my thoughts on the relations between religion and science in traditional China. This great congeries of issues can be explored in many ways, but one theme of a general kind calls for prior reflection. "Taoist" is a familiar term, and will perhaps seem to some readers too straightforward to pose methodological problems; but my own experience

---

This essay is best seen as the tentative effort of a generalist, offered in the hope of encouraging specialists in Taoism to replace it by an account more adequate to the needs of scholars in other fields. I am grateful to participants at two international conferences on Taoist studies (see nn. 2 and 4) and at the inaugural meeting of the Society for the Study of Chinese Religions for discussions of ideas I will present. I also thank James R. Bartholomew, Steven J. Bennett, Mark Elvin, Barbara Ruch, Michel Strickmann, and Dorothy Yep for aid not detailed in the footnotes, and the John Simon Guggenheim Memorial Foundation and the National Library of Medicine for support.

© 1978 by The University of Chicago.
All rights reserved.

VI

On the Word "Taoist"

and that of my colleagues suggest that in practice it is not so manageable. My theme will be the confusion in our understanding of Chinese science wrought by the frequent use of the word "Taoist" to denote nothing more specific than a frame of mind—nature-loving, perhaps, or mystical in a naturalistic way, or unconventional—in discussions that are meant to be about a religion—an association of persons who hold a body of beliefs.

I will demonstrate that such vagueness affects current discussions of the relations between Taoism and science (in which the roles of Taoism and Confucianism are often considered antithetical), that the equally vague results of those discussions have spread into general writings about China, and that all of this vagueness is related (as both a cause and an effect) to a lack of consensus about the most fundamental characteristics of Taoism—not unexpected in such a young discipline. I will then examine more closely two sorts of confusion which typically arise from the failure to ask whether a given instance of "Taoism" is sentimental, intellectual, social, or bibliographical.

The first example is the tendency, many centuries old, to regard as "Taoist" practices and beliefs which originated in popular religion and were very widely distributed. This often happens even in circumstances where no connection to Taoist organizations or writings can be demonstrated. The second example is the curious case of Ko Hung (283-343), whose modesty has failed to deflect hyperbolic assertions of historians about his stature as a Taoist and alchemist. Finally I will argue that a more satisfactory state of affairs will depend not on imposing a standard definition but on being explicit about which of the many senses of Taoism we are invoking in each instance.

DEFINITIONS AND THEIR LIMITS

In keeping with this last intention, and recognizing the limits of present knowledge and of my own understanding, I will not attempt to encompass all of Taoism in a single definition, though attempts of others to do so are cited below. Instead, I will attempt to use the much more specific terms "philosophical Taoism" and "religious Taoism" in a reasonably consistent way.

I do not mean them to correspond to the distinction between *tao chia* and *tao chiao*, known to every undergraduate who has dabbled in Chinese history. In a popular formulation, "the Chinese themselves sum up Taoism by dividing it into *Tao chia* and *Tao chiao*—the 'Taoist school' and 'Taoist sect.' The first category they restrict to partisans of the philosophy of Lao Tzu and Chuang Tzu. In the second they include all those groups that have taken immortality as their goal—alchemists, hygienists, magicians, eclectics, and, in particular, the members of the Taoist church."[1]

This neat distinction is the creation of modern historians. It is vague as a basis for synthesis and of little use in textual studies. "*Tao chia*" became current from the Han on as a bibliographic rubric, and in that capacity eventually came to cover works on alchemy, hygiene, magic, and religious ritual—everything in the imperial libraries connected with Taoism in any sense, however loosely. As a designation of persons, it was applied to ordained priests of "the Taoist church" at least through the Six Dynasties, and is even occasionally so used today. As for "*tao chiao*," before modern times it meant simply "the teachings of the Way." It was first applied in section 39 of the *Mo-tzu* to Confucian beliefs. By the Southern Dynasties it referred to Taoist teachings of every sort, not in contradistinction to the *Lao-tzu*, which in this sense it subsumed, but to Buddhism and Confucianism. As for the *Chuang-tzu*, although it was often read together with the *Lao-tzu* for its quietist ideals in the Han and afterward, it was not generally considered a canonic scripture until imperially sponsored for that purpose in the T'ang.

As I use it, "philosophical Taoism" has no sociological meaning. It refers to the content of the *Lao-tzu* and a few similar philosophical writings which bibliographers have classified with it. The philosophical Taoists were not a group, but a handful of authors scattered through history. I prefer not to apply the term to individual readers who used these books for moral or mystical inspiration, except in cases when I know a great deal about where this attitude was situated in their minds and careers. I also avoid the crippling assumption that the *Lao-tzu*, *Chuang-tzu*, and so on, had a fixed intellectual content independent of time and place. They meant one thing to their writers, another to their compilers, another to each reader, and quite another to moderns. None of these brings the same assumptions to them, and each finds different "original meanings" in them.

---

[1] Holmes Welch, *The Parting of the Way. Lao Tzu and the Taoist Movement* (Boston, 1957), reprinted under the title *Taoism. The Parting of the Way* (Boston, 1966), pp. 162–63.

VI

On the Word "Taoist"

By "religious Taoism" I refer to groups (also called "the established Taoist sects," or, better, "orthodox Taoism") whose liturgy was directed to the Tao as absolute divinity and to its emanations (*t'ien tsun*, etc.), which reveal or manifest it to man. The safeguarding and perpetuation of orthodox scriptural traditions depended on esoteric rites of transmission, specifying a line of predecessors never contaminated by people uninitiated into worship of the Tao. In addition to special objects of worship and spiritual genealogies, members of orthodox Taoist organizations, despite rivalries which sometimes led them to deny the legitimacy of each other's traditions, shared a recognition of Chang Tao-ling as the founder of true Taoism.

Although the *Lao-tzu* has been accepted by all Taoist sects as a central revelation, the religion was not scriptural in the same sense as Christianity. The spiritual orientations of orthodox Taoists gave meaning to the *Lao-tzu* text—a gnostic meaning very different from what outsiders found in it—to a much greater extent than the philosophy of *Lao-tzu* shaped the orthodox faith. There were too many revelations, accumulating century after century and winning attention away from the *Lao-tzu*, for Taoism to have been a religion of *the* book. Looking at records in the Taoist patrology, one might argue that individual faith was most decisively shaped by ritual and practice, which the scriptures served primarily to justify, prescribe, guide, and support. The canons can no more be read as pure philosophy, without reference to their use in religious activity, than can the written legacy of meditative Buddhism or Christian mysticism.

The term "religious Taoism" (but not the alternative terms) also refers to people initiated into a line of scriptural transmission which branched out of an orthodox group, whether or not these initiates took part in communal activities.[2]

---

[2] In "Taoism in the Lettered Society of the Six Dynasties" (paper presented at the Second International Conference on Taoist Studies, Tateshina, Japan, September 1972), Michel Strickmann emphasized that rituals which accompanied the transmission of texts are an important—although ambiguous—element in a definition of Taoism in the Six Dynasties.

This essay has been published in Japanese translation as "Bōzan ni okeru keiji: Dōkyō to kizoku shakai," in *Dōkyō no sōgōteki kenkyū* [Comprehensive studies in Taoism], ed. Sakai Tadao (Tokyo, 1977), pp. 333–69. In Strickmann's forthcoming "On the Alchemy of T'ao Hung-Ching" he has noted that the position accorded to Chang Tao-ling is another trait common to all orthodox groups. Although Strickmann deserves credit for putting together the elements of this definition, I am grateful for discussions with K. M. Schipper and Anna Seidel which helped greatly in the formation of my own understanding of this matter.

I do not mean to urge that the definitions I have given above be generally adopted; they are merely meant to clarify my uses of words. There has not yet been sufficient study of the Taoist literature to make any attempt to *fix* definitions profitable.

This may seem like a precise definition, but it is not. The ordination of the orthodox priest, which made him a member of the bureaucracy of gods and provided him with a roster of subaltern divinities on which he could call for help (*lu*), is unambiguous.³ Submitting ritually to the gods petitions written in the classical language is hardly less sure a sign of orthodoxy, for only members of the Taoist priesthood could rightfully initiate such documents. Even in Taiwan today, where the norms of traditional culture exert only the most vestigial power, this practice is usually avoided by popular priests.⁴

The ambiguity that begins just outside the small circle of ordained priests and their acolytes spreads and ramifies with great speed as we move outward in society. Priests of the popular religion avidly incorporated rites that did not require written petitions. In the eyes of the orthodox they were pretenders. But it is well to remember that this borrowing was part of a reciprocal process. The orthodox liturgy was built up through adapting and incorporating popular rites, a process that, like its inverse, continues today.⁵ In this respect Taoists were as catholic as those who shaped Buddhist sects and Confucian doctrines in the last two millennia. From the dynamic interplay between the Taoist and the village exorcist or medium the road runs downhill to the quack "adept" peddling a nostrum, or the alchemical confidence man

---

³ This is the situation in modern times, but in earlier Taoism a succession of *lu* was associated with different stages of initiation. This complicated matter has been sorted out in K. M. Schipper, "Some Remarks on the Function of the 'Inspector of Merits'" (paper presented at the Second International Conference on Taoist Studies).

⁴ K. M. Schipper, "The Written Memorial in Taoist Ceremonies." *Religion and Ritual in Chinese Society*, ed. Arthur P. Wolf (Stanford, Calif., 1974), pp. 309–24, esp. 309–10. Basic distinctions between orthodox Taoism and popular religion were drawn by Schipper (a Taoist priest as well as a sinologist) at the First International Conference on Taoist Studies (Bellagio, Italy, September 1968). They are summarized in Holmes H. Welch, "The Bellagio Conference on Taoist Studies," *History of Religions* 9 (1969–70): 107–36; see also ibid., 12 (1973): 392. The table presented on p. 125 of the summary does not accurately reproduce all of Schipper's main explanations. The statement that the priesthood of popular religion "was wholly unorganized" misses the point that there was no *single* organization to enforce orthodoxy, but various organized sects were recorded and can be seen today in Taiwan and expatriate communities. The association of orthodox Taoism solely with the Cheng-i (or Celestial Masters) sect, and folk religion with the Mao Shan sect, also misrepresents a complex situation.

⁵ The process has been described eloquently by Schipper in "Some Remarks on the Function of the 'Inspector of Merits.'" The blurring of distinctions between Taoist priests and popular exorcists in contemporary Taiwan, where the latter no longer need observe traditional constraints, has been documented by Michael Saso, who follows current laymen's practice in north Taiwan by referring to the latter as "'red-head' Taoists" (see his introduction to *Chuang Lin hsu tao tsang* [The Chuang and Lin clans' supplement to the Taoist patrology] [Taipei, 1973]).

VI

On the Word "Taoist"

who can recite imaginary canons at naive believers by the hour—a figure not peculiar to China, as Ben Johnson attests.

There are many figures in history and literature who call themselves Taoists for reasons of their own. Are they not entitled to that choice, and are we not bound to accept it?

To accept such claims at face value would be naive, but to ignore them would be equally naive. Sometimes they are merely a product of sinological oversight, as when the term *tao jen*, which in Six Dynasties texts refers to Buddhist priests, is literally translated "man of the Tao." In cases where they reveal nothing about collectivities or traditions they are powerful indices to individual motives. In one case the claim to be an adherent of the Tao may signal isolated but fervent devotion to the goal of transcendence (although written traditions emphasized the need for mutual support in the endeavor, and companions were seldom hard to find in traditional times); in another, an imagination caught by esoteric imagery; in another, possession of an arcane text of one kind or another received in a hazy transaction from a "remarkable person" (*i jen*), about whose background the writer says nothing; and in another, a sound knowledge of the sort of persona that helps one to succeed as a swindler. Vague terms paraphrasable as "Taoist" were free for the taking, just as the romantic association of "woodcutter" with rustic sagehood was frequently appropriated by retired civil servants who had no intention of touching any tool of forestry. Such terms always tell us something. Like the modern term "executive," without particulars they tell us very little.

PERPLEXITY

My own perplexity about the ways the word "Taoism" is used was first aroused by a type of argument that has become rather popular over the last half-century. Since Homer Dubs in 1929 claimed that abstract theory was ruled out by Confucian practicality, authoritarianism, and distaste for change, many writers have tried to explain what limitations of orientation or attitude made the Chinese incapable of developing systematic philosophy, especially natural philosophy of the kind that has become associated with the origins of modern science.[6]

More recently this discussion has taken a new turn as people have finally begun reading the enormous scientific literature that the

---

[6] Homer Dubs, "The Failure of the Chinese to Produce Philosophic Systems," *T'oung pao* 26 (1929): 96–109, esp. 108–9.

VI

Chinese were supposedly unable to write. They have found elaborate abstract theories of natural change, based on such concepts as yin-yang and the Five Phases that no one had bothered to understand before.[7] They have begun to reconstruct from historical records the large part science and technology played in traditional institutions, economy, and thought.

A number of Western scholars now see a race between traditional China and traditional Europe toward the scientific revolution that transformed man's view of nature and his place in it. Confucianism is now held responsible by some, and "bureaucratic feudalism" by others, for China's failure to win this race, despite what seems to have been an excellent head start—for which Taoism is given much of the credit.

Fifty years ago Taoism was still considered in part too irrational a mysticism and for the rest too degraded a superstition even to be mentioned in such discussions. Now it has been transfigured, made a milieu for objective and experimental science and technology. As Joseph Needham puts it, "Taoism was religious and poetical, yes; but it was also at least as strongly magical, scientific, democratic, and politically revolutionary."[8]

Anyone familiar with Needham's work will acknowledge his awareness of such differences as that between the metaphysical poetry of the *Lao-tzu* and the sacerdotal rites of the Celestial Masters. Yet at crucial points in his arguments that "the Taoists ... affirmed their science and their democracy at the same time"[9] the distinctions tend to blur: "... the Taoists show certain characteristic differences from any analogous groups in occidental history. They formed a much more organized element than the Cynics or the Stoics, and their combination of political anti-feudalism with the beginnings of a scientific movement has no parallel in the West."[10] The "organized element" can only be the orthodox Taoist sects. Neither Needham nor anyone else has produced persuasive evidence that they were politically anti-feudal or that they might be identified with the beginnings of a scientific movement. Needham's emphasis on "the enmity of the Taoists not only for Confucianism but for the whole feudal system" is pri-

---

[7] The best current understanding is conveyed in Manfred Porkert, *The Theoretical Foundations of Chinese Medicine. Systems of Correspondence*, M.I.T. East Asian Science Series, vol. 3 (Cambridge, Mass., 1974).

[8] Joseph Needham, *Science and Civilisation in China* (Cambridge, 1954–), 2 (1956): 35.

[9] Ibid., p. 103.

[10] Ibid., p. 129.

VI

On the Word "Taoist"

marily supported by his interpretations of the *Lao-tzu*, *Chuang-tzu*, and *Huai-nan-tzu*, the basic ideas of which are conventionally associated with early Taoist philosophy, not the organized religion.[11] As for the beginnings of a scientific movement, the theoretical and practical work of disparate individuals who may be called Taoist in one sense or another does not warrant generalizations about Taoism as either a religion or a philosophy. It remains to be proved through close study of each individual that these accomplishments were in some special sense due to Taoist connections or sentiments. It also has yet to be demonstrated that these associations and feelings formed a consistent pattern more significant for scientific accomplishment than that formed by the intellectual and social allegiances of equally important scientists who were in no sense Taoists.[12] As this example shows, many of Needham's general statements about Taoism are sociological in tone, but the more I ponder what social entity they may have referred to the more I am perplexed.

This vagueness is no cause for alarm in a "reconnaissance" (as Needham has called his project) of such vast scope, concerned largely with proposing "hypotheses for further research."[13] In view of the tentative nature of Needham's proposals about the connections of religion and science, and the enormous philosophic and historical difficulties they raise, one might expect sinologists

---

[11] Ibid., pp. 100–132. The quotation is from p. 100. This theme is clearly presented as hypothetical, and indeed the sources are susceptible of interpretations very different from Needham's. A passage in the third source, for instance, is described as "giving in enlarged form a picture of primitive collectivism" (p. 108). I am unable to find in the Chinese text any reference to collectivism in either the political sense (governance by all) or the social sense (common ownership of the means of production and distribution). Its theme is, to be sure, primitivity; it describes the happy and harmonious state of nature and man before this pristine simplicity was spoiled by human artifice (see *Huai-nan-tzu*, Ssu pu ts'ung k'an ed., 8:1a–1b; inadequate translation in Evan Morgan, *Tao the Great Luminant. Essays from Huai Nan Tzu with Introductory Articles Notes Analyses* [Shanghai, 1933; reprint ed., Taipei, 1965], pp. 80–81).

[12] Nor has Needham ever put forth a documented claim that Taoist scientists or technologists were predominant in either quantity or quality. Any list of major figures over the last 2,000 years would be drawn primarily from the scholar-official class for science and from artisans for techniques, with few Taoist connections demonstrable except in alchemy, often considered a heterodox art (see, for instance, the representative sample of twenty-nine men and women in *Chung-kuo ku-tai k'o-hsueh-chia* [Ancient Chinese scientists] [Peking, 1959]; the scope of the book includes engineering). In connection with this issue I have examined closely the careers of two great scientists whose major formative intellectual influences were Confucian in "Shen Kua" and "Wang Hsi-shan," in *Dictionary of Scientific Biography*, ed. C. C. Gillispie (New York, 1975), 12:369–93, and (1976), 14:159–68, respectively. The essay on Wang, revised to include data on the connections of science and Neo-Confucianism in the seventeenth and eighteenth centuries, will appear in Monica Croghan et al., eds., *Nothing Concealed* (Taipei, in press), vol. 2.

[13] Needham, 1:5 and 2:119.

to be much concerned with testing and refining them. Some caution there certainly has been, and, as other papers in this volume will no doubt indicate, the challenge has been taken up by a few. But the more common tendencies in the United States have been either to ignore the problem or accept Needham's hypotheses as proven. The first is no doubt a prudent course; the second an open-minded one. But both, particularly the second, remind us that vagueness about Taoism is typical, although by no means universal,[14] in American scholarship.

I will give only three examples, chosen from different sorts of writings. The first is from Holmes Welch's *The Parting of the Way* (1957), an original and commonsensical introduction to the philosophy of the *Lao-tzu* for a broad public, and incidentally a historical sketch of the Taoist movement from available secondary sources. We are told that "to a large extent the Taoists practiced experimental science. They were reluctant to alter their premises in the light of logic and experimentation, but they did at least experiment. They were ultimately responsible for the development of dyes, alloys, porcelains, medicines, the compass, and gunpowder. They would have developed much more if the best minds in China had not been pre-empted by Confucian orthodoxy," and so on, not a sentence of it (given the normal meanings of "to a large extent" and "ultimately") yet demonstrated to be true of any group that might conceivably be called "the Taoists."[15]

A second example is the introductory textbook *East Asia. Tradition and Transformation* (1973), probably the most discriminating synthesis of scholarship to date. The entire treatment of pre-Ch'ing Chinese science in this 969-page volume is as follows: "The many protoscientific discoveries and inventions in China were associated more with the nature-loving Taoists than with the scholarly Confucians. The promising beginnings of nature lore in China were never consciously rationalized and institutionalized like modern science in the West."[16] The first sentence is too vague

---

[14] For instance, the late Arthur Wright was sufficiently familiar with the excellent historical scholarship on Taoism of what might be called the School of Paris to write with salutary clarity on the subject (see his "A Historian's Reflections on the Taoist Tradition," *History of Religions* 9 [1969–70]: 248–55). The strength of European Taoist studies largely derives from the pioneering work of Henri Maspero and his colleagues at Paris, extended and refined in recent years by Max Kaltenmark, Rolf Stein, Schipper, and their associates, to the point that it provides a model of the critical approach which this essay seeks to further.

[15] Welch (n. 1 above), p. 134. It should be noted that Welch cautions the reader about the tentative nature of his generalizations (see "Foreword").

[16] John K. Fairbank, Edwin O. Reischauer, and Albert M. Craig, *East Asia. Tradition and Transformation* (Boston, 1973), p. 232. This is not exactly Need-

On the Word "Taoist"

to be translated into a testable assertion. The second is misleading when it implies that early Chinese science was not consciously rationalized and institutionalized in its own way, in a pattern very different from that of modern science but perfectly comparable with that of early Europe.

My final example is the attempt of a most intelligent scholar, writing for historians of religion, to restate the kernel of Needham's thesis: "In Joseph Needham's evaluation, the religious Taoists were scientists and activists who defied fate and any passive fatalism." Here the Taoist group is positively identified in a way that Needham has never attempted. The writer is obviously motivated by a desire for clarity. The resulting equation of religious Taoist and scientist is so clear-cut as to be unrecognizable to anyone familiar with the primary literatures of science and orthodox Taoism, which support only the vaguest generalizations about overlaps.[17]

The point of these examples is not to remind readers that life is short and our craft long. That truism could be illustrated as easily from my own writings.[18] The prevalence of vagueness in discussions about the relations of Taoism and science reminds us rather that little precision and consensus can yet be found in our more general understanding of Taoism.

BONES OF CONTENTION

Is Taoism one? Norman Girardot has recently pointed out the dissonance between two positions held by scholars of Taoism. The

---

ham's position. He believes that responsibility for sterilizing "the sprouts of natural science" is to be "laid at the door, not so much of the Taoists' complacent and conventional rival, social-minded Confucianism, as of the socio-economic system of feudal bureaucratism itself" (Needham, 2:162).

[17] Whalen W. Lai, "Toward a Periodization of the Taoist Religion" (essay review), *History of Religions* 16 (1976): 75–85, esp. 77, n. 7; compare Needham's discussion of "Taoism as a Religion" (Needham, 2:154–61).

[18] For instance, in *Chinese Alchemy: Preliminary Studies*, Harvard Monographs in the History of Science, no. 1 (Cambridge, Mass., 1968), esp. chap. 3, I wrote repeatedly of the great alchemical and medical author Sun Ssu-mo (alive 675) as a Taoist. Although I showed that early accounts of his Taoist and Buddhist connections were legendary, my understanding of the history of orthodox Taoism was too limited to specify in what sense he could be called a Taoist. Only one sense withstands scrutiny. In his therapeutic compendium *Ch'ien chin i fang* [Supplement to prescriptions worth a thousand] there is an extensive Canon of Interdiction (*Chin ching*). Its exorcistic formulas were plainly meant to be recited by a priest. They contain such phrases as "I am a libationer [i.e., priest] of the Celestial Masters (*Wu wei t'ien-shih chi-chiu*)" and "I am a son of the Celestial Master, dispatched by the Master" (Mei-ch'i ed. of 1307; reprint ed., Peking, 1955), 29:14a (p. 347) and 30:2a (p. 353). The canon is still learned from Sun's book by priests of the Celestial Masters sect in Taiwan today (private communication, K. M. Schipper 1971). It is unlikely that Sun would have had access to these formulas had he not been an initiated member of a Taoist order.

issue is "the relationship of the 'philosophical' nature of the *Tao Te Ching*, the *Chuang Tzu*, and other works such as the *Lieh Tzu* to the 'religious' Taoism of hygiene, liturgy, alchemy, and other subtraditions concerned with a type of soteriology or quest for 'immortality.'" One position "essentially holds that the Taoism of Lao Tzu and Chuang Tzu is an isolated phenomenon of 'pure philosophy' completely distinct from the goal of *hsien* immortality." The second "holds that an essential unity must be seen among the various historical forms of Taoism," without denying the distinctions between these forms.[19] Since Girardot has already mentioned leading European and American proponents of these two views, I will examine a few instances of the same contention in Japan:

To my mind the case for the underlying unity of Taoism has been most eloquently stated by Yoshioka Yoshitoyo: "... in the simplest axiomatic sense, Lao-tzu is the embodiment of the 'Tao'; the *Tao te ching* represents the teachings of the 'Tao.' The masses assimilated this artless logic and transformed it into action. This was the origin of organized Taoism.... Generally speaking, the relation between orthodox Taoism (Dōkyō) and the Chinese people, and between orthodox Taoism and Lao-tzu and the *Tao te ching*, has been of this kind. If we ignore this fundamental interrelationship, and merely consider superficially the diverse forms of Taoist belief, it will be impossible to reach a true understanding."[20]

As the proponents cited by Girardot indicate, on the whole the first position has been adopted by sinologists whose study has concentrated on the pre-Han philosophic classics and on views of the Taoist traditions recorded in historical sources, and the second by those who have gone on to become familiar with the scriptures preserved in the Taoist patrology (*Tao tsang*). It may be that as thorough study of this literature becomes the rule in Taoist studies the issue will be laid to rest. It is equally likely that the dichotomy

---

[19] Norman Girardot, "Part of the Way: Four Studies on Taoism" (essay review), *History of Religions* 11 (1972): 319–37, esp. 320–24.

[20] Yoshioka Yoshitoyo, *Eisei e no negai. Dōkyō* [Taoism. The aspiration toward eternal life], Sekai no shūkyō [Religions of the world], vol. 9 (Kyoto, 1970), pp. 23–24. This work is summarized in Lai. The original Japanese does not imply that what came to be called the *Tao te ching* was the sole teaching of Lao-tzu. It was the oldest revelation by that divine figure, but there were many others. See, for instance, the texts in the *Tao tsang* listed in K. M. Schipper, *Concordance du Tao-tsang. Titres des ouvrages*, Publications de l'Ecole Française d'Extrême-Orient, no. 102 (Paris, 1975), p. 67, and the Tun-huang scripture reproduced, translated, and discussed in Anna K. Seidel, *La divinisation de Lao tseu dans le Taoïsme des Han*, Publications de l'Ecole Française d'Extrême-Orient, no. 71 (Paris, 1969).

VI

On the Word "Taoist"

will be translated into more refined distinctions (for instance, greater awareness of the diversity of concepts that today tend to be lumped into the notion of *hsien* immortality).

Is Taoism to be defined by sentiments alone, or must institutions be considered as well? Even in Japan, where the study of Taoist literature has been more thorough—and better organized—than elsewhere, there is no agreement. A catholic definition, in which organization plays a due part, has been given by Kubo Noritada: "Taoism ... was founded on a variety of ancient popular beliefs; was centered about immortality lore; incorporated the lore of philosophic Taoism, the Book of Changes, yin-yang, the Five Phases, divination and apocalyptic prognostication, astrology, and so on, as well as shamanistic beliefs; and was given religious forms modeled on the style and organization of Buddhism. Taoism is a religion of benefit in this world, the chief goal of which is eternal life, exempt from aging."[21]

Kimura Eiichi, a leading interpreter of the *Lao-tzu*, has constructed his definition very differently:

There are cases where various religions were unified into one standard and major religion for a people.... When such varied racial faiths were not quite unified into one organized body, but had prominent features peculiar to and common among all of them, a certain name could be given to the common features to indicate the major religion of that people.... Judaism for the Jewish people, Hinduism for the Indian people, Taoism for the Chinese people and Shintoism for the Japanese people are the major racial religions. They are closely related to the living customs of their respective society *whether individuals are ardent or conscious believers in the religion or not*.... Needless to say, however, a national religion does not spread to another race, though it colors the life of all the members of the homogeneous society from which it springs.

Here the essence of Taoism is a congenital sentiment that need not even be conscious. This view has survived documentation by Japanese historians of the spread of Taoism to Korea and Japan. Kimura does not, in fact, use systematically the distinction between orthodox Taoism (the Taoist sects) and popular religion.

---

[21] Kubo Noritada, *Kōshin shinkō no kenkyū* [Studies in *keng-shen* beliefs], 3 vols. (Tokyo, 1961–69), vol. 1. Although immortality is usually stressed by modern scholars as the end of individual Taoist practice, it is only one of several ways to express the chief goal of Taoism. Taoists also thought of this goal as union with the Tao—godhead immanent in the cosmic order and absolute beyond it—or as appointment to the celestial hierarchy. An outsider might see the goal as harmony with the social as well as the divine order, but a functionalist sociological view tends to lose sight of the core of religious striving. One might say that the practices and beliefs of Taoism (including immortality lore) conditioned the imagination and thus prepared the individual to embody the perfect order which the Tao implies.

His definition is easily understandable if it is referred to the latter.[22] Despite the virtues of this definition—for instance, its attention to the fundamentally religious impulse of Taoism—it could hardly be employed to study the relations between scientific and religious activities in China.

The great student of the Taoist patrologies, Ōfuchi Ninji, has avoided these difficulties by defining not Taoism itself but what underlies its manifestations:

> Insofar as we use the word "Taoism" (dōkyō), and insofar as we attempt to deal with Taoism over a period of hundreds and thousands of years, we must be presupposing something common and fundamental that has remained unchanged over time. . . . If we cannot find a single system that can be objectively delineated, we have no choice but to explore . . . more subjective territory, namely, the hearts and the minds of the Chinese people. Taoism, in my view, cannot be explained except by reference to the timelessly unchanging, realistic, and optimistic sentiments of the Chinese people who, concerned solely with reality and placing their trust entirely in man's present life, unashamedly regard man's happiness in this world as the ultimate value.[23]

This characterization conveys the common foundation of Confucianism, folk religion, and certain characteristically Chinese types of Buddhism as well as that of Taoism. Popular belief shaped them all.

To sum up, an important reason for the prevalent vagueness about the relations of Taoism and science is that the field of Taoist studies is still too young to have settled even the most general questions about its content and scope. Those questions

---

[22] The quotation is from Kimura Eiichi, "Taoism and Chinese Thought," *Acta asiatica* 27 (1974): 1–18, esp. 2–3; my italics. The passage immediately preceding the one cited occurs with somewhat different wording in an earlier essay of the same title by Kimura in Japanese; there it is introduced by the statement, "First and foremost there is the question of popular religions and world religions" (see his "Dōkyō to Chūgoku no shisō" [literally, "Taoism and Chinese Thought," but given in the English table of contents as "The Position of Taoism in the History of Chinese Thought"], *Tōhō shūkyō*, no. 38 [1971], pp. 1–20, esp. 2). The secondary literature on Taoism in Japan is massive, but a good deal of it is about popular beliefs for which the role of Taoism in transmission to Japan is undocumented. An excellent survey which pays a good deal of attention to Taoist liturgy and organization is Shimode Sekiyo, *Dōkyō. Sono kōdō to shisō* [Taoism in thought and action], Nihonjin no kōdō to shishō [The Japanese people in action and thought], vol. 10 (Tokyo, 1971). On Taoist institutions in Korea, where the penetration of the religion was deeper and more abiding, the standard historical account is Yi Nung-hwa, *Han'guk togyosa* [A history of Korean Taoism] (Seoul, 1956), in literary Chinese. An important recent study is Kubo Noritada, "Chosen no dōkyō" [On Taoism in Korea], *Tōhōgaku* 29 (1965): 118–31.

[23] Ōfuchi Ninji, "Dōkyō no keisei" [The formation of Taoism], in *Shūkyō* [Religion], Chūgoku bunka sōsho [Chinese culture series], no. 6, ed. Kubo and Nishi Junzō (Tokyo, 1967), p. 32. This work of popularization contains essays by leading authorities on every aspect of Taoism, Buddhism, and popular religion, with some interesting articles on religion in recent Chinese history.

315

VI

On the Word "Taoist"

will be settled not by protracted debates on abstract issues such as those I have just raised, but by thorough and critical studies of the *Tao tsang* and other sources. With that in mind, I will return to issues that bear directly on science.

TAOISM, CONFUCIANISM, AND SCIENCE

On a certain level there is no reason to disagree with the observation that an undeniable tension in Chinese scientific and technological viewpoints was due to the opposition of Taoist and Confucian values. "Confucian" is a defensible one-word code for the hierarchic, bureaucratic, and bookish values that in traditional times were regularly invoked against change (and also, lest we forget, for change). The rhetoric of the *Lao-tzu* and *Chuang-tzu* was adopted by various people on one or another margin of society who wished to justify their receptiveness to novelty, or who found it esthetically satisfying to contemplate nature and man's relation to it, or who wanted others to think of them as men of wisdom, or who simply found conventional stuffiness laughable. Whatever else they may have meant, "Taoist" and "Confucian" in the popular imagination were vague clumps of sentiments, as fuzzy and as handy as "counterculture" and "conservative" are today. The study of such clichés certainly deserves a place in the history of Chinese sentiments.

If, on the other hand, we want to understand what efforts of human beings formed the doctrines, mental sets, and practices of Taoism and Confucianism, made a place in society for them, and modified them in changing circumstances, then studying their currency as universal clichés becomes a very minor issue. Generalizations about people who accept a certain doctrine have no significance for social history unless such people can be shown to act as a group, or at least to identify themselves as a group.

By "Taoist" a sinologist may mean a mystical author; a hereditary priest, whether or not ordained by the Celestial Master or the head of another orthodox sect; a monk; a lay member of a sect dedicated to worship of the Tao and its emanations; an initiate, whether isolated or a member of a coterie; any priest, operator, healer, medium, shaman, or supporter of popular liturgies; anyone who took seriously or practiced occult disciplines, even fakes and swindlers who merely claimed mastery of them; anyone who lived a noncomformist life outside Buddhist circles; or anyone who harbored anti-feudal feelings as certain historians define them today (over the past few years "Legalist" has begun to supplant "Taoist" in this last usage).

The term "Confucian" is used indiscriminately, sometimes even by specialists, to refer to a master of state ceremonial, a recognized teacher of Confucian doctrines, a philosopher who contributed to the elaboration of these doctrines, anyone who attempted to live by Confucius's teachings, any member of the civil service regardless of whether he lived in accordance with Confucius's teachings, any educated person regardless of official ambitions, or any conventional person (since it was conventional to quote Confucian doctrines in support of conventional behavior).

If we were to bring all the possible "Confucians" together, we would encounter everyone in traditional China who had the slightest claim to social or intellectual standing. All those so-called Taoists would make up a group just as motley and probably a great deal larger. The overlap between these two groups would defeat any attempt to generalize about their differences—unless, of course, we were to open Pandora's box by treating them, come what may, as mutually exclusive.[24]

What sense can we make of the statement, "Taoists were more friendly toward science or technology (or democracy, or revolution) than Confucians"? There are many ways in which the claim might be translated into a form that historical evidence can confirm or refute. One might, for instance, contrast Taoist priests with recognized teachers of Confucian texts, or initiates whose spiritual orientations were demonstrably shaped by Taoist scriptural traditions with thinkers explicitly committed in some specific sense to the ideals of Confucius and his followers. In neither case is the claim that the Taoists favored science (et cetera) proven by the evidence that has been presented for it so far, and prosopo-

---

[24] They are understood on the whole as mutually exclusive by historians in the People's Republic of China, where it is administratively feasible to keep Pandora's box closed. I might add that the identification of Confucianism with convention seems to do no harm in purely political history from the sixteenth century on, when orthodox Taoist organizations had ceased to play an important role. I find appropriate to the subject of Frederic Wakeman, Jr.'s excellent volume, *The Fall of Imperial China* (New York, 1975), his use of "'Confucian' . . . to denote the political and moral orthodoxy that prevailed after the T'ang period" (p. 17, n. 2). I would put the final crowding of Taoism and Buddhism out of the state orthodoxy considerably later than the T'ang, but earlier than the events described in Wakeman's book. More common in introductory histories is a blurring of the Confucian-Taoist distinction: "The man in power was usually a Confucian positivist, seeking to save society. The same man out of power became a Taoist quietist, intent on blending with nature around him. The active bureaucrat of the morning became the dreamy poet or nature lover of the evening" (Fairbank et al. p. 49).

On the Word "Taoist"

graphic studies of this kind have not been published. If, on the other hand, we translate it into oppositions between the scholarly and the nature-loving, or between the conforming and the antifeudal, we may be left with something not very far from "educated individuals who hold unconventional sentiments are more inclined to value activities unconventional for the educated than are educated people who hold conventional sentiments." That is probably not quite a tautology, but it is sociologically vacuous and historically not very stimulating.

TAOISM AND POPULAR RELIGION

In one study after another disciplines and concepts connected with transcendence have been called "Taoist"—the Taoist concept of immortality, Taoist amulets, Taoist breath disciplines, Taoist alchemy, and so on.

In what sense are they Taoist? Let us take the idea of *hsien* immortality and the arts of breath control as examples. Did the authors of the *Lao-tzu* invent them? Hardly. Although they and the writers of the *Chuang-tzu* and the *Kuan-tzu* were the first to leave writings about breath control that we can read today, they appear to be discussing established practice. It is generally recognized that the notion of immortality developed to the point reflected opaquely in the *Chuang-tzu* (earliest parts from late fourth century B.C.?) and clearly in the *Yuan yu* (late second century B.C.?) and the *Lieh-tzu* (fourth century A.D.?).[25]

Were thought about immortals and the practice of breath control subsequently restricted to people who were committed in a special way to the *Lao-tzu*? No. Was breath control practiced, or

---

[25] For an authoritative summary of current understanding of *hsien*, see Ying-shih Yü, "Life and Immortality in the Mind of Han China," *Harvard Journal of Asiatic Studies* 25 (1964–65): 80–122. David Hawkes's translation of the *Yuan yu* in *Ch'u Tzu. The Songs of the South. An Ancient Chinese Anthology* (Oxford, 1959), pp. 81–87, follows his understanding of it as "a Taoist's answer to *Li Sao*."

Priority in description of breath-control techniques is another issue which cannot be resolved in the light of present understanding, since all the allusions cited are ambiguous and are interpreted in very different ways. The strongest case for the practice of physical disciplines as reflected in the *Chuang-tzu*, *Lao-tzu*, and *Lieh-tzu* was made by H. Maspero in *Le taoïsme*, Mélanges posthumes sur les religions et l'histoire de la Chine, vol. 2 (Paris, 1950), pp. 201–18. There has been very little publication of high scholarly quality to carry this work further.

But then there is Mencius's famous passage about his *hao-jan chih ch'i* ("floodlike *ch'i*" in D. C. Lau's rendering; 2A.2.11). Wei-ming Tu has suggested that this is Mencius's attempt to add or counterpoise a dimension of moral cultivation to established breath-cultivation techniques (private communication). Unlike the sources cited above, Mencius (no Taoist) was describing his own practice. In short, it is premature to regard the special association of breath control with early Taoist philosophy as proven.

immortality believed in, only by Taoist initiates? There is every reason to believe that, even before the first Taoist sects originated and right up to the mid-twentieth century, immortality and breath control were taken seriously by numerous people in every segment of Chinese society. Both still can be found in the few habitations of Chinese where tradition maintains some semblance of life. Did breath-control techniques, or other arts of long life and immortality, originate in the Taoist religious organizations? We have yet to see persuasive evidence that *anything* originated in the Taoist religion outside of a particular way of embodying Godhead in the Tao and its emanations, and a number of rites devoted to these divinities. Orthodox Taoism is built largely upon beliefs and practices adapted from popular religion. As for popular religion—the communion of the ordinary people of China with their great celestial bureaucracy and the spirits of their dead—we understand it little better today than did most scholars of ancient times, who casually dismissed it under the labels "vulgar beliefs" and "folkways."

The issue is not whether *hsien* immortality or breath disciplines had Taoist connotations in the minds of certain Chinese, but whether such beliefs and practices reliably signal Taoist influence. The alchemical or medical practices of T'ao Hung-ching (456–536), the systematizer of the Mao Shan Taoist sect, could certainly be called Taoist to the extent that they were associated with him, regardless of their origin or prevalence in non-Taoist circles.[26] What makes them Taoist in this sense is their context and the special character he gave them, not their nature. To transfer the label "Taoist" to another alchemical or pharmaceutical context without good cause is not sound method.

When Taoist initiates performed certain techniques (for instance, physical and breathing disciplines) which do not pass any of the above tests, it is worth asking whether they were doing them as Taoists per se or simply as Chinese in a certain time and place.[27] We do not, after all, assume that rice was Taoist because Taoists ate it.

---

[26] The connection between Tao's religious and medical activities has not been explored in depth, but Strickmann has studied T'ao's alchemy in connection with his establishment of the Mao Shan sect of Taoism.

[27] See *Taoist Yoga. Alchemy and Immortality* (London, 1970), by the learned Buddhist adept Lu K'uan Yü. It translates the work of a writer born in 1860 and describes practices often encountered outside esoteric circles in present-day Taiwan and Southeast Asia. No evidence of Taoist origin or particular association is given. The discussion of "the Taoist school" in the same author's *The Secrets of Chinese Meditation* (London, 1964) is equally vague about what makes it a school.

On the Word "Taoist"

Why did ancient authors habitually associate occult practices and conceptions with Taoism without being concerned with the special character of the linkage?

Chiao Hsun (1763–1820), the great classicist and mathematician, provides a clue in his argument for the extravagant hypothesis that books perish because bibliographic classification is sloppy: "I have observed that, although the Buddhists and Taoists passed through prosperity and decline, their writings have survived. This is surely not a matter of personal efforts alone. Because both schools have established clear schemes of classification, they have been able to preserve their books fervently over the generations, so that although they might be destroyed they would not perish."[28]

In other words, the compilers of the successive Taoist patrologies provided rubrics for classifying books of all kinds that had been collected and stored in Taoist monasteries and temples. These books soon would have perished had there been no shelter but that provided by the whims of individual collectors.[29] The Taoist collections were quite eclectic. They preserved many important works of medicine, materia medica, geomancy, and other sciences. It was primarily the printed *Tao tsang* that preserved the neglected *Mo-tzu* until in modern times it began to attract the attention of philologists. I have never seen any sign that the compilers of the *Tao tsang* considered the *Mo-tzu* Taoist.[30] It was enough that it was useful.

Even more to the point, the Taoist librarians collected records of a multitude of beliefs and practices created by the illiterate majority and by literate people who were in touch with the folk milieu. We have yet to see proof that more than a fraction of those who wrote on such subjects as breath control and immortality were connected with Taoist organizations, or were even aware of the special tenets of Taoism; but certain Taoist sects were the only large organizations (outside the central government, of course)

---

[28] Chiao Hsun, *Kuo shih ching chi chih*, end of ch. 3 in *Ming shih i-wen chih, pu pien, fu pien* (Peking, 1959), 2:932.

[29] The selection of Taoist texts freely available in traditional times depended as much on literary beauty as on intrinsic importance. There was, of course, a great bias toward books useful in self-cultivation. Important aspects of the religious Taoist literature—spiritual and magical exercises based on elaborate and fantastic image-forming meditations, and the entire basis of Taoist ceremonial—have remained almost unknown to the Chinese general reader.

[30] A simple test of orthodox Taoist attitudes is the priest Po Yun-chi's *Tao tsang mu-lu hsiang chu* [Bibliography of the Taoist patrology, with detailed annotations, completed 1636]. Although Sun Ssu-mo's medical writings elicit a complimentary note, the *Mo-tzu* passes without comment, T'ui-keng t'ang ed., 4:30b–31a, 33b.

motivated to collect and preserve quantities of such "heterodox" writings.

I do not mean that Taoist eclecticism was a mere matter of librarianship. Nor was it a matter of Taoism's inherent logic. Most Taoists through history—initiates of early sects, later priests conducting rituals of renewal and laying the dead to rest in the villages and towns—did not regard the *Mo-tzu*, the Travels of King Mu [*Mu t'ien-tzu chuan*], the construction of special water clocks to time meditation, and so on, with special interest.[31] Groups primarily oriented toward individual salvation (for themselves and their powerful patrons) were willing to ransack every current belief and practice. When those groups, particularly the Mao Shan sect and the Ch'üan-chen sect after it, classified their books, they gave structure to everything they borrowed and created.

There was no Taoism, Ōfuchi has argued—no community of interest or consciousness of shared conviction among early sects now considered Taoist—until it was created by the classification of scriptures. Ōfuchi has shown that from the beginning of such classification by Lu Hsiu-ching (406–77) and T'ao Hung-ching (456–536) a primary motive was to assert the paramount spiritual status of the compiler's own tradition.[32] Another aim was to preserve old religious traditions against encroaching Buddhism. A third motive was to assert in the spiritual realm the high self-estimate of the older Wu gentry as northern refugees with Celestial Masters sect traditions became politically ascendant over them. It is thus understandable that the Celestial Masters tradition, despite its greater antiquity and popularity (especially but by no means only in the north) was not even represented in the tripartite *san tung* classification of the early patrology.[33]

Although rivalry fueled the formation of the Taoist patrologies, the result over a long period was a stock of beliefs and practices

---

[31] *Mu t'ien-tzu chuan* (*Cheng-t'ung Tao tsang*, vol. 137; *Ch'üan-chen tso p'o chieh fa*, vol. 988) (reprint ed., Commercial Press).

[32] Ōfuchi Ninji, "*Dōzō no seiritsu*," *Tōhōgaku* 38 (1969): 49–57, esp. 50–54; translation by Leon Hurvitz in "How the *Tao-tsang* Took Shape" (paper for the First International Conference on Taoism). A good deal of the same evidence is scattered through Ch'en Kuo-fu, *Tao tsang yuan-liu k'ao* [Researches in the history of the Taoist patrologies], rev. ed., 2 vols. (Peking, 1963). See also the paper by Strickmann cited in n. 2 above. Lu was associated with the Ling-pao tradition and T'ao with the Shang-ch'ing, i.e., Mao Shan.

[33] The basic Celestial Masters scriptures were tacked on as one of four ancillary groups of texts (*ssu fu*). The motivations of the southern aristocrats have been explored in detail by Strickmann in "Taoism in the Lettered Society of the Six Dynasties" (see n. 2 above).

On the Word "Taoist"

upon which all the sects that worshiped the Tao could draw. Thus a willingness to adapt elements of popular liturgy, of Buddhist organization and scriptural form, and of the state bureaucratic ritual, was central to the formation of Taoism. This borrowing of orthodox Taoists represented only half of a complex and ill-understood dialectical relationship. For instance, Rolf Stein has investigated with his customary penetration the movement of Taoist liturgies into popular religion in early times. In the process he has shown how useless recorded accusations of heterodoxy and degenerate practices are in our endeavor to untangle established Taoist institutions from their imitators and heretics. He has accumulated an impressive array of examples in which the stigma "lewd cult" was directed against orthodox Taoist groups by orthodox rivals.[34]

Stein's discovery seemingly renders hopeless the distinction between Taoists and their popular competitors. To the contrary, it provides us with a powerful light by which we can read the documents critically and reconstruct the dynamics of institution-building (as well as doctrine-building) in the transitional milieu of Taoism.

It is scarcely surprising that the distinction between Taoism and popular religion was of negligible interest to conventional members of the educated elite. They read about popular beliefs and practices in books and collections of anecdotes that stressed Taoist connections. Outsiders were not concerned that the Taoists did not originate certain disciplines and rites, were not their main sponsors, and in many cases had no reason to alter them and give them a specifically Taoist flavor in use.

"Popular religion," as I have indicated, was not an acceptable cubbyhole to historians and other educated people trying to make sense of beliefs very much at odds with their own rationalist humanism. It could be all the more easily rejected as a cubbyhole because "Taoism" provided an alternative.

Taoism has usually been considered heterodox by pedants (although it is not clear whether the majority of the educated elite would have agreed with them before the end of the T'ang period). But there are heterodoxies and heterodoxies. Taoism had organization, well-defined literary traditions (with a certain number of its

---

[34] Stein, "Taoïsme religieux et religion populaire (du IIe au VIIe siècle)" (paper for the Second International Conference on Taoism), and "Etude du monde chinois: Institutions et concepts," *L'Annuaire du Collège de France* 67 (1967–68): 411–15; 68 (1968–69): 453–57.

canons circulating in the lay world, where they were admired as fine writing), and imperial recognition. Far from being politically revolutionary, orthodox Taoist sects after the Han period played no active role in rebellions, messianic or otherwise, and never represented rebellion as desirable (it was because certain "Taoist" ideals were so thoroughly diffused among the people that rebel groups freely adapted them to their own purposes). Instead, the Taoist religious organizations consistently sought the favor of the temporal powers, provided support for the government, and modeled relations with the gods on the usages of the imperial bureaucracy.[35] "Taoism" was an epithet that came readily to the minds of the literati, but there was no incentive to be fastidious in its use.

THE CASE OF KO HUNG

Nor has modern scholarship always been more fastidious. An example which bears on the study of early science is the general estimate of Ko Hung (283–343) as a major figure of the early orthodox Taoist religion and as "the greatest alchemist in Chinese history."[36] Let us examine the evidence with due care.

Ko claims in his Inner Chapters of *Pao-p'u-tzu* to have received several texts from one Cheng Yin, whose teacher had been Ko's great-uncle Hsuan. Ko Hsuan was reputedly the first person in South China to receive them (from the northerner Tso Tz'u). Ko Hung devotes two chapters to the alchemical procedures he had been taught. In both he assures his readers that he could never afford the ingredients to prepare them. Nor is there evidence that if Cheng Yin ever carried out an alchemical preparation Ko witnessed it.[37] Ko states his intention to prepare the Divine

---

[35] Anna K. Seidel, "The Image of the Perfect Ruler in Early Taoist Messianism: Lao-tzu and Li Hung," *History of Religions* 9 (1969–70): 216–47. As for the Han, when the picture is not as clear as in later periods Seidel notes that the Celestial Masters sect "did not attack the established political power" (p. 227).

[36] Needham (n. 8 above), 2:437; Yuan Han-Ch'ing, *Chung-kuo hua-hsueh-shih lun-wen-chi* [Collected essays on the history of chemistry in China] (Peking, 1956), p. 180. Needham recently has modified this characterization to "the greatest alchemist of his age, and the greatest Chinese alchemical writer of any age" ([1976], 5, pt. 3:79). Note that the editors of the Ssu-k'u Catalogue state flatly that the Inner Chapters "are in their discourse purely Taoist" (*Ssu k'u ch'üan shu tsung mu t'i yao*, Kuang-tung shu-chü ed., 1868, 146:42b). There has been some confusion about Ko Hung's dates, but see N. Sivin, "On the *Pao p'u tzu nei p'ien* and the Life of Ko Hung (283–343)," *Isis* 60 (1969): 388–91.

[37] Ko denies performing alchemy in *Pao-p'u-tzu nei p'ien*, P'ing chin kuan ts'ung-shu ed., 4:2a, and 16:1b. Needham's assertion that Ko achieved some minor elixirs with which he prolonged his life is not supported by the source he cites, nor by any reliable source known to me (Needham, 5, pt. 3:82.)

VI

On the Word "Taoist"

Medicine after his philosophic writings have been completed.[38] He consistently phrases his previous alchemical interests as manifested by searching for unusual writings and canonic formulas rather than by joining those who were performing the Great Work.[39]

Ko stood at a watershed. He was an initiated adept seeking to transcend his mortal limitations. In this restricted sense he can be considered "part of a long line of Taoist masters concerned with magic, medicine, alchemy (the T'ai-ch'ing and San-huang line)." But attempts to connect him with large-scale Taoist organizations (for which salvation was a communal matter) have been based on tenuous evidence indeed, mostly on marriage connections and his descendants' involvements.[40] He does not even seem to have known that the Celestial Masters or any other contemporary Taoist sect existed in his time. The expectation of an impending new order, at once cosmic and political, that brought together

---

In ch. 4 Ko adds, "My teacher Master Cheng was a disciple of my great-uncle Hsien-kung [=Hsuan], and received [several canons on elixirs of immortality] from him; but [Cheng's] family was poor and he did not have the wherewithal to buy the ingredients." In 16:1a, speaking of the Medial Canon of the Yellow and White (*Huang pai chung ching*), concerned with the preparation of silver and gold, Ko states, "Master Cheng said that he had tried to prepare [these formulas] on Mount T'ung in Lu-chiang with Master Tso, and that they all were successful." This Medial Canon was not the same as the scriptures concerned with the elixir of immortality mentioned in ch. 4. Whether Cheng is believable may be assessed in the light of the fact that he convinced Ko that Tso Tz'u had performed many thaumaturgical marvels (2:4b, 5:6a, 12:5b, 15:6b, and 18:3b). Cheng was modest about his own feats in which Ko Hsuan and Tso Tz'u were not involved. In any case, Ko was acquainted with the experiment at Lu-chiang only by hearsay. Kaltenmark states explicitly that "Ko Hung admits that he never undertook any experiments," but makes Ko "after Wei Po-yang, the greatest theoretician of alchemy" (*Lao tseu et le taoïsme* [Paris, 1965], p. 168; *Lao Tzu and Taoism*, trans. Roger Greaves [Stanford, Calif., 1969], p. 131). Kaltenmark later appears to qualify his emphasis on Ko's theoretical eminence (*Lao Tzu and Taoism*, p. 132). I contend that Ko's chapters on alchemy contain very little of theoretical consequence; their value lies in the practical information he transmits from earlier sources. Ko's limited understanding makes some of his formulas indecipherable, but the operations described in the rest make sense chemically (see Sivin, *Chinese Alchemy* [n. 18 above], pp. 40–47). For a complete but often unreliable translation of the *Nei p'ien*, see James R. Ware, *Alchemy, Medicine and Religion in the China of A.D. 320* (Cambridge, Mass., 1966). It is keyed to pages of the edition cited in this footnote.

[38] Ko, 4:17a.

[39] Ibid., 4:2a, 4a; 16:1a; *Pao p'u tzu wai p'ien*, P'ing chin kuan ts'ung-shu ed., 50:7b, trans. Ware, p. 15.

[40] For a thorough and judicious account of Ko's family connections and his associations with Taoist textual traditions, see Max Kaltenmark, "Religions de la Chine" (Rapport sur les conférences), Ecole pratique des hautes études, V^e section, sciences religieuses, *Annuaire* 77 (1970–71): 125–27. I would add that the only scripture taught to Ko which was clearly of Taoist origin and concerned primarily with matters other than alchemy and cosmology was the *San huang nei wen* (*Nei p'ien*, 4:1b, 19:3a). This was probably but not certainly part of the *San huang ching* and was devoted to disciplines for individual immortality (see Kaltenmark, "Notions sur quelques grands sutras taoïstes" [paper for the Bellagio conference, 1968], and Ch'en Kuo-fu [n. 32 above], 1: 71–81).

believers in every early sect, never entered his consciousness.⁴¹ He acknowledged the authentic knowledge and spiritual authority of no one but his teacher. The spread of the Celestial Masters into the south did not reach discernible proportions until after his time, and his direct influence on the form Taoist institutions finally took seems to have been negligible—although his vision of a Taoism designed for the needs and tastes of the aristocratic adept did turn out to be the wave of the future.⁴²

The Inner Chapters are anything but the writings of a Taoist man of wisdom or organizer for his disciples or for other initiates. This book is a vast trove of commonplaces and hearsay about popular beliefs in which Ko's few incontestably Taoist texts play an essential but small part. Its goal is not to catalogue, synthesize, or provide a handbook of techniques. It is rather a dialogue in which Ko hurls scattershot against a skeptical anonymous interlocuter.

The Inner Chapters are a one-issue book. Ko seeks to convince his questioner, and thereby his readers, that immortality is a proper object of study and is attainable—not only by the ancients but in his own time, not only by a destined few but by anyone with enough faith to undertake arduous and dangerous disciplines. The devotion that Ko calls for implies wholesale acceptance of legends, myths, tales of prodigies, magical beliefs, religious faiths—practically every belief current in the popular imagination of Ko's time and the inverse in almost every sense of what "fundamentalist Confucian" humanists considered worthy of thought (but then they were no longer setting the intellectual style). The only notable exceptions to Ko's credulity were what he ridicules as the notions of heterodox and uninitiated self-styled Taoists

---

[41] Ko managed to copy off the titles of the books his teacher owned, and among them we find the *T'ai-p'ing ching*, associated with the major sect of Northeast China and a source of ideology for the Yellow Turban rising of A.D. 184. It was not among the texts Cheng allowed Ko to read (*Nei p'ien*, 19:2b–3a), and Ko does not refer to specific T'ai-p'ing beliefs. Ko's couple of citations from the *Ling pao ching* (17:4a, 5a) do not prove that he had been initiated into knowledge of it. In *A Gallery of Chinese Immortals* (London, 1948), pp. 60–61, Lionel Giles translates an account of Chang Tao-ling's "Taocratic" state in late second-century Szechwan, supposedly from Ko's *Shen-hsien chuan*. This would indicate that Ko was at least familiar with early Celestial Masters organization. But the ascription of this book (along with many others) to Ko is most unlikely; even if it were his, the description is not part of Chang's biography in early versions of *Shen-hsien chuan* (e.g., that in *Yun chi ch'i ch'ien*, Cheng-t'ung Tao tsang ed.), 109:19a–21a), but only in spurious recensions concocted out of the *T'ai-p'ing kuang chi* (cf. reprint of the latter [Taipei, 1968], 8:32a–33b; see *Ssu k'u ch'üan shu tsung mu t'i yao*, 106:46a–46b).

[42] For instance, the *Pao p'u tzu nei p'ien* is not mentioned in the foundation texts of Mao Shan Taoism. See the citation lists in Yoshioka Yoshitoyo, *Dōkyō keiten shiron* [Studies in the history of the canonical Taoist literature] (1955; 2d ed., Tokyo, 1966), pp. 350–92.

On the Word "Taoist"

(*su tao-shih*), too ignorant of authentic arcana to overcome their skepticism about Ko's enthusiasms. He represents them as literate and owners of books, and thus not socially outside his pale.

Ko's sketchy, forced, and contradictory understanding on many points is remarkable when compared with the internal consistency and assured tone of Taoist texts of the time.[43] It is all the more remarkable because Ko is discussing very widespread beliefs. Whom is he trying so desperately and with so inadequate a command of his materials to convince? Hardly the people in the fields and marketplaces around him, for it was their own familiar lore of magic, gods, and spiritual power, bolstered with what he had learned of Taoism from the north, that he was trying to persuade his readers to take seriously. Who were the skeptics?

Ko Hung, "marquis of Kuan-chung, entitled to support from two hundred households of Chü-jung town,"[44] was writing for people of his own quality. Some of the small aristocracy, both southerners and the northern immigrants who were beginning to settle among them, knew a little of Taoism; it was they he searched out and discredited. Few, no doubt, cared to know much about the vitalistic, animistic, divinity-centered world view of the peasants whose toil supported them.

This view of Ko as an obsessed bookman and indiscriminate lore-collector may seem abrupt so soon after Liu Ts'un-yan's recent essay on the viewing of tuberculosis microbes through compound microscopes by Taoist priests in the twelfth century, in which he found occasion to elevate Ko Hung to the rank of Taoist priest.[45] But the image of a priest, a doer in the community of the faithful, is not what the credulous and labored preachments of the Inner Chapters suggest. Ko's style was rather that of a pedantic purveyor of occultism to the upper class. I can only think of him as the Alan Watts of his time.

Why should Ko Hung's quixotic attempt to turn the old southern aristocracy into dropouts have been taken with such deadly seriousness for so long? Because the alternative is to insist on delving into the social and historic circumstances in which his

---

[43] This point has been made by Michel Strickmann in *Encyclopædia Britannica*, 15th ed., s.v. "Taoist Literature": "For all his charm, however, the contradictory opinions that the author in his enthusiasm lets slip raise doubts as to his real understanding of the beliefs he so lustily defended." This article, Strickmann's "Taoism, History of," and Anna K. Seidel's "Taoism" are by far the best available short surveys of these topics, in point of learning, critical approach, and clarity.

[44] *Pao p'u tzu wai p'ien*, 50:11a, trans. Ware, p. 20.

[45] Liu Ts'un-yan, "The Taoists' Knowledge of Tuberculosis in the Twelfth Century," *T'oung pao* 57 (1971): 285–301.

books were written. It is premature to expect clarity about the changing social and historic character of religious Taoism, but steady efforts in this direction are obviously called for.

CONCLUSIONS

I have drawn attention to the frequent confusion between many things that "Taoist" can mean and have suggested that this confusion often mires us down when we try to comprehend the historic character of these phenomena and their complex interplay. I have also examined the frequent failure to distinguish between Taoist and popular beliefs and practices that are recorded in Taoist literature. This neglect has led to wasted effort in studying the evolution of Taoism.

Few of the sorts of confusion I have discussed are the creation of sinologists; most were prevalent centuries ago. To some extent the confusion reflects the complexity of the historic circumstances. A great deal of additional ambiguity was generated in traditional times by the social prejudices of people who happened to write about religion and by a general reluctance to interpret philosophic and religious texts critically, to think of them as documents produced by subjective observers in historic predicaments. It is perhaps a bit early in the development of modern studies of Chinese religions to expect everyone involved in these subjects to work out a fresh understanding unconstrained by traditional limitations of viewpoint.

What can be done now, in the present state of religious studies, to make the exploration of Taoism's connections with science more fruitful?

Although carelessness about definitions is my theme, the last thing I would propose as a remedy is that we all adopt a single operational definition of "Taoist," "Confucian," "popular religion," and so on, and apply them with merciless rigor. There is no single definition of each term that can encompass the diverse historical questions that we might want to ask of the sources. A study of the interaction between popular and Taoist liturgy might demand great precision of definition, limited finally by the ambiguity of the documents. A study of Taoism as reflected in vernacular literature might accept as Taoist everything the writers called Taoist.

My point is a modest one, inspired by a saying of Confucius himself that can hardly be improved upon as a guide to critical

VI

On the Word "Taoist"

research: "When you know something, to know that you know it; when you don't know it, to know that you don't know it: that's knowledge."[46] I propose merely that as our learning proceeds we remain aware, and keep our readers aware, of the ambiguity in our sources and of what definitions we are applying as we interpret them.

Surely there is nothing wrong with considering Taoism religious, poetical, magical, scientific, democratic, and politically revolutionary, so long as we remind our readers that these descriptions apply to a variety of phenomena which are Taoist in very disparate senses, and that such a view of Taoism is not an assertion about any collectivity which interacted with other collectivities to shape science or other aspects of Chinese history.

It is, I believe, simple and feasible, when we speak of something as Taoist (or Confucian), to be explicit about the sense in which we so consider it and the criteria by which we so judge it. If we are concerned with orthodox Taoism and the only available sources may or may not be Taoist, it is a simple matter to maintain the distinction between what we are sure about and what we are not. There is enough blurring beyond our critical control without adding more needlessly. When we use the writings of people who were Taoist in a perfectly definable sense, it is a simple matter to specify what that sense is. It is a simple matter, when we do not know what private motives and social situations inform a writer's assertions about reality, to emphasize our ignorance. We may be specialists in the religious aspects of Taoism and thereby feel relieved of all obligation to explore in depth the political commitments or scientific interests of the people we study; but the people we study were more than disembodied Taoist consciousnesses.

It is not a simple matter, but it is essential nevertheless, to make certain that our historical reconstruction of traditional Chinese religions adequately represents their diversity, their unity with the rest of what people did and thought, and their complex textures, which span every level of spiritual experience.

---

[46] *Analects* 2.17. *Chih* refers to understanding and recognition of significance as aspects of knowledge, not to objective factual knowledge isolated from the act of understanding and evaluating.

GLOSSARY OF CHINESE CHARACTERS

Ch'en Kuo-fu, Tao tsang yüan-liu k'ao 陳國符、道藏源流考

Cheng-i 正一

Cheng-t'ung Tao tsang 正統道藏

Cheng Yin 鄭隱

Chiao Hsun 焦循

Ch'ien chin i fang 千金翼方

chih 知

Chin ching 禁經

Chuang Lin hsü tao tsang 莊林續道藏

Ch'üan-chen 全真

Ch'üan-chen tso p'o chieh fa 全真坐鉢捷法

Chung-kuo ku-tai k'o-hsüeh-chia 中國古代科學家

hao-jan chih ch'i 浩然之氣

hsien 仙

Hsien-kung 仙公

Huang pai chung ching 黃白中經

i jen 異人

Kimura Eiichi 木村英一

Ko Hsüan 葛玄

Ko Hung 葛洪

Kubo Noritada 窪德忠

Kuo shih ching chi chih 國史經籍志

Lieh-tzu 列子

Ling-pao 靈寶

lu 籙

Lu Hsiu-ching 陸修靜

Mao-shan 茅山

Ming-shih i-wen chih, pu pien, fu pien 明史藝文志、補編、附編

Mo-tzu 墨子

Mu t'ien-tzu chuan 穆天子傳

Nishi Junzō 西順藏

Ōfuchi Ninji 大淵忍爾

Pao-p'u-tzu nei p'ien 抱朴子內篇

Po Yün-chi, Tao tsang mu-lu hsiang chu 白雲霽、道藏目錄詳註

San-huang 三皇

San huang nei wen 三皇內文

san tung 三洞

Shang-ch'ing 上清

Shen-hsien chuan 神仙傳

Shimode Sekiyo 下出積與

ssu fu 四輔

Ssu k'u ch'üan shu tsung mu t'i yao 四庫全書總目提要

Sun Ssu-mo 孫思邈

T'ai-ch'ing 太清

T'ai p'ing ching 太平經

T'ai p'ing kuang chi 太平廣記

VI

On the Word "Taoist"

T'ao Hung-ching 陶弘景

Tao tsang 道藏

t'ien tsun 天尊

Tso Tz'u 左慈

Wu wei t'ien-shih chi-chiu
吾為天師祭酒

Yi Nung-hwa, Han'guk togyosa
李能和, 韓國道教史

Yoshioka Yoshitoyo 吉岡義豐

Yüan Han-ch'ing, Chung-kuo hua-hsüeh-shih lun-wen chi 袁翰青, 中國化學史論文集

Yüan yu 遠遊

Yün chi ch'i ch'ien 雲笈七籤

VII

TAOISM AND SCIENCE

中歲頗好道
Devoted to the Dharma *(tao)*
in my middle years . . .
Wang Wei

The Problem
Was Taoism in some sense responsible for—or did it at least further—the development of the sciences in China? Whether there was some such relationship will concern anyone trying to understand religion or thought about Nature in China. The character of this influence may also cast light on why the technical subcultures of China and Western Europe diverged so decisively in early modern times.

Attention in Chap. VI to the consistent vagueness in scholarship about Taoism was an unavoidable preliminary to historical questions. The purpose of this essay is raise some of them.

Points of Transition
People who think about science without having practiced it often assume that it progresses by a steady accretion of knowledge. That this belief is so widespread testifies to the enduring influence of Francis Bacon, who argued in the days when modern science was being invented that people ought to do it bit by bit, in groups, according to a rational division of labor. But that is not an adequate description of what scientists do. There is a larger rhythm in which, as knowledge accumulates, it becomes more and more evident that old patterns are inadequate to encompass newly discovered phenomena. Eventually a new pattern makes sense of otherwise anomalous knowledge. Much previous knowledge may be quite irrelevant to this new pattern. After the transition, it sets the standard until the need for a still more comprehensive pattern can no longer be ignored.

A transition of this sort has taken place in studies of Taoism. The view of Taoism and its evolution prevalent among historians until the 1970's gave rise to a good many hypotheses and opinions about the relations of Taoism and science. As the old conventional wisdom has been replaced by a fundamentally different understanding of Taoism, no one has reassessed these claims about a link with science. It is time to ask whether they fit the emerging new framework, or are extraneous to it.

2

Two Histories of Taoism
The first history of Taoism originated centuries ago among orthodox Chinese authors. Historians there and elsewhere continued to flesh out its structure until the late 1960's. The convulsions within China in the last century and a half, and metamorphoses in the rest of the world's relations with it, are reflected in the fine texture of this complex of interpretations, but it is not difficult to summarize in broad outline. Its view of Taoist practice was based on a few readily available texts that, as Anna Seidel put it, "enjoyed the esteem of Confucian schoolmasters." When tracing the vicissitudes of the religion it did not look beyond the prejudiced accounts in the official histories. Not surprisingly, it has been largely replaced by a second history that mines the enormous collections of Taoist scriptures and a far wider range of historic sources.

History 1. Very early attempts to placate Nature were shaped by the mystical practices of shamans to form a philosophy that sought "the union of the individual with an impersonal natural order," unlike Western mysticisms that strove for oneness with a personal deity. Taoism resembled Confucianism only in that calling it a religion was a matter of definition; those of philosophers and religious scholars tended to diverge sharply. The most prominent Taoists were Lao-tzu 老子, variously dated from the sixth to the third century, and Chuang-tzu 莊子, somewhat more closely placed between 399 and 295. Although a number of other authors were included in the "Taoist school," descriptions of key Taoist ideas were almost entirely based on the writings of these two.

Somehow out of this school, in the first and second centuries A.D., evolved more than one "secret society with strong Taoistic tendencies." They fomented a great peasant uprising in the first century A.D. Late in the second century Taoism metamorphosed, possibly under "barbarian" Buddhist influence, into "an organized popular religion." As the Han dynasty collapsed, this church even governed a large piece of West China as an autonomous theocracy. A leading British authority summed up the usual judgment: "It is hardly possible to dignify with the name of religion such a strange medley of magic, legend and gross superstition; and one cannot believe that its scriptures were regarded very seriously by any large section of the community."[1] Taoism was often at the root of rebellion, and some scholars

---

1. Giles 1935–1937: 1. Since most of this essay is devoted to Joseph Needham's hypotheses, my usage of the words "Taoist" and "Taoism" reflects his own. When clarity is advisable I use other terminology. References in the form "II, 123" or "V.3, 76" are to Needham et al. 1954- . I acknowledge with gratitude the counsel of Timothy Barrett, Derk Bodde, Kenneth J. DeWoskin, David Keightley, Terry Kleeman, G.E.R. Lloyd, Victor Mair, Joseph Needham, and Donald B. Wagner. This paper was written in Need-

VII

TAOISM AND SCIENCE

affirmed that "we always find the Taoists with the party opposing the literati."[2]

Taoists borrowed monasticism from Buddhists, competed against them for imperial patronage, split into sects that competed with each other, but failed to build "an organized church." They remained sunk in degeneracy, the intellectual and moral standards of their communities low, until the last "Taoist pope" was ousted in 1927.

There was what one might call a last gasp of respectability in Neo-Taoism, a short-lived movement of the third and fourth centuries of brilliant intellectuals who devoted to the Taoist classics the deep study that in earlier centuries would have been channelled into those of the Confucians. Aside from that, the various enthusiasms of cultured gentlemen for immortality, alchemy, and escapist "pure conversation" were, if not perceptibly related to the popular tradition, at least "Taoistically inclined" in ways that no one felt the need to explain.

The transition. This first history was not at all coherent, and (in hindsight) could not have been maintained without large gaps. The lacunae existed because historians of late imperial China did not find the religions of the masses a suitable topic for exploration. Their overviews were cobbled together out of incidental accounts and biographies that had found their way into the Standard Histories for reasons that had nothing to do with a desire on the parts of ancient historiographers to document an organized religion.

The massive Taoist Canon *(Cheng-t'ung tao tsang* 政統道藏*)*, printed ca. 1477 for the use of religious communities, was reprinted for commercial purposes in 1924-1926. It quickly found its way to centers of learning around the world (there were already incomplete copies in Japan and Paris). That is not to say that the nearly fifteen hundred treatises it contains, and still more scriptures in reprints of smaller scope, were promptly digested. With respect to the question of Taoism and alchemy, for instance, I believe that by 1950 there was only one person in the

---

ham's East Asian History of Science Library. When the first version was presented at the Third International Conference of Taoist Studies, Unterägeri, Switzerland, in September 1979, he was commentator. That draft carried a dedication to Needham and his colleagues marking the twenty-fifth year since the first volume of *Science and Civilisation in China* was published. This final version commemorates three great explorers of Taoism who are no longer with us, Joseph Needham, Anna Seidel, and Michel Strickmann.

2. Weber 1922. This summary is drawn from the best textbook survey of East Asian studies before 1970, Reischauer & Fairbank 1958: 72–76, 122, 137–141, and the most authoritative anthology of philosophy, Wing-tsit Chan 1963: 138–177. These were unadmiring but benign accounts; those of European missionary scholars (on whom Weber and others depended) written in the days when imperialism was a good word were a great deal less tolerant. See, for instance, Wieger 1917.

world both trained as a scientist and widely and critically familiar with the collection, namely Ch'en Kuo-fu 陳國符.³ By 1968, however, Japanese historians had accumulated a great deal of monographic research that implied new and complex patterns. A few French scholars, brought up in the Parisian sociological tradition, had begun looking at Taoist sources in new ways. Well before 1950 Edouard Chavannes, Paul Pelliot, and Henri Maspero were taking them seriously as religion, as varieties of individual self-cultivation, and as documents of a social history quite unlike that of conventional scholarship. The Japanese and French work had already rendered History 1 obsolete, but did not replace it. Writers on Taoism elsewhere largely ignored their publications.

Between 1968 and 1979 a series of three international conferences brought together a total of thirty or so scholars from Japan, China, Europe, and the United States, one of them an ordained Taoist master, a few of them specialists, and a majority who had been working on one or another margin (including two historians of science). This encounter of diverse research experience and insight realigned into new patterns a great deal of what earlier had never quite made sense. Taoist studies gradually became a new presence in Sinology, with the usual apparatus of specialities from journals to Internet sites. Since it normally takes at least twenty years for research breakthroughs to engender a real consensus, and to osmose from the conference site to the undergraduate survey, the old prejudices are still vigorous. But the contours of a quite different history have for some time been visible from the research frontier.⁴

---

3. Chen 1949 is a remarkably useful study of the history of the Taoist canons.

4. This overview does not express a consensus, but my own reading of one that is forming amongst those immersed in the primary sources. It is based on the post-1968 literature, on personal communications from most of the specialists in China, Japan, and elsewhere, and on my study of the primary sources. For histories of Taoism that reflect current scholarship see Seidel 1990 and Robinet 1991 (the latter up to the Yuan). There is no such history in English, although Seidel 1974 and Strickmann 1974 are excellent as early overviews of modern Taoist studies, and Baldrian-Hussein 1987 provides an up-to-date introduction. Seidel 1989–1990 is a thorough and magisterial essay on Western scholarship, updated by Verellen 1995. Boltz 1987 is a detailed guide to late Taoist literature. Teiser 1995 lists Western publications on popular religion, very broadly defined. Particularly useful monographic studies include Seidel 1969 for the Han, Fang Shih-ming 1993 for the second century, Mollier 1990 for apocalyptic movements, Mather 1979 for early court Taoism, Bokenkamp 1983 and Bell 1988 on the Ling-pao tradition, Strickmann 1981 and Robinet 1984 for the Shang-ch'ing movement, Verellen 1989 for the end of the T'ang, Chin Chung-shu 金中樞 1974 for the late Sung, Zürcher 1980 for the early influence on Buddhism, various papers in Strickmann 1981–1985 for the adaptations of Tantrism, and Sakade 1977 and Akizuki Kan'ei 1986 for recent Japanese studies (Fukui

History 2. Although the philosophical and religious meanings of the *Lao-tzu* and *Chuang-tzu* are debated with as much gusto as ever, their historical significance has changed fundamentally. As I have noted in Chap. VI, in the first couple of centuries after their compilation readers did not take them to be exemplars of a single philosophy. There is no reason to believe that any part of either book was written much before 300 B.C. The *Lao-tzu* was probably compiled late in the third century, and the *Chuang-tzu* as late as the late second century.

The two books did not provide a philosophic basis for the early Taoist movements, which either ignored the *Lao-tzu*, reinterpreted it in terms of cult practices that had nothing to do with metaphysics, or were fixed on the mythical figure of Lao-tzu as a savior. Those groups that treated the book as a revelation revered it alongside a multitude of other revelations, some of them "revealed" by Lao-tzu after his transfiguration as a (not the) divine emanation of the Way (T'ai-shang Lao Chün 太上老君). Writings of the early Taoist movements likewise tend either to ignore or to scoff at the *Chuang-tzu*. It was rather the noble dabblers of the third century who concentrated their "studies of the mysteries" *(hsuan-hsueh* 玄學) on the *Lao-tzu*, the *Chuang-tzu*, the Canon of Supreme Mystery, and the Book of Changes. The canonic prestige of the *Lao-tzu*, the *Chuang-tzu*, and a few similar works rose later, primarily because of their official recognition by T'ang emperors. They claimed that the god Lao-tzu, whose surname was the same as their own, was an ancestor.

Finally, if we are simply interested in these two books as distinct presences in intellectual life, they had little traceable influence on philosophy before the first century B.C., and did not stand out afterward, except for their great literary influence. In the first place, few historians would place the compilation of either long before the Han. Between the mid third and the late second century, quietist ideas became part of a rich eclectic stew, reflected in a succession of books from *Lü shih ch'un-ch'iu* 呂氏春秋 on. The meat in this hodgepodge was a new theory of the state in which it became a replica of the cosmos and a simulacrum of the human body.[5] A new Confucianism, which discarded the humanism of the founder and his immediate successors for a broad range of doctrines cooked in that very stew, furnished the state with its first orthodoxy in 135 B.C. Important additions to it continued until the end of the first century (see Chaps. I and III).

Kôjun 1983 offers a somewhat different group of themes). See the provocative discussion of current problems in Fukui Fumimasa 1995. Current Chinese views, still affected by official disapproval of Taoism as "superstition," are represented in the collective volumes edited by Jen Chi-yü (1990) and Ch'en Ku-ying (1992-).

5. Sivin 1995.

VII

6

As that stew became the common diet, ideas derived from the *Lao-tzu* and *Chuang-tzu* were no longer the property of Taoists, but were simply part of what was in everyone's bowl. A few books of the time were partial to quietist ideas, and some, such as *Huai-nan-tzu* 淮南子, drew heavily on the two literary masterpieces. But even in the third century the label "Taoist" does not tell us very much. Han-fei-tzu 韓非子 displays similar affinities, but he was also in the direct line of teaching descended from Confucius, and served as a high minister of the Ch'in regime. Historians blithely call the book named after him Taoist, Legalist, eclectic, and so on, according to fad and personal taste. Such isms do more to invite confusion than to shed light.

In short, from the Han on one can identify certain ideas as Taoist to the extent that they echo those in books conventionally assigned to that bibliographic category, but embodying such ideas in some subclass of society called "the Taoists" is more likely than not to be a vacuous exercise. The best corrective for that misleading old habit is to ask "exactly what individuals does this proposition refer to?" When, as often happens, that question has no answer, rephrasing the proposition is the candid solution.

One of the most important new elements in this second history is the understanding that popular religion was the common stem out of which all religious phenomena grew. It was not a folk phenomenon, but was universal in the sense that it united elite and commoners as the form of local collective life. Officials were sometimes expected to persecute it, and ideologists despised it as a threat to the government's control of correct thought, but when they went back to their villages they were likely to take part in, and help pay for, the festivals they shared with everyone else.

Worship of local gods was a part of every community's shared activities. Officials distrusted the potential of communal organization for heterodoxy, and in didactic statements were unwilling to take seriously any culture but that of their own class. In their writing most of them simply ignored the religious character of popular rites, acknowledging them only as "vulgar practices" *(su* 俗), and their practitioners as "wizards" *(wu* 巫, a term we will explore below). The heirs of the old pedants, Nationalist and Communist party ideologues, tried to expunge it, and encouraged scholars to label it "feudal superstition" *(feng-chien mi-hsin* 封建迷信).

Taoism, a tradition of the literate minority, was, for those who drew the distinction, more respectable than popular religion. This was partly because of the high moral ideals that it professed, moving away from the popular view of gifts as central in asking favors of the gods. It was partly because Taoism did not threaten the state's authority to define what should be worshipped, unlike the creativity of ordinary people whose cults peopled the pantheon. It was also because, like Buddhism and the state cult, certain Taoist rites incorporated an elite fondness for display of wealth. As Erik Zürcher has put it with regard to the upper-class cults of the Six Dynasties, "the strict observance of the complicated rules governing the life of the Taoist adept, the very expensive drugs and the frequent and equally expensive banquets and purificatory ceremonies must, as far as laymen were concerned, have remained the privilege of the happy few who had both the leisure and the financial means to fulfil the demands of the Taoist way of life."[6]

Historians today use "Taoism" as a cover term for a number of religious movements with diverse aims: organization and maintenance of theocratic communities of believers (the Way of the Celestial Masters or T'ien-shih 天師, end of the second century), missionary activities to save people from the impending cataclysm (among other early traditions, the Divine Spell or Shen-chou 神呪 movements, recurrent from the fourth century on), collective access through ritual to the power of the Way (the Numinous Treasure or Ling-pao 靈寶 movement, shortly before 400), support for the state in return for patronage (the first theocracy in 215, and several exemplars from 425 on), individual self-cultivation leading to immortality and appointment to the celestial bureaucracy (the Supreme Purity or Shang-ch'ing 上清 movement, ca. 500), tantric "thunder magic" traditions that won imperial patronage in the eleventh century (Divine Empyrean or Shen-hsiao 神霄 Way), and quasi-Buddhistic monastic striving for spiritual perfection (the Way of Complete Perfection or Ch'üan-chen 全眞, twelfth century), to mention only the best-known varieties.

These traditions were built atop and interacted with popular belief, but their ritualists distinguished themselves from those of local cults. Popular religion simply incorporated the ideals and social relations of a place; everyone belonged. Taoists were initiates, whether en masse or individually. In its first centuries the Celestial Masters movement trained and initiated whole populations, but by perhaps 500, only educated initiates had access to its canons, its liturgy, and its practices. While popular religion remained local in its focus even as the worship of certain gods spread widely,[7] each Taoist movement saw itself as universal. Taoists organized gods, local and regional, into a bureaucratically structured pantheon of which they themselves were a part. New movements tended to claim that their

---

6. This quotation comes from Zürcher 1959: 290. Appendix VI is a study of Buddho-Taoist conflict considerably ahead of, and generally ignored in, its time. Strickmann entitled the analysis of initiation fees in his path-breaking study of the Supreme Purity movement "The Free Enterprise of the Spirit" (1977).

7. For this process of diffusion see, e.g., Dean 1993, Kleeman 1994.

heavens and their emanations of the Tao were superior to those previously known—all of them superior to those of the popular gods.

Taoist masters were hereditary and highly trained, and held registers of initiation (lu 籙). The highest of them were ordained. Popular masters were generally self-selected. Instead of the oral formulas by which operatives of popular religion implored the gods, Taoists ceremonially submitted written documents in the classical language, following forms derived from those of the imperial government. Taoist movements regularly defined themselves by reference to bodies of revealed scriptures, and like the government were much occupied with questions of orthodoxy and heterodoxy; none of this had any meaning for popular religion with its oral traditions. Some popular masters, particularly in late imperial China, were literate and owned books, but writing was not fundamental to their work. Such distinctions have blurred as part of the social tumult of the late twentieth century.

Popular belief was enmeshed in the ongoing reality of the everyday world. Taoist movements were born anticipating that an imminent catastrophe would end the world, and that only initiates would survive. Although the popular master's access to the divine world was often achieved through trance and possession, the Taoist master's rites did not depend on letting a god take over his body. The authority of the ordained master came instead from no less than membership in the bureaucracy of the gods, which implied immortality. He did not seek favors of the gods, as the popular master did; he was one of them, and could issue orders.

The Taoist maintained that authority by elaborate forms of self-cultivation that maintained communication between the outer pantheon and the thousands of gods within his own body. Many of these techniques that were oriented toward immortality, such as alchemy and ingestion of rare natural drugs, began as popular practices among the southern aristocrats. A precursor of the individually-oriented Supreme Purity movement adopted them in the fourth century. The Numinous Treasure movement later competed for believers by drawing on the ceremonial resources of Buddhism.

Taoist masters gave their reverence many liturgical forms, directing it through various intermediaries ranked above the popular pantheon to the unnameable and ineffable Way itself. Because that was its object, Taoism is often called "China's indigenous higher religion."[8]

Although the Taoist movements were millenarian, with the possible exception of their cloudy role in the uprisings of the second century, they remained politi-

---

8. For further discussion, see Chap. VI, p. 307.

cally conventional. When their writings take up political issues, they consistently support the established order, and gratefully accept the state's support. Like Buddhist sects, when not drawing on the state's largesse, they absorbed their share of the government's recurring distrust for unofficial foci of popular esteem. Rebels generally appropriated symbology, as we would expect, from their own popular faiths rather than the specialists' scriptural religions.

The distinction between practitioners and liturgies cannot obscure the considerable integration of Taoist and popular worship for nearly two thousand years. From the T'ang on, as the autonomous Taoist communities disappeared, priests gradually became higher ritual specialists employed by local cults to perform essential ceremonies more powerful than those of the popular masters. In the Northern Sung, as the popular pantheon and the structures that Taoists had built on top of it began to merge, certain regimes extended their recognition not only to Taoist movements but to local gods. This was not a matter of popularizing Taoism, but, as Schipper has put it, of "the upgrading and emancipation of local power structures" on terms shaped by Taoist doctrines. By the Yuan, Taoism was furnishing "the organizational framework for all vocational and associational non-official bodies." These services made the movements of the time viable, but gradually moved the non-monastic Taoists out of their own physical institutions into popular temples and other public spaces.

Several important ingredients of the first history of Taoism play no significant role in the second. The most obvious are the imperial offerings to Lao-tzu recorded in A.D. 166, and the "studies of the mysteries" of "Neo-Taoism." The state cult was not linked in any perceptible way to any individual or group that can be called Taoist, although it reflects popular and imperial worship of the legendary sage not as a philosopher but as a god. What is original in "Neo-Taoism" has turned out mainly to be adaptations of Mâdhyamika Buddhism. Some scholars believe that a new ingredient, the early Han Huang-Lao 黃老 intellectual fashion, tied together the philosophy of the *Lao-tzu* and the religious movements of later centuries in some unspecified way. This movement has inspired much vague discussion and tentative links to recently unearthed manuscripts, but testable hypotheses still await some concrete and systematic research.[9] The early influence of Buddhism has turned out to be limited in extent and type, generally mediated not by religious specialists but by laymen. Here too, as Zürcher once said, Taoism and Buddhism are "two branches springing from a single trunk."

Finally, as I have argued in Chap. VI, individual spiritual techniques of immortality, including alchemy, are not Taoist in any fundamental sense. They originated in the popular milieu, and continued to be widely practiced by laymen. Immortality is an important theme in Taoist writings over the centuries, but no more so than in secular writings on occult topics. Again it is an accident of bibliography that the literature on various arts of long life and eternal life was most fully preserved in Taoist abbeys. To automatically prefix "Taoist" to "alchemy," a habit of many Sinologists, adds an increment of confusion but none of meaning.

Findings. For the purpose of this investigation, we need not pause over the vicissitudes of the Taoist movements, at least two of which survive, much changed through acculturation. Of the considerable differences between these two histories, three bear particularly on attempts to link Taoism and science:

1. Clarifying the differences between popular religion and the Taoist movements has dispelled confusion about what practices were peculiar to the latter.

2. Appreciating that the aims of Taoist masters were squarely religious rather than a degenerate form of philosophic inquiry obviates much speculation about whether Taoist masters were studying Nature, doing research, carrying out experiments, and so on. We are now attentive when they describe their methods in terms that have nothing to do with study or rational inquiry, much less systematic experimentation. Taking seriously their own assertions that they are striving for union with godhead, we can acknowledge that they used others' technical knowledge and practice toward that end, but rarely added to it. Because they kept written records and artisans did not, historians have given them credit for innovations that were more probably borrowed.

3. Understanding the First Neo-Confucianism allows us to recognize that many quietist ideas that were originally far from conventional became, from the Han on, part of the furniture of conventional minds. It is foolhardy to assume that they imply any unconventional conviction, or association with "the Taoists."

The Role of Joseph Needham

Among the many claims posited for a linkage between Taoism and science, those by Joseph Needham stand out. He is the only scholar who has offered hypotheses substantial and rich enough in implications to serve as a starting point. If we are finally ready for a comprehensive look at how science and Taoism intersected, it is because, for the first time, Needham has imposed order on a large portion of the primary and secondary literature. The Taoist connection is a recurrent issue in a massive work that is, after all, not an attempt at a definitive history, but a preliminary reconnaissance published over forty years and still in process.[10]

It is salutary to remember, now that there are nearly a thousand historians of Chinese science,[11] how few of its sources had even been read outside China by

1950. Beginning with the first volume of *Science and Civilisation in China* in 1954, Needham proposed a first comprehensive pattern connecting Taoism and science. My concern here is to explore his use of the words "Taoism" and "Taoist," to determine what his interpretations imply, and to assess their bearing on the evolution of science. His assumptions and hermeneutics were widely shared, although never set by others in so broad a framework. Thus when I discuss details of analysis or interpretation, "Needham" is shorthand for "Needham, his collaborators, the eminent authorities whose opinions he cites, and the many scholars who accepted his findings without testing them independently." I need hardly add that the present inquiry depends from start to finish on hindsight.

Needham's ideas about linkage were largely formed before 1960. Original though many of them were, they reflect the conventional wisdom about Taoism at the time he wrote. They also reflect the readership toward which he chiefly aimed his book, namely scientists with humanistic leanings like himself. They then formed the main audience for writing on the history of science. They expected technical sophisticated narratives of a grand march away from error toward objective knowledge. Needham was not alone in urging a nuanced but still positivistic view. Lynn Thorndike's massive *History of Magic and Experimental Science* (1923-1958), which argued that the two had grown up together, strongly influenced him. But most of the discipline-builders rejected such ideas, and dismissed the idea that non-European science was worth taking seriously. Needham's willingness to see religion as a positive force made his account additionally unconventional without challenging in principle the positivism of the time.

Science and Which Taoism?

Needham's understanding of the relation between Taoism and science is not set out concisely or in one place. A large piece of the picture is found in the 1956 volume of *Science and Civilisation in China* devoted to the history of thought. He fleshed it out in subsequent volumes down to 1986 (when the last two he wrote for the series appeared), but did not alter it in any fundamental way.[12]

---

9. Yates 1994: 144, studying MSS excavated in 1972, questions the identification of many Ma-wang-tui finds with "Huang-Lao Daoism."

10. Taoism is a rarer theme in volumes of *Science and Civilisation in China* by other contributors. The term does not occur in the indexes of vols. V. 9 and VI. 1, or in the part of vol. V. 6 not written by Needham.

11. See the worldwide survey of specialists in Sivin 1988.

12. II, 33-164. Needham also wrote part of vol. V. 6 long before it appeared in 1994.

Many of Needham's views cannot be compressed beyond a certain point without losing their essential texture. I will accept the risk of recapitulating them a little more amply than has been done before (although I cannot avoid oversimplification). I am asking where and when in Chinese society Needham finds Taoists whose "complex and subtle set of conceptions . . . lies at the basis of all subsequent Chinese scientific thought."[13]

This question devolves naturally into two others that call forth less vague replies: First, are these Taoist precursors of science related to each other by any plausible definition of "Taoist," or at least by a coherent range of definitions? This is tantamount to asking whether the label is being applied in a way that illuminates the origins of science, or whether conversely a range of definitions so wide as to convey no information is being used to back up an initial assumption that every shaper of science is a Taoist. Second, how conclusive is Needham's evidence that the thoughts and acts of these Taoists, whoever they may be, were essential to the evolution of science?

Needham's use of the designation "Taoist" is far from consistent. Indeed it would be gratuitous to assume that he is striving for consistency. I will demonstrate, however, that a coherent and testable general notion of Taoism, as it influenced science, emerges from his writings. I will argue that when his evidence is critically evaluated and reinterpreted in ways consistent with today's fuller understanding of Chinese religion, its bearing on the course of science is negligible. As a check on this conclusion, I will survey the careers of a number of great contributors to science, technology, and medicine, inquiring whether Taoism played a role in the sum of their achievement different from that played by other systems of belief. Finally it will be appropriate to ponder what the outcome implies for the history of science.

Various Taoists

The historical variety of Taoisms. Needham's pro-scientific Taoists were agnostic naturalists, mechanists, materialists, experimentalists, empiricists who

---

My discussion centers on the extended essay in Vol. II (1956) of *Science and Civilisation in China,* and on the conventional wisdom as exemplified in Needham's studies. In forming my argument, however, I have drawn on all his publications. I should acknowledge the extremely limited extent of my own understanding of Taoism before 1970; see, for instance, Sivin 1968: 157n26. I began research for this and the preceding chapter in the late 1970's to overcome the confusion in thinking about the subject that I detected in my earlier work. That this confusion was commonplace did not make it acceptable.

13. I, 95.

avoided preconceived ideas, democrats and heterodox collectivists. They appear in Needham's overview of philosophy as the Confucians' "mortal enemies the *tao-chia* 道家 (Taoists), whose speculations about, and insight into, Nature, fully equalled pre-Aristotelian Greek thought, and lie at the basis of all Chinese science." "The philosophy of Taoism . . . developed many of the most important features of the scientific attitude . . . Moreover, the Taoists acted on their principles, and that is why we owe to them the beginnings of chemistry, mineralogy, botany, zoology and pharmaceutics in East Asia."[14]

The ideal Taoists in the first half of the last quotation, then, are what Sinological custom has called philosophers, not priests: *tao-chia*, not *tao-chiao* 道教, to use a modern distinction. His discussion depends largely for evidence on the *Lao-tzu*, *Chuang-tzu*, its late imitation the *Lieh-tzu* 列子, and eclectic compilations such as *Lü shih ch'un-ch'iu*, *Kuan-tzu* 管子, and *Huai-nan-tzu*. In 1956, one must note in passing, the last three were still conspicuously ignored by most students of Chinese thought. There was even less agreement than there is today about what ism to assign them to.

After this initial flowering, a few additional "Taoist" philosophers emerge in the Wei and Chin periods: Wang Pi 王弼 (226–249) and other "revisionists," Pao Ching-yen 鮑敬言 (early fourth century?) as a representative of the "radicals,"[15] and Ko Hung 葛洪 (283–343), "the greatest alchemist in Chinese history," as a representative of the strengthened "experimental traditions of ancient Taoism."[16] Finally, "there was a second flowering of true Taoist philosophy" in the *T'ien-yin-tzu* 天隱子 of Ssu-ma Ch'eng-chen 司馬承貞 (ca. 700, mentioned but not discussed or cited), the Book of the Gatekeeper (*Kuan-yin-tzu* 關尹子), probably of

---

14. II, 33–164, esp. p. 162; p. 1; p. 161; "the physics of magnetism" joins this list on p. 493.

15. Needham calls Pao "the most radical thinker of all the medieval Chinese centuries," but gives no evidence that his critique of the segregation of power and wealth is wedded to proposals for fundamental change in the structure of government or society (which I gather is what "radical" means). A later discussion of Pao in terms of class struggle argues—on an inadequate foundation of evidence, I believe—that his ideas were anarchist (Uchiyama 1965). Pao is unknown except for the dialogue in *ch*. 48 of *Pao-p'u-tzu wai p'ien* 抱樸子外篇. Needham allows for the possibility that the dialogue and Pao himself may be fictitious, or that Pao may have lived much earlier than Ko Hung. Later Needham proposed tentatively that Pao was Ko's father-in-law Pao Ching 鮑靚 (V. 3, 76). I believe that the last two possibilities are ruled out by the fact that Ko refers to him at the beginning of *ch*. 48 as "Pao sheng 生 Ching-yen," not a term for a senior.

16. II, 432–441. For a more qualified estimate of Ko's alchemical achievement, and of the extent to which he can be considered an experimentalist, see Chap. VI, pp. 323–324.

the Southern Sung, and the Book of Transformations (*Hua shu* 化書) attributed to T'an Ch'iao 譚峭 (tenth century).[17] As for the Taoists who "acted on their principles," they are clearly not drawn from these *tao-chia*. I will return to them.

In later writing Needham considerably broadened the scope of proto-scientific Taoism. The hiatus between the early Taoist writers and those of the Wei and Chin has been filled with those who carried on an intrinsic Taoist "artisanal element ... The *fang shih* 方士, of whom we hear so much between the -5th and +5th centuries, were certainly in general Taoist, and they worked in all kinds of fields (apart from divination and incantation) as star-clerks and weather-forecasters, men of farm-lore and wort-cunning, leeches, irrigators and bridge-builders, architects and decorators, metal-winners and smiths, above all, alchemists."

The distinction between alchemist and Taoist tends to blur, as in this elision: "If then we may take Ko Hung as fairly representative of all the early medieval Chinese alchemists, some clear conclusions may be drawn about their beliefs ... The maintenance of weight on cupellation was therefore for the Taoists not the only, or the main property ... which entitled a gold-looking substance to be called gold."

Needham finds the identification of alchemist and Taoist plausible because most alchemical literature either survives in the *Tao tsang* or has "Taoist connections." What connects alchemy, the saleable skills of the *fang-shih,* and Taoist philosophy is apparently not the propinquity of their literary remains but certain notions held in common: the naturalism, empiricism, anti-feudalism and related attitudes, and the willingness to take seriously magic and science, which he tells us repeatedly were inseparable in early times.[18]

Needham believes that both the "Taoist Church" founded in the second century A.D. and what used to be called the "Neo-Taoism" of the third and fourth centuries undermined the scientific attitudes bequeathed by their predecessors and negated the experimental impulse of the *fang-shih*. Here is how he puts it: "How could it have come about that the high philosophy (at one and the same time scientific and mystical) of the Taoist fathers ... was transformed into a theist and supernaturalist religion, heavily laden with superstition, and not without an element of conscious mystification?" "The strangest transformation of all was that which converted Taoist agnostic naturalism into full-blown mystical religion and ultimately theist trinitarian theology, Taoist proto-scientific experimentalism into fortune-telling and rustic magic, Taoist primitive communalism into a way of per-

---

17. II, 442–454.
18. V. 2, 9, 70; V. 2, 1; e.g., II, 57.

sonal salvation, Taoist anti-feudalism into equalitarian secret societies of anti-foreign or anti-dynastic tendency." This change was not so much the doing "of the Taoists' complacent and conventional rival, social-minded Confucianism, as of the socio-economic system of feudal bureaucratism itself."[19]

In brief, "the entire development was fundamentally the working up of an indigenous opposition system to Buddhism. First, political Taoism was sent underground. . . . Then Confucian feudal bureaucratism allowed no outlet for the scientific energies potentially present in the Taoist philosophers and the shamanist magicians. Thought thus being sterilised and experimental techniques despised, the shamans, from the +1st century onwards, found their living being taken away from them by the new foreign religion of salvation from India. . . . But now, with half-conscious resource, the Taoists copied theology, sutras and discipline to such good effect that for many centuries they were able to hold their own in the form of an organised religious institution."[20] There is clearly a discrepancy between the scientific value of the great Taoist scriptural collections and the negative role assigned to the "shamanist magicians" who made them.

The transition was not only fateful but final, and reinforced by later history. The "Neo-Taoists" Wang Pi, Hsiang Hsiu 向秀 (ca. 221–ca. 300), and Kuo Hsiang 郭象 (d. 312) were probably not "quite so mystical as Feng Yu-lan represents them, since there is a tradition that Hsiang Hsiu at least practiced alchemy." As a result of their commentaries on the *Lao-tzu* and *Chuang-tzu* "the whole Taoist system was emasculated for continued existence in, and adaptation to, a milieu in which the Confucian conventions were dominant." Note the assumption that anyone who practiced alchemy was not very mystical.[21] Whatever "the whole Taoist system" may mean, the *Lao-tzu* and *Chuang-tzu* coexisted with a Confucian state orthodoxy for some time after 135 B.C. In the era of these "Neo-Taoists" the power of "Confucian conventions" was hardly in evidence, as emperors increasingly ruled under the aegis of Buddhist or Taoist movements.

Before returning to test the influence on science of Needham's "Taoism" as I have just summarized it, it is advisable to pause over two special characteristics of his historical account, namely the absence from it of popular religion, and his interpretations of imperial cults.

Popular religion. The word "Taoist" is conventionally used for a great variety of popular beliefs and practices—notions about gods and their relationships with

---

19. II, 154, 162.

20. II, 161.

21. II, 433. I have argued (1976) that the aims of alchemy were squarely mystical.

humans, liturgical rites, sacerdotal functions, disciplines for self-cultivation, magical and divinatory techniques, and so on. There is no reason to believe that *tao-chia, tao-chiao,* Neo-Taoists, etc., originated these aspects of popular culture. Nor did they use them more often than did masters untrained in the worship of the Tao and its emanations, pious but otherwise conventional gentlemen hoping for an appointment to the celestial civil service, honest businessmen who followed the trade of physiognomy or geomancy, or itinerant rainmakers. Rolf Stein has described the "ceaseless dialectical movement of coming and going" by which Taoism, like the other high traditions of China, drew upon and contributed to the forms of popular culture (1979).

Some Sinologists, especially in Japan, are aware that these cultural artifacts far transcend the organized religious movements. They speak of them as manifestations of "popular Taoism," which they treat as an entity distinct from "popular Buddhism." This species of Taoism, for good reason, has not been adopted by critical scholars elsewhere. It implies that the masses, incapable of creating their own religion, depended upon Chang Tao-ling 張道陵 and his progeny for it. The notion that everything worthwhile began as a grant to the commons from a legendary founder is so entrenched in traditional culture that this form of it lingers on despite the historical evidence accumulated against it.

Needham usually avoids the term "popular Taoism" and the cast of mind behind it.[22] He belongs instead to the larger company of those in the West for whom any generally diffused belief or practice that cannot readily be sorted into such categories as "Confucian" or "Buddhist" must be Taoist without further qualification. A few examples from *Science and Civilisation in China* illustrate this point. The universal themes depicted in the renowned silk painting from Tomb 3 of Ma-wang-tui (crow in the sun, toad and rabbit in the moon, etc.) are "Taoist myths and legends." A court lady is said to have been "a specialist in Taoist sexual techniques," although the Ming source does not mention Taoism, but merely speaks of the sexual "techniques of the Yellow Lord and the Pure Girl" *(Huang-ti su-nü chih shu* 黃帝素女之術*)*.[23] In the *Pao-p'u-tzu nei p'ien* 抱樸子內篇 "strong Taoist influence on alchemy is apparent from the use of charms and amulets,

---

22. For an exception see V.6, 231.

23. V. 3, 21–22; see also IV. 1, 31, 91; V. 3, 39; V. 2, 260. The idea of Huang-ti beliefs as in some sense particularly Taoist apparently accounts for the mental leap, but a few pages later Needham judiciously uncouples the two rubrics: ". . . the study of Huang (Ti) and Lao (Tzu), which can mean either Taoism or the art of immortality . . ." (p. 52). The name of Huang-ti was not a monopoly of Lao-tzu, Chuang-tzu, the *fang-shih,* or the various sects that worshipped the Tao.

Taoist magic and ceremonies," and so on.[24] Finally a discussion of "how far Taoist ideas penetrated into Japan" is concerned with such matters as "the Taoist theory of the 'three corpses *(san shih* 三尸*)*' in the body, . . . the Taoist respiratory exercises *(fu ch'i fa* 服氣法*)*," and *"pai shu* 白朮, that characteristically Taoist longevity medicine."[25] All of these examples are drawn from the popular milieu. The terms were as familiar to literari with a taste for the occult as to Taoist initiates. "Taoist magic and ceremonies" is another matter, but when their components are identified, they too regularly turn out to be parts of popular religion.

Even being Confucian in various senses does not rule out being Taoist, even though the two were "mortal enemies." A section on conduits and canals in the "K'ao kung chi 考工記" section of the canonical *Chou Li* evokes the remark "One should not fail to note the extremely Taoist character of these maxims."[26] A passage in which Mencius uses an astronomical example is "of distinctly Taoist flavour" because "he is criticising the scholars of his time for forcing facts and going against nature."[27] In 1959 Needham described the inscription on a stele erected in Fukien by Cheng Ho 鄭和 and his admirals in 1432, although composed by a civil servant on behalf of a "Confucian" state cult at imperial behest, as "an inscription of sailors' gratitude to a Taoist goddess." Twelve years later he called it a votive offering "to the Buddhist-Taoist goddess of the sea."[28] Here too the issue, I suggest, is popular belief.

---

24. V. 3, 106.

25. IV. 1, 91; V. 3, 175–177. Sinologists often make the first two links. The last is not obvious, but atractylis is a macrobiotic plant established in conventional medicine long before the cited works were compiled. Atractylis, it seems, is Taoist without qualification because immortality is Taoist without qualification. This is not the only case in which a drug used universally is considered Taoist because it figures in immortality legends, alchemical practice, etc. For examples, see III, 642, and below, p. 18.

26. IV. 3, 256, note a. This usage has hoary antecedents. For instance, in his translation of the *Li chi* 禮記, James Legge says of Confucius' reference to *"ta tao chih hsing* 大道之行 (lit., 'the practice of the great Way')," "this sounds Taoistic," and he speaks of "the Taoistic period of the primitive simplicity" (1885: I, 364–365).

27. III, 196. In this passage (4B.26) Needham reads *ku* 故 first as "cause and effect" and then as "phenomena." The "Taoist flavour" and concern with scientific reasoning disappear in the more literal translation of D.C. Lau: "In talking about human nature people in the world merely follow former theories *[ku]*. They do so because these theories can be explained with ease. . . . In spite of the height of the heavens and the distance of the heavenly bodies, if one seeks out former instances *[ku]*, one can calculate the solstices of a thousand years hence without stirring from one's seat" (Lau 1970: 133).

28. III, 557; IV. 3, 523.

VII

State Cults. A number of important imperial cults were based on Taoist structures of meaning and legitimation. Needham mentions many of them, alongside others more doubtfully Taoist. A general view of the relations of imperial Taoism to science, medicine, and technology, and to other Taoisms, does not emerge. Here are a few examples:

1. The Han Martial Emperor (Wu-ti), we are told, was in contact with "the shamanistic strain of Taoism" but, because he was "active and ambitious," brought about "the very triumph of Confucianism." Eight years later Needham writes of "the time of Taoist dominance under Han Wu Ti."[28]

Lack of clarity about the ideological commitments of the Martial Emperor and his immediate predecessors is universal in modern studies. Historians in the Han and after emphasized the conflicts between Confucian and Taoist philosophic doctrines. They were not inclined to consider how easily these doctrines (and rituals related to them) could be, and how frequently they were, used side by side as adjuncts to state power. Needham's reconnaissance of science and material culture has encouraged interest in the Western Han. Perhaps it will motivate someone to take a fresh look at the promiscuous use of ideology by the Martial Emperor and his predecessors—not only that derived from the *Lao-tzu* and the search for immortality, but the assortment of conceptions, slogans, symbols, and rites that historians by force of habit label Confucian.

2. Needham notes that in 401 the Grand Progenitor of the Northern Wei (T'ai-tsu, first emperor, r. 386-409) "established a professorship of Taoism *(hsien jen po shih* 仙人博士) . . . and a Taoist workshop for the concoction of medicinal preparations." The primary source does not mention Taoism, although it remarks that the emperor "was fond of the words of Lao-tzu, and recited them without tiring." The Erudite, or professor, taught a number of texts on attaining immortality through taking drugs. The workshop was for preparing them. The imperial enthusiasm, the source tells us, soon ebbed and the project ended. We cannot decide on the basis of this evidence that the emperor's employment of specialists in immortality techniques was in any way related to his fondness for chanting the *Lao-tzu* (more often admired as a source of salvation than of mysticism). These two Taoisms in turn do not imply his involvement in any other Taoism, such as the religious movement that briefly turned the rule of his successor into a "Taoist theocracy" under the master K'ou Ch'ien-chih 寇謙之.[29]

3. Needham's survey of alchemy discusses the Taoist involvements of the Per-

---

28. I, 106, 108; III, 581.

29. II, 441; *Wei shu, 114*: 3048. Needham quotes Ware 1933: 224. See *Wei shu, 113*: 1973, for the date. On K'ou and court Taoism see Mather 1979, esp. p. 107, and Yamada 1995.

fected Ancestor (Emperor Chen-tsung of the Sung, r. 997–1022). This emperor's attempts to attract recluses from 998 on, and the series of revelations vouchsafed to him beginning in 1008, are "nonsense" that "may seem as meaningless as it is amusing" unless "one realises that it was exactly at this juncture that a number of important military inventions were made." In an otherwise carefully documented summary, Needham explains this by "his Taoists . . . consulting with their military colleagues," but backs this with no evidence at all.

He suggests further that the court fabricated messages from heaven as an alternative to "military might." Imperial use of Taoism was based on a policy of "theocratic mystification." Needham's monograph on astronomical clockwork declares that "Chen Tsung was particularly devoted to Taoism, so it is not surprising that he took a great interest in the astronomical equipment made by Han Hsien-fu" (and paid for it). Needham does not suggest that Han was connected with any Taoist movement, so this must be another instance of Taoism as enthusiasm for technology, quite compatible with Taoism as mystification.[30]

4. Finally there is the intricate case of the Excellent Ancestor (Emperor Hui-tsung, r. 1101–1125), who inaugurated the Divine Empyrean cult. Orthodox historians of the last thousand years have not explained this grand act of patronage as mystification. After all, they see sincere Taoism as merely one more sign of self-indulgence in the ruler they hold morally responsible for the fall of the Northern Sung. Strickmann, who unlike them has studied the primary sources in the Taoist Canon, sees it in part as an expression of Chinese religiosity counterposed to the Buddhism of the Khitans who threatened the empire from the north.[31]

Needham has taken a particular interest in this period, and his work has already played a part in prompting more specialized research. The issue is broader than religion alone. ". . . Hui Tsung . . . consistently supported the reformers . . . except between 1107 and 1112 . . . Now one of the most striking features of the period was the alliance between the reformers and the Taoists, counterbalancing the strict Confucian orthodoxy of the conservatives. . . . The reforming party of the Sung were bureaucratic scholars who broke away from the typical Confucian ideas and were prepared to ally themselves with Taoist science and technology. It was highly significant, for example, that Wang An-shih 王安石 (1021–1086), and again in 1104 Ts'ai Ching 蔡京 (1046–1126), included mathematics and medicine among the subjects which could be offered in the imperial examinations." Exactly what tie Needham had in mind between Taoism and these examinations, offered as qualification for specialized but in no sense unorthodox civil service functions,

---

30. V. 3, 183–184; Needham, Wang & Price 1960: 70. For the context of the revelations see Wu Yiyi 1989.

31. Strickmann 1978.

he did not explain. Thanks to the historical survey of medical qualifying examinations by Needham and Lu we know that they were not new in the Sung period, and that association with Taoism (unless one makes physicians "Taoists" by definition) had not previously distinguished their use. A footnote in a different volume mentions another strikingly pertinent point—that "the Reforming Party became associated with Taoism" only after Wang An-shih's lifetime![32]

The relation between Taoism and science was not merely a matter of social and political expediency. In discussing the elevated status of Taoism Needham asserts an intellectual link: "All this should be viewed in conjunction with the background of interest in natural phenomena and mechanical invention which led us at an earlier stage to call Hui Tsung's court an 'entourage of virtuosi.'" He is inspired to compare the Excellent Ancestor to Charles II, founder of the Royal Society. Earlier he likens the ambience of the court to that of Rudolf II at Prague, the great patron of astrologers and alchemists.[33]

The claim, then, is that scientists, mechanicians, and Taoist adepts (explicitly related to the Divine Empyrean movement) were allied against the orthodox Confucians. This union was not a mere marriage of convenience between the heterodox, but was based on a genuine overlap of convictions.

Now what evidence does Needham offer for this view? He lists the characters he considers central to the Taoist side of the alliance. All are familiar from older conventional accounts, which present them as living testimony of the Excellent Ancestor's Taoist leanings. The orthodox historiographers often thought of this enthusiasm as (to quote Kracke's paraphrase) a "tendency toward mental instability" that "led to definite unbalance in his later years."[34]

Neither in those discussions nor in others I have cited earlier do particulars of cooperation between the reform group and Taoists against "strict Confucian orthodoxy" appear. A single anecdote bears the main burden of illustrating a working alliance. It deserves examination.

The great escapement-regulated astronomical clock that Needham and his colleagues have reconstructed was designed, built, and documented under the supervision of Su Sung 蘇頌 over the years 1086–1094. The political circumstances of this project were in many respects diametrically opposite to those depicted for a similar feat proposed a quarter-century later, in the Excellent Ancestor's time.

---

32. Needham, Wang & Price 1960: 124; Needham & Lu 1963; Strickmann 1978: 335n15; Needham 1954– : I, 138. On Hui-tsung's Taoist associations see also Miyakawa 1975 and 1976.

33. V. 3, 190–191; IV. 2, 501, note d.

34. Needham, Wang & Price 1960: 124–125; Kracke 1953: 25.

Su was one of the several unprecedented polymaths of his century. By 1094, when he presented to the Emperor his book on the clock, he had risen to the top of the civil service to become Vice Director of the Secretariat and Chancellery (i.e., Grand Councilor, 1092). This was the period in which the more extreme enemies of Wang An-shih were extirpating his adherents (1085–1093). Su was thus the tool of a group that most principled conservatives avoided. He himself seems to have held no factional grudges. No one has argued that he was in league with Taoists of any sort.[35]

Now for the anecdote. In the Excellent Ancestor's time, Wang Fu 王黼 (1079–1126), a powerful official remembered mainly for his sycophancy and his complicity in the fall of the Northern Sung, proposed building an astronomical clock to replace that of Su. This was a technological step in the campaign of Wang, the prime minister Ts'ai Ching, and other remnants of the reform period, to wipe out every trace of the faction which had wrecked their careers a generation earlier. Whether Wang's clock was superior to its predecessor has yet to be determined; it is interesting that the only contemporary evaluation we have likens its design to one built three and a half centuries earlier.[36]

Needham counterposes the two projects: "Su Sung's clockwork was associated with the Confucian Conservatives—Wang Fu was one of the Taoistic Reformers." He explains in a note that Wang "was closely associated with a number of Taoist adepts and certainly acquainted with some of their arts, but this did not prevent him from making an adventurous, and somewhat unscrupulous, career in the official bureaucracy." In *Heavenly Clockwork,* published five years earlier, the nuance is a little different: "It is quite clear that he disposed of the talents of certain more obscure Taoist technologists . . ."

To back this claim Needham cites Wang's memorandum of 1124. The only relevant portion says "In [1102] I chanced to meet a wandering unworldly scholar at the capital, who told me his family name was Wang and *gave me a Taoist book which discussed the construction of astronomical instruments in detail."*

The phrase the translation of which I have italicized is *"mien ch'u su shu i* 面出素書一," more literally, "he took out in front of me one book written on unbleached silk."[37] Nothing in the story hints that a Taoist produced it. With due

---

35. See Franke 1976: 969–970 and, for more detail, Weng Fu-ch'ing 翁福清 1986.

36. There is a great deal of information on the mechanized astronomical instruments of the Sung period in the *Sung hui yao* 宋會要 that has not yet been used by historians of science. This source states definitely (LIII, 2: 15a, p. 2151b) that the Wang Fu clock was completed, a matter that is not resolved in *Heavenly Clockwork,* 2: 15a.

37. Needham 1954- : IV. 2, 500–502; Needham, Wang, and Price 1960: 118–119, 125. The anecdote is found in *Sung shih, 80*: 1906.

sagacity, Needham does not *assert* that the unworldly scholar was either a Taoist or the author of the book.

In Wang Fu's statement, then, we are left with his chance meeting with a man of unknown allegiances who gave him a book of ideologically undetermined character more than twenty years before he based his proposal on it. This is not an unusual tale, but rather the most common way of suggesting that a book is of more than mortal origin.

Where was Wang's obscure band of Taoists in 1124? This question can be answered on the assumption that, here as so often elsewhere, Needham is using the word "Taoist" to mean "technician."[38] Wang's association with Taoists would thus amount to making use of mechanicians to design his clock. That workers skilled in manual arts would be essential to this job hardly needs to be proven. In this view, the generalization about reformers and Taoists becomes a statement that Wang and others of his faction used technicians when technicians were needed, as indeed civil servants—orthodox, heterodox, reforming and reactionary—had usually done before them. The weight in this equation of other Taoisms—of "Taoists" who were not just technicians—remains unknown.

One can make a case, although it has nothing to do with science, and Needham did not make it, for a political alliance between individuals in the "New Policies" group and the most powerful representatives of orthodox Taoist movements in the Excellent Ancestor's court. The latter included Liu Hun-k'ang 劉混康 of Supreme Purity, and Lin Ling-su 林靈素 with his epochal Divine Empyrean revelations. The emperor supported them—at least up to a point. When Lin arranged the downfall and execution of the previous Taoist favorite, Wang Tzu-hsi 王仔昔, no infusion of imperial grace saved the latter.[39] At the same time, the leading members of the emperor's government were adherents of what had once been Wang An-shih's faction. These two groups (both riven by internal rivalries) were thus joined in access to power, and we know that they cooperated to secure

---

38. Wang Fu's long memorial, excerpted above, is accorded only a one-sentence summary in *Sung hui yao, loc. cit.*: "Previously Fu memorialized that he had obtained a *fang-shih's* book on armillary spheres." The historiographic official (an occupational group not fond of drawing fine distinctions among the heterodox) is equating *fang wai chih shih* 方外之士 and *fang-shih* 方士.

39. On Liu, see *Mao shan chih* 茅山志, S 304, TT 153–158, *3:* 3a–4: 7a; on Lin, Strickmann 1978; and on Wang Tzu-hsi, *Sung shih, 462:* 6a, *T'ieh wei shan ts'ung-t'an* 鐵圍山叢談, *5:* 9b–10a, and Chin Chung-shu 1974: 301. Needham has linked Lin with medicine by calling *Su Shen liang fang* 蘇沈良方, a collection of prescriptions and theoretical discussions by Shen Kua and Su Shih 蘇軾 (1037–1101), "a conflation of their writings under the supervision of the Taoist Lin Ling-su" (V. 3, 193). Lin did not perform this service. Lin's preface (to *Liang fang*) merely says that he paid for printing an MS in his possession.

it. What else joined them—principle, vision, intentions toward the use of science—we do not know.

Neither of these perfectly plausible types of association—administrators' use of technicians for engineering tasks, and the sharing of imperial favor by Taoist favorites and Ts'ai Ching's feuding clique—casts light on our central issue, namely the consequences for science of Taoism as an imperially sponsored cult.

Was either alliance opposed to Confucian orthodoxy? Recent studies of the New Policies *(hsin fa* 新法*)* generally differ from the perspective of the 1950's in two pertinent respects:

1. Critical historians have abandoned "Confucian orthodoxy" as a criterion for distinguishing the enemies from the adherents of the New Policies. The reasons are both general and specific.

Considered generally, the term has scant inherent meaning. Through most of political history both sides of a power struggle usually flaunted their orthodoxy, freely redefining it. "Confucian orthodoxy" was a necessarily amorphous area of discourse *because* groups competing to hold, control, or define legitimate access justified their opposed claims by reference to established usage.

In the New Policies period, assertions of orthodoxy were frequent and fervid on both sides. Both factions represented, and of course modified, trends founded in the formal Confucian schools of the time. Wang An-shih was one of the greatest shapers of orthodoxy in Chinese history. He requiring those preparing for the examinations to master a single set of classical interpretations, namely those in his own commentaries.[40] Wang's reforms gave the bureaucracy enhanced authority; in return he expected more conformity from civil servants. This combination of authority and orthodoxy became part of his enduring legacy (see Vol. I, III 3-4).

2. Specialists in the politics of the Northern Sung period see the group around Wang An-shih as heterogeneous. Its members ranged from sincere and innovative reformers to unscrupulous conventional careerists taking what for a time was the obvious route to power. As they diverged, the New Policies ceased to be an effective program of reform by the time Wang retired in 1076. The restoration of its promoters in the Excellent Ancestor's time was so completely under the control of those avid for power that it is often called the "post-reform period."

The dominant figure of the Excellent Ancestor's reign was Ts'ai Ching. "Under him the political persecution intensified, corruption increased, and the

---

40. J. T. C. Liu 1959: 24–29, 88–89. These mandatory interpretations, which were promulgated in the schools, were composed by Wang, a son, and a disciple. See the references in Liu and the prefaces in *Lin-ch'uan chi* 臨川集, *wen chi* 文集, *15*: 147–149.

government administration deteriorated in many ways." Ts'ai, as prime minister, was deeply involved *ex officio* in the Excellent Ancestor's Taocratic extravaganzas, and was involved in distaff politics in which enthusiasm for Taoists played a role.[41] The evidence has yet to appear that, during the last quarter-century of his life, reform meant more to him than a slogan useful in accumulating power and personal wealth.

To sum up the case of the Excellent Ancestor, Needham's picture of an alliance between Taoism and political reform blurs as we scrutinize the evidence. Taoists did not support reform in this instance any more than in the earlier ones.

Labels for Taoists

A necessary last step before examining the links between Taoism and scientific attitudes is to look at terminology. The Sinological faith of the 1950's in the clear distinction between *tao-chia* and *tao-chiao* as philosophy and religion has turned out to be mistaken. Those terms, and others that routinely have been translated "Taoist," turn out to have much more diverse meanings in the historical sources.[42] I will summarize the results of research on several such terms, and on *fang-shih* and *wu*, two words that scholars often relate to Taoism without considering their meaning in context. Each discussion begins with a literal translation of the term. "*Tao*" may refer, in various contexts, to "the Way" or "a way, an art or method."

Tao-chia 道家, "masters of the Way *(or* with a way)." This term in early writing consistently refers to books rather than people.

Ssu-ma T'an's 司馬談 (d. 110 B.C.) "On the Essentials of the Teachings of the Six Schools" (discussed in Chap. IV) first associated this term with philosophy. He argued at length that the doctrines of the *tao-chia* were best because they included the best points of the others. Ssu-ma's Taoism was much more comprehensive than that of the *Lao-tzu* and *Chuang-tzu*, because he drew the boundaries of his description from the Springs and Autumns of Master Lü (ca. 239) or some other eclectic work like it.

In the "Treatise on Bibliography" of the History of the Former Han, *tao-chia* is a bibliographic rubric. We can assess its historical value from the assertion that the Taoists, like the proprietors of nine other classifications, were descended from bureaus of the royal Chou government. This class of books comprises thirty-seven

---

41. Liu 1959: 10; Chin Chung-shu 1974, esp. p. 296.

42. The most useful summary is Sakai & Fukui 1977, with an English summary in Fukui 1995. See also Chap. VI above.

works. The titles of the lost majority do not reveal what they had in common.

Over the centuries the *tao-chia* class of books also came to incorporate an indiscriminate range of treatises on religious, occult, and legendary topics. This was not the result of a sustained, conscious effort to define *tao-chia* as an ideal type and collect books accordingly. Books became "Taoist" because conventional people often read them, or because imperial patronage of one sect or another had enriched a palace library.

Ko Hung, that great enthusiast of immortality (ca. A.D. 335), contrasts the contents of two of his books, Inner Writings and Outer Writings. The first, which "belongs to *tao-chia*," includes not philosophy but "immortality, medicine, spirits and prodigies, transformations, self-cultivation, longevity, exorcism and the avoidance of calamity." The Confucian *(ju* 儒*)* content of the second is far from classical: "success and failure in human relations, good and bad in worldly affairs." The same use of *"tao-chia"* without philosophical overtones for those practicing occult disciplines can still be found in a twentieth-century treatise on charm healing written for anyone who cared to buy it.

Beginning in the Six Dynasties, members of Taoist organizations used *"tao-chia"* to distinguish themselves from their Buddhist counterparts. As late as the Ming it referred to those who practiced techniques, i.e., to what had earlier been called *fang-shih*.[43]

In short, when modern historians went hunting for Chinese parallels to European philosophy, they were bound to find them. They read Ssu-ma T'an's partisan argument carelessly and ignored large ranges of meaning elsewhere that did not suit such a high-minded enterprise.

Tao-chiao 道教, "teachings of the Way." Every philosophic tradition had a Way of its own. This term, like those that follow, was available to anyone. In *Mo-tzu* 墨子 39 (fourth century B.C.) and many later historic sources, it refers to the teaching of the Confucians. It was not much used in the Han. Buddhists used it rather widely beginning just after the Han to refer to the teachings of their own *dharma*. Authors associated with Taoist movements, engaged in anti-Buddhist polemics from ca. 480 on, began using it to distinguish their doctrines from those of their rivals. "Before modern scholarship," as Seidel puts it, the term "was never used to distinguish Taoist philosophy from Taoist religion but to differentiate the Taoist tradition from Confucianism and Buddhism." That, we have seen, was also the case for *tao-chia*.[44]

---

43. *Shih chi, 130:* 7–14; *Han shu, 30:* 1729–1732; *Pao-p'u-tzu wai p'ien, 50:* 9b; Anon. 1930: 16; Ch'en Kuo-fu 1949: 259.

44. Seidel 1989–1990: 229. Sakai & Fukui 1977: 432–433 cite a number of Buddhist

VII

Tao-jen 道人, "man of the Way (or with a way)." Hsun-tzu 荀子 (ca. 250 B.C.) a successor of Confucius, in a passage concerned with the foundations of good government, uses the term for people who abide by his version of the Confucian Way. In the *Chuang-tzu*, two occurrences mean "a person who has attained oneness with the Tao" and one means "a sycophant." In the *Lun heng* the word means "one who follows a discipline," a *fang-shih* (five occurrences). With the advent of Buddhism, as a simple extension of the Han meaning, *tao-jen* often designated Buddhists. By the first half of the third century, it was used for that purpose in translating Indian sutras.⁴⁵

Tao-shih 道士, "gentleman with a way." Beginning with a famous passage from the mid second century B.C. in which "an ancient *tao-shih*" advises on nourishing the vital principle, this term simply meant someone who followed a way, a discipline. From the Chin period on, this broad sense of *tao-shih* to some extent replaced *fang-shih*. Thus Ko Hung writes of "dilettantish practitioners" *(ch'ien-po tao-shih* 淺薄道士), homeless "vulgar practitioners" *(su tao-shih* 俗道士) and "common practitioners" *(fan-yung tao-shih* 凡庸道士) of low-grade esoteric disciplines. Like *tao-jen*, the term was used for Buddhist monks until the Southern Dynasties. Then, again via polemics, *tao-shih* came to mean Taoist masters in contradistinction to Buddhist *tao-jen*.⁴⁶

Tao-shu 道術, "techniques of a way or discipline." The passage that begins the final chapter of the *Chuang-tzu* equates "those who cultivate some method" with "those with a *tao-shu*," and uses *"tao-shu"* for all the "hundred schools." Sakai and Fukui have shown that, by the Han, conventional authors were commonly using *tao-shu* for "the methods of the former kings and sages." Thus the skeptical Discourses Weighed in the Balance (completed A.D. 70/80) speaks of Tung Chung-

---

examples of *tao-chiao* from the early third century to the beginning of the fifth. Ku Huan's 顧歡 biography quotes the first extant Taoist usage from his lost *I hsia lun* 夷夏論 (a polemical treatise, ca. 480); *Nan Ch'i shu*, 54: 932, 934. It is interesting that both treatises in the *Tao tsang* that include the word *tao-chiao* in their titles, the *Tao-chiao i shu* 道教義樞 (ca. 700?), and the *Tao-chiao ling yen chi* 道教靈驗記 (after 905), also borrow heavily from Buddhist literary forms in order to argue for the superiority of Taoist religion. Recently some Japanese specialists have argued that we cannot speak of a Taoist religion until initiates began using this term to designate themselves in the fifth century (Fukui 1995: 8; Kobayashi 1995: 26–27). I would prefer to define it by its doctrines, practices, and organization rather than by its polemical stances.

45. *Hsun-tzu*, 21. 31–34, trans. Knoblock 1988–1994: III, 103–104; *Chuang-tzu*, 12. 86. On Buddhist usage, Sakai & Fukui 1977: 446.

46. *Ch'un ch'iu fan lu* 春秋繁露, *16. 7*: 23b; *Pao-p'u-tzu nei-p'ien*, *15:* 4a; *4:* 1a; *15:* 7a; Ch'en Kuo-fu 1949: 258–259. On Ko Hung see Chap. VI, pp. 323–327.

shu's outline of neo-Confucian orthodoxy as a "treatise on *tao-shu*, in which he repeatedly spoke of portents as due to failures of government," at the same time affirming that "in Chung-shu's book he neither turned his back on the Confucians nor contravened Confucius." Sakai and Fukui find Huang-Lao enthusiasts, like those of other schools, using the term. Stein has demonstrated that after the Han it still often referred to conventional literati and Buddhists.[47]

Fang-shih 方士, "gentleman who possesses techniques, technician." When Ch'en Kuo-fu argues that *"fang-shih"* is equivalent to *"fang-shu shih"* 方術士 and Yü Chia-hsi 余嘉錫 equates it with *fang-chi chih shih* 方技之士, they are substantially agreeing. Both phrases might be translated literally "gentleman who possesses techniques." Wang Ping 王冰, the eighth-century annotator of the Plain Questions of the Inner Canon of the Yellow Lord, glosses *fang shih* literally as "a gentleman who clearly comprehends an art *(ming wu fang-shu chih shih* 明悟方術之士).[48]

I have discussed the movement in and out of fashion of various terms that used to be uniformly translated "Taoist." Discussions of technicians also use a shifting vocabulary. The treatises devoted to *fang-shih* in the Standard Histories refer to their arts as *fang-chi* 方伎, *fang-shu* 方術, and *i-shu* 藝術. All the technicians (*fang-chi*) mentioned in the first "Treatise on Bibliography," that of the Han History, are physicians. Other masters of technical arts are scattered under such designations as "diviners of propitious days" (*jih-che* 日者) and "recluses" (*yin-i* 隱逸).

The chapters on technicians of the later Histories include wonder-workers, astrologers, physiognomists, diviners, imperial favorites, initiated Taoists, a Buddhist patriarch, a man famed for his wealth, architects and other artisans, and an assortment of doctors in and out of the Imperial Medical Service.[49] This diversity affirms the ambiguous but generally low status of the *fang-chi* label. There is no reason to consider it more than a catchall for people who made their mark in arts of which gentlemen were not encouraged to know more than a smattering.

---

47. *Chuang-tzu*, 33.1. Watson 1968: 362 renders *chih fang shu che* 治方術者 as "those who apply themselves to doctrines and policies," which is not translation. Graham 1981: 274 renders it as "who cultivate the tradition of some formula." The context does not validate so narrow a sense for *fang*. Sakai & Fukui 1977: 441–443; *Lun heng*, 29: 7a, 4a. For the contexts see Forke 1907–1911: I, 466 and 84. Forke translates *tao-shu* as "magical arts," but the issue is plainly omens. Unlike Forke, I accept Sun I-jang's 孫詒讓 emendation of the phrase *pu-chi K'ung-tzu* 不及孔子 to *pu-fan K'ung-tzu* 不反孔子, cited by Huang Hui 黃暉 1938: 29: 1163–1164 and Liu P'an-sui 劉盼遂 1957: 571. Stein 1963: 39.

48. Ch'en 1949: 258–259; Yü 1958: 682; *Huang-ti nei ching su wen* 黃帝內經素問, 3: 67. 2.

49. Sakade 1978: 627–628.

VII

One can only agree with Yü Chia-hsi that *fang chi* is a general term for technical skills, which need not involve magic—even, I would add, in the broad acceptance of "magic" that includes divinatory powers.

Unlike the terms discussed above, there is a more elusive dimension to *fang-shih* and *wu*, which requires further scrutiny. For our purpose I will pay attention mainly to the use of the term *fang-shih* by historiographers. No two used it identically, and other sorts of writers used it in other ways. Despite considerable evolution, to a first crude approximation several criteria for the use of *"fang-shih"* appear general from the Han at least through the Sung:

1. The *fang-shih* usually belonged to the tiny privileged segment of the population who could read books and leave records. The writings we have, not a random sample, are of high literary quality. Early stories about technicians often have them confounding philosophers. The *fang-shih* usually came from a family that we know held official rank, even in periods when such rank was normally hereditary.

2. The *fang-shih* himself did not usually hold high rank in the regular civil service. If he did, it tended to be obtained irregularly, most often as an imperial gesture. Someone who reached a high post through a conventional career, although he might have considerable mechanical skill, scientific knowledge, or mastery of the occult, was not often called a *fang-shih*. Chang Heng 張衡 (78-139), astronomer, cosmologist, inventor, "patron of the art of yin and yang," was, in the words of Ngo Van Xuyet, essentially a *ju* literatus, and so merited a biographical chapter all his own in the History of the Later Han.[50]

3. The *fang-shih* did not strive for the personal goals that the well-born expected of their own kind. He usually held conventional moral and political opinions, if we can rely on the record, but the stigma of inappropriate technical enthusiasms, however faint, is commonly visible. Someone in a conspicuous position of orthodoxy, regardless of technical expertise, was not considered a *fang-shih*.

In the eleventh century we see emerging the label of "literatus-physician" *(ju-i* 儒醫), which elevated doctors given it far above the *fang-shih*. For instance, Sun Ssu-mo 孫思邈 (alive 673), although not related to any Confucian lineage, eventually became one of the models for aspiring scholar-physicians because of his emphasis on philosophic studies and ethical standards as preparation for medical practice. In his own dynasty he was a mere *fang-chi*. That is how he was listed, at least, in the Old History of the T'ang (945), a book rather faithful to the contemporary record. When the New History was compiled in the interest of better coverage and stylistic improvement (1060), Sun was promoted to "recluse." Finally, in the famous sixfold classification of early doctors in the Introduction to Med-

---

50. *Hou Han shu, ch.* 59; Ngo 1976: 71.

icine of 1575 he figured prominently among the scholar physicians.[51]

4. The *fang-shih* had powers only rarely seen in the orthodox literatus—to foresee the future, to arrogate to himself the shaping and transforming powers of natural process *(tsao hua* 造化*)*, and so on. At the same time descriptions of him never limn the full humanity, the mastery of the social Way, of the more conventional great.[52]

What pattern do these criteria reveal?

This is a category that originates not in some set of objective criteria but in the eye of the beholder. Ingenious craftsmen, diviners, physicians, thaumaturges, seekers after immortality, monks: I have found none who says "I am a *fang-shih.*"[53]

Fang-shih is not a social grouping toward which people align themselves, but rather an imputation of aims, powers, or behavior that the literatus biographer may admire or despise, but does not share. People become *fang-shih* in the eyes of others because of what they have done, not where in society they were born. Outside of technical skills and an identity that is nonconformist in a specific way (while usually conformist in other ways) they may or may not have anything in common with other *fang-shih.*

In what way, then, are they nonconformist? Here we benefit from the systematic and thoughtful researches of Ngo Van Xuyet into *fang-shih* in the Later Han. Ngo points out that "just as Confucian literati were more or less imbued with esoteric literature . . . the *fang-shih* were versed in the canonic books. No clear demarcation can be established in this period." Texts had long been the basis of claims to authority and, for the elite, livelihood. The state orthodoxy of the Former Han based on the Confucian writings aimed precisely to enforce the boundaries between respectable and disreputable texts, that is, those that furthered, and those that potentially obstructed, careers. The platitude always used for this purpose is "The master never spoke of marvels, prodigies of strength, anomalies, or

---

51. *Chiu T'ang shu*, *191*: 5094–5097; *Hsin T'ang shu*, *196*: 5596–5598; *I hsueh ju men* 醫學入門, *shou chuan* 首卷, pp. 54a–80b; Hsieh Kuan 1935: 51a–51b.

52. Ngo 1976: 66.

53. The opening section of T'ao Hung-ching's 陶弘景 preface to his *Pen-ts'ao ching chi chu* 本草經集注 (ca. 500) seems to use *fang-chi* as a self-reference. This preface is universally accepted as genuine, and indeed the later part appears to speak with T'ao's voice. But the first four characters, *"Yin-chü hsien-sheng* 隱居先生, the Hermit Master," are not the way a Six Dynasties author would have referred to himself. At least the initial thirty-nine graphs (including the phrase about *fang-chi)* are a prologue by someone else, perhaps a disciple. This curious point might be easier to resolve were it not that what corresponds to the beginning is missing from the very early Tun-huang MS (p. 1).

spirits." Fang-shih were those whose authority was based on the wrong books.⁵⁴

By the Later Han that orthodoxy had crumbled, and any sort of text might be useful to anyone. Those who hoped to preserve the dying order, marginal in a new society, needed such labels to mark those who should have remained on their side but had given in.

Ngo's insights suggest a pattern in the use of "fang-shih" in conventional historiography. The word acknowledges that someone claims exceptional skill in a technique or discipline, which implied spiritual penetration of the cosmic order. This claim might or might not be true. In either case it denies his entitlement to the conventional status for which his birth or rank would otherwise qualify him. "Fang-shih" is thus an epithet that denies social authority based on skill, although the latter may imply cosmic power.

This description only holds for one kind of author. The few early writings that come from the popular milieu indicate that *fang-shih* could be a positive term. An interpretation of a revelation in the Canon of Great Peace, for instance, says that the word refers to the leader of the community who will carry its message to "a lord who has virtue and power."⁵⁵ Without further multiplying distinctions, we can move on to the corollary.

The patterns outlined above tell us nothing about the concrete data that led someone at a certain time and place to decide that someone else was a *fang-shih*. The word does not express a fact, not even a fact about behavior. I have demonstrated for Sun Ssu-mo that even the most soberly stated accounts of *fang-shih* tend to resolve under close scrutiny into scattered crystals of history in a matrix of legend. As DeWoskin puts it with understatement, the term typically "was somewhat akin to 'others,' and did not attach to any readily definable tradition."⁵⁶

What facts underlie a given use of this epithet can only be determined case by case through the study of evidence. Some of those labeled *fang-shih* were no doubt members of politically marginal groups, while others (T'ao Hung-ching, Sun Ssu-mo, and I-hsing 一行 among them) passed their lives as pets of monarchs. Some *fang-shih*, we already know, were initiated masters of an orthodox sect. Some were brilliant technicians with average political and religious beliefs. But what little we know with fair certainty, once we have teased the facts away from the biasses and stereotypes, does not support the assumption that a *fang-shih* was a proto-scientist, and at one with the masses.

---

54. Ngo 1976: 64; Analects 7. 21.

55. T'ai-p'ing ching 太平經, 39. 50: 68–69. Cf. Kaltenmark 1979: 40.

56. Sivin 1968: 81–144; DeWoskin 1983: 6. On sources for the study of *fang-shih* see also DeWoskin 1981.

Wu 巫, "mediums." Lin Fu-shih's 林富士 thorough study of *wu* in Han dynasty records and artifacts reminds us that, although the character appears in the oracle script of the second millennium B.C., experts on the period do not agree on what it meant then. In the philosophical writings of the late Chou some *wu* served rulers and others pursued occupations allied to divination and exorcism, but all seem to have served the gods. They continued to be possessed by gods in the Han.[57]

The senses that *"wu"* has accumulated over the centuries—a dancing ritualist who brings the gods down to earth, a medium of either sex, a curer, one who behaves wantonly, and so on—are related closely enough to suggest not scattered meaning but sparing and fitful curiosity about a single phenomenon. In view of the scant detail in the documents about who *wu* are and what they do, I submit that the word, like *"fang-shih,"* is an epithet, not the name of an occupation. Its linguistic function is like that of the English word "superstition," which labels beliefs that the speaker doubts are religious and is not interested in understanding.

Wu is a garden-variety literary term, patronizing or disapproving, for mediums and other sorts of ritualist that the author could not be bothered to distinguish from them. Men of letters and officials most often applied it to the masters who performed the priestly functions of popular religion, often as mediums. They also imposed the word indiscriminately on the much rarer shamans (who instead of being possessed take spirit journeys on behalf of their clients), trance healers, Taoist priests, and others. Its use generally reveals less about the person it points to than about the attitudes of him who points.[58] To take a statement about a *wu* as evidence that someone was a specialist spirit medium, or belonged to another particular occupational category, misses the point.

Application to science. For Needham and most of his secondary sources of a generation or more ago, a Taoist is a Taoist. He does not need to keep several species distinct, since all embody a single ideal with clear positive consequences for science. I will give a few examples in which one or another hallmark of a Taoist evokes the wider complex of ideas (naturalist, empiricist, etc.) conducive to technical endeavor.

1. In his discussion of flood control Needham tells us that Chia Jang 賈讓 (fl. 6 B.C.), a great engineer, "was in fact an advocate of the channel expansion theory, a Taoist in hydraulics, who believed that the great river should be given plenty of

---

57. Lin 1988: 14–22. Lin seems to consider servants of the gods a separate category, but is not clear on this point. Chow 1978 is much more limited in value.

58. Historical and anthropological studies of popular religion in a a few localities indicate that its rituals are commonly mediumistic, but the breadth of this generalization remains untested.

room to take whatever course it wanted." Chia was a "Taoist" because he did engineering. Needham reasons that Chia's reluctance to confine the river was an instance of "non-interference," *wu-wei* 無為, a notion that harks back to the *Lao-tzu*. By perfectly analogous reasoning he could be labelled "anti-Taoist," "Confucian," or "Legalist," according to taste. His preferred plan, involving "the wholesale resettlement of the populations of prefectures bordering the river," was anything but *laissez-faire*.

2. That is not the only leap from technology to high philosophy. Another example appears in the section on the fenestrated rudder: "the device was probably quite empirical in origin . . . but it is not at all too fanciful to suppose that some medieval Taoist sailor, finding that his work was eased and that his ship sailed better, was fully content to follow the principle of *wu-wei*, and letting well alone, recommended the arrangement to his friends" (p. 656).

If Taoist necessarily implies *wu-wei*, however, *wu-wei* need not imply Taoist. Elsewhere Needham reminds us that "although the concept of *wu-wei* was emphasized particularly by the Taoists, it was part of the common ground of all ancient Chinese systems of thought, including the Confucians." This excellent point, unfortunately, does not recur often enough to avoid frequent confusions.[59]

3. A group of Taoists turns up in a palace laboratory at the end of the first century B.C.: "the *Han Kuan I* 漢官儀, a book on the Han bureaucracy written or published by Ying Shao 應劭 in +197, says that Wang Mang's coins were called *Pai Shui Chen Jen* 白水眞人, i.e., 'White-Water Adepts', a distinct indication of the role of his Taoist alchemists in the 'adulteration' of the bronze." Is "Taoist" meant to convey more than "alchemist" alone would do?

The Chinese text says nothing about Taoism *or* alchemy. It merely remarks that Wang changed the coinage, and that the new coins carried the words *huo ch'üan* 貨泉 (specie currency). The *Han kuan i* continues, "these words when dissected yield *pai shui chen-jen* [lit., 'white water immortals']. This was an omen that [the Han] would be restored under Shih-tsu [i.e., Emperor Kuang-wu]." In other words, Wang's enemies, intent on restoring the Han dynasty, used "dissection" *(fan* 反*)*, a popular method of divination, to forecast success. They split the character *ch'üan* into its component parts, which can be read separately as the words *pai* 白 (white) above and *shui* 水 (water) below. Similarly, but more arbitrarily, they divided *huo* more or less diagonally into *jen* 人 (man) and *chen* 眞 (real, realized). *Chen-jen*, literally "realized immortal," was a most propitious word. The point of this exercise in dissection is that the restorationists identified the "white-water immortal" with the future emperor. Wang's own coinage, they asserted, was

---

59. IV. 3, 234–235, 656; II, 563.

VII

TAOISM AND SCIENCE 33

predicting his defeat. The passage is about political propaganda, not alchemy.[60]

4. Needham observes that Taoists made instruments "for bringing the rays of the sun to a focal point," that is, burning lenses. He quotes two passages from Discourses weighed in the Balance. One says "that the *Chi Tao chih Chia* 伎道之家 (Taoist technicians) do it," and the other "this is the climax of Taoist learning and a triumph of their skill."

One might retranslate the first phrase to reflect the syntax. According to Han word order, *"tao"* cannot describe the type of technician. *"Chi"* (skill, technical), which comes first, tells what kind of *tao*. The phrase means nothing more specific than "masters who possess technical ways," i.e., "technicians."

Needham took the English of the second phrase directly from Alfred Forke, half a century earlier. Forke read into the text the idea that the learning was Taoist. One might read the text (about a marquess who made artificial pearls) literally as "[the success of the technique is due to] the perfection of acquired skill on the part of those who practice this art *(tao-shih)*, the application of ideas by ingenious people."[61]

5. Needham's tendency to link technical skills to Taoism without evidence (echoing Forke and many others) is not confined to discussions of the Han.

In reviewing the social backgrounds of technologists, he observes that "in view of the close association between Taoism and technical arts in ancient China, one would expect to find more Taoist inventors in the middle ages than have so far made their appearance . . . on the whole the Buddhists were more illustrious as technicians in these times." Among the exceptions to this most significant assessment is the "Taoist swordsmith" Ch'i-wu Huai-wen 綦毋懷文 (mid-sixth century). His identification with the co-fusion method of steelmaking, according to Needham, provides the clearest link of Taoism to ferrous metallurgy. The source states only that Ch'i-wu "put at the service of the Exalted Progenitor (Emperor Kao-tsu) the arts that he commanded."[62] No evidence indicates that Ch'i-wu was a Taoist. *"Tao-shu"* implies nothing more than Ch'i-wu's technical arts. The only ones mentioned are swordmaking and reasoning according to the color associations of the Five Phases, both of which were widely practiced among those uneducated in the *Chuang-tzu* and uninitiated into the worship of the Tao.

Needham describes a carriage of ca. 340 with a mechanism that animated "a

---

60. V. 2: 218; *Tai-p'ing yü lan* 太平御覽, *835:* 6b–7a.

61. 道士之教至，知巧之意加也. IV. 1, 111–112, repeated IV. 3, 677. Lun heng 論衡, *16.* 47: 694, *2. 8:* 71, tr. Forke 1907–1911: II, 350 and I, 378. For *chiao* as "acquired skill" see *Kan-Wa daijiten* 漢和大辭典, item 13212, sense 1.2.

62. 以道術事高祖. IV. 2, 34; *Pei ch'i shu, 49:* 3b; cf. Needham 1958: 26.

large wooden figure of a Taoist ... with its hands continually rubbing the front of the Buddha" as well as "ten wooden Taoists ... continually moving round the Buddha." A footnote is undecided "whether there was religious syncretism here, or whether the Taoists were supposed to be paying homage to a superior religion, or whether even the phrasing does not simply mean followers of the Buddhist Tao." Since the text reads *"mu tao-jen* 木道人" and speaks of the Hunnish patron's surpassing devotion to the Buddhist teachings, the third alternative is the only plausible one. I have noted that *"tao-jen"* routinely referred to Buddhists at the time. The passage is not evidence for a Taoist presence in technology.[63]

In Needham's writing the hiatus between the *tao-chia* of ca. 300 B.C. and the *tao-chiao* that began nearly five hundred years later is solidly filled with Taoist activity in relation to early technology, medicine, and science, especially in the technical efflorescence of the Han. Some of the actors are "Taoist *fang-shih.*"

Fang-shih enter this picture because they both personify the Chuang-Lao philosophies and complement them. Taoism as an organized religion unites the "Taoist school of ancient philosophers" with "magic-scientific accretions ... around a nucleus of primitive shamanism." On the one hand, the interest of the *fang-shih* in techniques puts them on the Taoist side of the Confucian-Taoist antinomy. On the other, their magic connects Taoism with "the most primitive sorcery of the North Asian peoples." This sorcery is of course shamanism. "The Chinese had a word of their own for shamanism, ... namely, *wu* 巫." Needham's extensive discussion of "Shamans, *Wu*, and Fang-shih" devotes only one sentence to the latter: "The only remaining important term is *fang-shih* 方士, which some like to translate as 'gentlemen possessing magical recipes'—we think they were just straight magicians." Since Needham also consistently describes the *wu* as magicians and thaumaturges, *"wu"* and *"fang-shih"* appear, in the absence of a conclusive assertion to that effect, to be synonyms or close to it. But does *fang* always imply magic? We have seen that it does not.[64]

In considering anyone called a *wu* from the Middle Period of history on to be Taoist, Needham follows his chief source, the turn-of-the-century pioneer J. J. M. de Groot, for whom "Taoism may ... be defined as Exorcising Polytheism, a cult of the gods with which Eastern Asiatic imagination has filled the Universe, con-

---

63. IV. 2, 159–160; *T'ai-p'ing yü lan, 752:* 3a.

64. I, 117; II, 132, 134. It is pertinent that the distinction between magic and religion, after close scrutiny by cultural anthropologists, has proved to be indefensible. Tambiah 1968 is the classic demonstration. Anthropologists increasingly use the rubric "symbolic behavior" to cover rites of all kinds, whether the appeal to the gods is explicit or tacit. My own research into popular healing in traditional China fully supports this view. There is no point in splitting hairs over, e.g., whether magic includes divination.

nected with a highly developed system of magic . . ." Both scholars believe that the *wu* were gradually "incorporated in the Taoist system." Needham paraphrases de Groot to the effect that "the *wu* were by no means always on good terms with the ruling authorities, . . . so that the magical as well as the political-philosophical aspects of the Taoist system drove it inevitably into general opposition to the government." Finally "the *wu* aspect of Taoism was driven underground, and tended to take the form of those secret societies among the people which in later centuries played such an important part in Chinese life."⁶⁵

Finally Needham, even in his latest writings, follows Sinological habit in finding Taoists in two other unconventional corners of Chinese society. First, he typically translates a sentence in T'ao Hung-ching's 陶弘景 preface to the Canon of Materia Medica with Collected Annotations (ca. A.D. 492): "There are also things mentioned in the writings of the Taoists *(hsien ching* 仙經) as necessary for their techniques . . ." *Hsien ching* means nothing more than "canons of immortality," and immortality was a very general pursuit at the end of the fifth century.

This elision misses an important point. T'ao is one of the few indubitable Taoists we have considered. He also practiced alchemy, and studied techniques of immortality. But what he did in his edition of the Canon was to *reject* the traditional arrangement of drugs by their use in seeking immortality, and to arrange them by origin instead. He contrasts those who use drugs for immortality with those who use them primarily for therapy, the aim of his book.

Needham also enumerates, among authors of sixteenth-century books on edible wild plants, Chou Lü-ching 周履靖, "a Taoist naturalist," and Kao Lien 高濂, "another Taoist naturalist." Chou edited a collection of works for high-minded gentlemen living in retirement, but an autobiographical sketch does not indicate involvement in any Taoist activity. Kao was a Hangchow poet, and compiled several tractates on cultivation of the vital *ch'i*. Elsewhere Needham qualifies his description, writing of Kao as one of "a group of scholars, largely of Taoist inspiration, who sought no office and lived in seclusion, cultivating plants and writing about them in order to console the heart in bad times and nourish the spirit." Needham presents no evidence that the "Taoist inspiration" consisted of anything more specific than seclusion. What does probing the influence of seclusion on the evolution of natural history tell us about the relation of the latter to Taoism?⁶⁶

Let me sum up this necessarily extensive look at "Taoists" who neither contributed to philosophic Taoism nor were members of a Taoist movement. They,

---

65. II, 137–138; de Groot 1892–1910: VI, 931.

66. VI. 1, 246, 349–350; *Pen-ts'ao ching chi chu* 本草經集注, p. 4. See the biographic data on Chou in Goodrich 1976: 658b, 1279b, and on Kao in Ch'ang Pi-te 1976: 892a.

VII

rather than the philosophers and priests, are the central *actors* in the concrete interplay between Taoism and science exemplified in *Science and Civilisation in China*. The book depicts their contribution as strong and positive, in contrast to the obstacles posed by the Taoist religion. It explains such Taoists' contributions to science by their magic, which in early times was supposedly akin to science, and by their manual operations, contrasted to the prejudice of Confucians against using their hands. These two themes are intimately related: "If I believe that by taking a wax statue of [someone] and sticking pins in it I can cause him evil, I am adopting a belief for which there is no foundation, but I do at any rate believe in the efficacy of manual operations, and science is therefore possible."[67]

This classic British empiricist stance can lead to odd conclusions. Is the wizard conducting his own rituals, a Taoist in the abstract, a better scientist because he dirties his hands than the imperial astronomer, a Confucian at the office, working out new algorithms to compute events in the sky? And what sort of history of science will emerge from generalizations based on abstract Taoists while ignoring all but a handful of initiated ones?

Residual Taoists. I have argued so far that Needham conceives of Taoism in relation to science largely in three ways: first, as a set of attitudes toward Nature and its apprehension derived from the *Chuang-tzu* and other early classics, without reference to scientific inquiry; second, as the work of "technicians" *(fang-shih, tao-jen,* etc.) without, in most cases, reference to their individual attitudes; third and less frequently, as popular beliefs and rites. Most of the variety of undefined references to Taoism and Taoists scattered through Needham's voluminous writings can be understood with reference to one or the other of these senses. Others do not clearly fit any identifiable species of Taoism, so that their bearing on the Taoism-science relationship is especially nebulous. I will give a few examples.

1. Of several "Taoist authors" writing on the formation of minerals from *ch'i* 氣 exhalations, the only one about whom anything whatever is known was King Chien-p'ing 建平王 (fl. 444) of the Liu Sung Dynasty. His biography does not hint at a connection with the *Chuang-tzu,* Taoist sects, popular observances, etc.[68]

2. Learning from experienced artisans and craftsmen "had been a long tradition in Taoism, as witness the story of Pien the Wheelwright . . . and Liu Tsung-yuan's old gardener . . . In Thang, Han Yü 韓愈, though so Confucian a scholar, had written a famous essay on what he had learnt from the mason Wang Chheng-fu 王承福." Of the three "Taoists" who learned from artisans in these anecdotes, one

---

67. Needham 1969: 162.

68. III, 638–639; *Sung shu, 72:* 2a–6a.

was an unimaginative ruler in the familiar *Chuang-tzu* story. The wheelwright was the "Taoist," and the lord his straight man. The second and the third can hardly be called Taoists. Han (768–824) was the most ostentatiously orthodox Confucian of his time. William Nienhauser remarks of Liu Tsung-yuan 柳宗元 (773–819) that "although he took an interest in Buddhism, especially in the Ch'an school, which was then popular in South China, he was a servant of the state at heart." Elsewhere Needham describes him merely as "a Thang writer of naturalistic interests." None of the three who learned from technicians is linked to Taoism.[69]

3. In pre-Han legends of taming the waters, "the unsuccessful, or at any rate disapproved, irrigation engineers are identical with the corps of legendary rebels ... It was therefore quite natural that some of them should have become heroes for the Taoists in later ages, since the Taoists opposed feudalism root and branch, urging a return to the collectivist golden age. Consequently it is of great interest that Chhü Yuan, in the *Li Sao* and especially the *Thien Wen* odes (c. -300), strongly takes the part of Kun," saying that he "met with failure through no fault of his own."[70] No "Taoist" other than Ch'ü Yuan 屈原 is mentioned in this connection. Needham does not present Ch'ü's otherwise unknown credentials as a Taoist. This association may be a matter of Ch'ü's unconventionality, or of his fascination with mythical cosmography.

4. Yang Hsiung 揚雄 (53 B.C.–A.D. 18) "was devoted to astronomy and used to discuss it with the Taoists. He made an armillary sphere himself. An old artisan once said to him ..." Needham evidently understood *huang-men* (literally, "yellow gate") as "Taoist." Taking the word in its established Eastern Han sense of "the gates to the private quarters of the emperor and his women," one might read " ... was devoted to astronomy. He asked an old artisan about it who was making an armillary sphere at the Yellow Gates. He said ... ."[71]

Summary of Findings. Despite a few cases such as the examples just given that do not fit any pattern, Needham portrays Taoism as an essentially threefold influence on science:

1. The "Taoist" literary works shaped science because they embodied a set of attitudes which prefigure, or are in principle identical to, those of the modern sci-

---

69. IV. 3, 85 note a; II, 577; Nienhauser in Liu & Lo 1975: 569.
70. IV. 3, 250; also II, 115ff.
71. 揚子雲好天文。問之于黃門作渾天老工。曰。 *T'ai-p'ing yü lan*, 2: 11a–11b, or *Ch'uan Hou Han wen* 全後漢文, *15*: 2a; Needham, Wang & Price 1960: 129; Bielenstein 1976: 24 for *huang-men;* for a translation of the whole passage, Pokora 1975: 114–115, item 114.

entist. They are distinctive because they were opposed to attitudes encouraged by Confucians.

2. "Taoist" popular religion shaped science through its immortality disciplines, magic, etc. (to which Needham gives more attention than communal rituals): "since in their beginnings magic, divination and science were inseparable, we cannot be surprised that it is among the Taoists that we have to look for most of the roots of Chinese scientific thought."[72]

3. The "Taoist" *fang-shih* shaped science because they contributed to technical inquiry.

We have seen that once Needham has identified an individual in principle with any of these three Taoisms, a concrete linkage with Taoist activity becomes unnecessary.

Taoism and Scientific Attitudes

I have argued that the least ambiguous linkage of science asserted by Needham is to the tradition of the *Chuang-tzu,* the *Lao-tzu,* and other books customarily shelved with them. How are the links formed? How strong and clearly articulated are they? In other words, can one assume that all early philosophic Taoist authors were agnostic naturalists, mechanists, materialists, experimentalists, empiricists who avoided preconceived ideas, democrats and heterodox collectivists?

Empiricism. In the interest of concision I will sample only two of these characteristics, namely empiricism and its corollary, the observation of Nature without preconceptions. These are central, but the reader will do well to keep in mind that they are closely tied to the others. In a more rigorous discussion it would be necessary to examine them all.

We are told that "'Cognoscere causas' [understand the cause] . . . became the motto of the Taoists. Through all the convulsions caused by the substitution of feudal bureaucratism for feudalism at the time of the unification of the empire by Chhin Shih Huang Ti, they continued to pursue it." In the sources, curiously, this motto remains unrecorded. Needham sees a consistent message running through the philosophical writings enumerated earlier: that correct Taoist practice requires regard for "what may be ascertained about causes and intrinsic principles in Nature." This is how he puts it: "*Wei* 爲, then, was 'forcing' things, in the interests of private gain, without regard to their intrinsic principles, and relying on the authority of others. *Wu-wei* 無爲 was letting things work out their destinies in accordance with their intrinsic principles. To be able to practise *wu-wei* implied learning from Nature by observations essentially scientific." Needham considers

---

72. II, 57. Here Needham follows Thorndike 1923–1958.

Taoists opposed in this respect to Confucians. Of the latter he asserts that "their rationalism was limited to human society and did not even admit that the world of Nature was worth theorising about at all."[73]

This strongly stated opposition will be discussed later with respect to individual careers of scientists (pp. 49–56 below). First it is necessary to weigh the evidence for the idea that, as an aspect of *wu-wei*, Taoist philosophers urged their readers to learn about principles, to theorize through observation of Nature.

Needham's discussion of empiricism is not documented from the *Chuang-tzu* or *Lao-tzu*. Chuang-tzu's contemporary Shen Tao 慎到 (ca. 300 B.C.) provides the earliest statement: "As regards the people who protect and manage the dykes and channels of the nine rivers and the four lakes, they are the same in all ages; they did not learn their business from Yu the Great, they learnt it from the waters."

Different authorities classify Shen Tao as a Legalist, a Taoist, and an Eclectic. His work has survived only in fragmentary form. Since we have no context for this sentence, both its relation to Taoist thought and its bearing on empiricism are murky. For instance, although the fragment speaks clearly to one empiricist theme, learning from experience of Nature, it appears to contravene another, the cumulative and progressive character of experiential knowledge.

A close translation would differ in detail from Needham's free paraphrase: "Those who control the waterways, shoring up with bundles of sticks, dredging and closing ruptures, even among the barbarians, [use techniques] so much alike that they are practically identical. They learned them from the water, not from Yü the Great," the legendary king who tamed the universal flood. This laconic utterance would seem to be about methods, not principles. In reflecting on whether it is Taoist, a preamble to this passage in the earliest ample collection of the fragments, that of 1578, provides food for thought: "Hsu Fan inquired of Master Shen 'Where do laws originate?' Master Shen said 'Laws do not descend from heaven, or emerge from the earth. They issue from the human realm, in harmony with human hearts and minds. That's all.'"[74] We can hardly take that bit of dialogue as a verbatim recording, nor can we be sure it is ancient. It tells us how readers in the Ming or earlier understood the original fragment. It is anything but a philosophical Taoist reading. No ancient reading of the passage in the style of

---

73. II, 55, 10, 71, 94 (also 395).

74. [許犯問於慎子曰。法安所生。慎子曰。法非從天下，非從地出，發於人間，合乎人心而已。] 治水者，茨防決塞。雖在夷貊，相似如一。學之於水，不學之於禹也。II, 73; Thompson 1979: 271, fragment 68. See on the same page of Needham the quotation from *Kuan-yin-tzu*, 5. 20b, similarly paraphrased to speak of "those who can think."

VII

Chuang-tzu is recorded.

The extended citations from pre-T'ang works that bear Needham's burden of proof are also not quite about empiricism. One, from the *Huai-nan-tzu,* is part of a long passage that argues for remaining aware of the origins of social order in deeply-rooted human tendencies that even the harshest punishments cannot overcome. It criticizes "the draconic Legalists" for turning their backs on these foundations and thus inventing disastrous methods of government. Needham reads the conclusion as "So therefore, rather than begging or borrowing fire, you had better take a burning mirror, and rather than drawing water from other people's wells, you had better dig one yourself."[75] He then remarks by way of interpretation "finally, go to Nature and not to Authority, make your own fire and dig your own well." A more literal translation would read "therefore begging for fire is not as good as using a firemaker, and depending on someone else to draw water is not as good as digging a well." The point of the passage is not observing while undistracted by authority, but building institutions on the basis of what is inherent in human beings. The fire one starts oneself and the water one draws from one's own well are not more natural than those received from other people. Rather, they represent direct access to the sources of fire and water.

The second text, from the Springs and Autumns of Master Lü (ca. 239 B.C.), "may be considered one of the finest affirmations of the ancient Taoist technologists against the politicians and sophists of their time," but it mentions neither Taoists nor politicians. Its theme is the need to recognize the limits of one's knowledge, and especially to avoid false analogies, a matter stressed as much by Confucius as by Chuang-tzu. It ends, in Needham's translation, "the Sage follows (Nature) in establishing social order, and does not invent principles out of his own head." There are no principles, no Nature, and no technologists in the Chinese.[76] It might be translated literally and in context as "the sage accords [with the actual circumstances] when creating institutions, rather than conforming to his own inclinations."[77]

As these examples indicate, one may venture to question the significance and Taoist credentials of the pre-T'ang evidence Needham brings to bear on the issue of empiricism.

In view of the crucial role that Needham allots to scientific observation in

---

75. 是故，乞火不若取燧，寄汲不若鑿井. A *sui* can be a mirror, lens, or drill.

76. 聖人因而興制，不事心焉。

77. *Huai-nan-tzu, 6:* 10b; *Lü shih ch'un-ch'iu, 25: lun* 5. 2. 1643; II, 73, translates the whole chapter.

Taoist philosophy, one is surprised to find him saying that "the Taoists never developed a systematic theoretical account of Nature, analogous to that of Aristotle. The Yin and the Yang, the various forms of *chhi*, the Five Elements [i.e., the Five Phases], were insufficient for the task assigned to them," so that the Taoists' progress was mainly in "all practical technology." These two sentences appear to be asserting that certain Taoists attempted to construct a systematic theoretical account of Nature on the basis of inadequate concepts, and failed. At the same time, Needham does not identify Taoists who were making such an attempt.[78]

Nor does he consider obvious objections. First, systematic theoretical accounts of Nature did appear in the last three centuries B.C., not from Taoists but from literati. Second, how are we to judge adequacy "for the task assigned to them" without specifying what that task was, and perhaps who assigned it? If we are to insist on the criteria of modern science, Aristotle's account, devoid of mathematical abstraction, is just as inadequate. If we are prepared instead to take seriously the expectations *of Chinese in the last three centuries B.C.,* we must admit that they did not find these concepts at all inadequate. We also find them, not only in philosophic treatises but in the emerging technical literature, able to deal abstractly with deep theoretical issues. They did not attain a single, tightly organized consensus, although we find their modes of explanation converging in such matters as Five Phases associations and their articulation with yin-yang explanation. If we move back from Aristotle and look at all of Greek natural philosophy in the same period (before he drowned out his rivals), we find among the schools at least equal diversity.

Nor apparently is Needham convinced that such a failure would prevent the transition to modern science. He notes that "failure to develop adequate scientific terminology was characteristic of medieval European science, and this was one of the limiting factors which the upsurge of the Renaissance swept to one side."[79] Note that Chinese and Greek "limiting factors" were not things or phenomena, but rather *failures* of something to happen. Pondering that logical difficulty, we are left with the unanswered question of why the Chinese "limiting factor" was not "swept to one side" as well.

Avoidance of Preconceptions. The scientific empiricist not only observes Nature, but does so without the preconceptions that blind others to the actual phenomena. In this respect too Needham considers Taoism and Confucianism opposite: "the observation of Nature, as opposed to the management of society, requires a receptive passivity in contrast to a commanding activity, and a freedom

---

78. II, 84.
79. IV. 3, 403, note d.

VII

from all preconceived theories in contract to an attachment to a set of social convictions." Now "what was the main motive of the Taoist philosophers in wishing to engage in the observation of Nature? There can be little doubt that it was in order to gain that peace of mind which comes from having formulated a theory or hypothesis, however provisional, about the terrifying manifestations of the natural world surrounding and penetrating the frail structure of human society.... This distinctively proto-scientific peace of mind the Chinese knew as 'ching hsin 靜心.'"[80]

Needham then quotes the Writings of the Gatekeeper (Kuan-yin-tzu) on the sage's freedom from the obsessions that lead others to be possessed by demons: "for every day the sage faces the facts of Nature, and his mind is untroubled."[81] A literal translation might be "for every day the sage is responsive to the totality of phenomena (wu 物), his mind stilled," i.e., in a state of incipience.

On the same folio of the source, an epigram reminds us that for the Gatekeeper wu are not facts: "When one recognizes that there are no wu in the mind one knows that there are no wu in wu. Recognizing this, one knows there are no wu in the Way. Thus one is neither moved to emulate exceptional conduct nor to admire subtle and penetrating discourse [and so on]." In another place the book tells us that "of the myriad phenomena (wu) in sky and earth, not one is my own wu. Although wu are not my own self, I have no choice but to respond to them. Although I am not my own self, I have no choice but to cultivate my self. Although [I] respond to wu, there is no such thing as wu. Although [I] cultivate my self, there is no such thing as my self ... The Way is one; that is all there is to it. It cannot be approached by ordered progression."

Wu in this late book, obviously much influenced by Buddhism, are phenomena in the mystic's sense, mere appearances, individually of no greater interest to the sage than images are to the mirror. Sage and mirror are "responsive" in the sense that they reflect phenomena. The presence or absence of theoretical preconceptions is not an issue.

Another key passage that Needham cites from Writings of the Gatekeeper to illustrate "the old theme of the necessity of being without partiality or preconceptions" turns out to be equivocal evidence. He translates "...the sages were taught by the myriad things, and in their turn taught the worthies, who taught the people. But only the sages could understand the things (in the first place); they could unify themselves with natural principles, because they had no prejudices and pre-

---

80. II, 57, 446–448, 63.

81. 日應萬物。其心寂然。

conceived opinions." A plain English version of the phrase after the semicolon might be "only the sages were at one with the phenomena, and therefore devoid of self."[82] There is nothing in the Chinese about natural principles, prejudices, or preconceived opinions. It is about not rational inquiry, but the intuitive oneness with the world around them that furnished sage rulers with patterns for social institutions. What sages have overcome is not presuppositions but the illusions of selfhood.[83]

Needham presents a short passage from *Huai-nan-tzu* as an injunction to seek a unitary principle in Nature: "He who is of an intelligent nature is not terrified by any of Nature's operations; he who is wise by experience is not disturbed by any strange phenomena. The sage infers the far from the near, and concludes that the myriad things are based upon a single principle." More literally, the part of the source that corresponds to the last sentence says "Thus one who has attained sagehood understands what is distant from what is close at hand, so that all the diversity (of the phenomena) becomes one."[84] Again the word "principle" has no Chinese counterpart. Again the sage ruler seeks, not to draw a conclusion from study, but to merge himself with the world in all its particularity, social and physical, by a leap of intuition or illumination. He does so, this chapter asserts, in order to create a balance in the political order that can be achieved only by harmony with the cosmic balance.

Needham sees meditative activities as perfectly compatible with investigations aimed at the public welfare: "Thus 'emptying the mind' did not mean emptying it of that true natural knowledge which Chuang Tzu contrasted with the false knowledge of feudal social distinctions, but rather emptying it of distorting memories, prejudices and preconceived ideas, so that true practical knowledge might flourish and all abundance come in its train. The absolute justification of this complex of thought is seen in the great inventions of ancient China, as, for example, the use of water-power." But where do we find promises of material abundance and exhortations to hydraulic innovation in the *Chuang-tzu?*

The closest Needham comes to giving an answer is in the famous *Chuang-tzu* story of the old man who carries water out of his irrigation well by hand, refusing to use a simple dipping machine because "those who have cunning devices . . .

---

82. 唯聖人同物，所以無我。

83. II, 67, 447; *Kuan-yin-tzu* 5. 18b–19a; 9. 31b–32a; 3. 10a. For *chi-jan* see the Great Commentary to the Book of Changes, A. 9. This literal translation cannot convey the metaphysical depths of this passage, e.g., the distinction between the "I" and the "self," both written with the same Chinese word. Cf. Steininger 1953: 40.

84. II, 66; *Huai-nan-tzu, 8:* 4a–4b. 故聖人者由近知遠，而萬殊爲一。

have cunning hearts." This and similar anecdotes elsewhere Needham explains by "the popular feeling that whatever machines or inventions might be introduced it would be only for the benefit of the feudal lords; they would either be weighing-machines to cheat the peasant out of his rightful proportion, or instruments of torture with which to chastise those of the oppressed who dared to rebel." How a swape *(kao* 槔*)* could be used in either way is not obvious. Chuang-tzu's "anti-technology complex" sits uncomfortably with Needham's claim that the philosopher advocates "emptying the mind" as a means to prosperity for all.[85]

Needham acknowledges the contradiction when he translates two excerpts from Writings of the Gatekeeper that mix "magic, experimentation, bodily culture, and the invulnerability complex...suggesting that techniques should be used for the understanding of Nature *rather than* for benefiting human society." But these passages too are about attaining spiritual power and charisma rather than with either scientific investigation or social benefits of technology.[86]

What I have discussed in this section is the picture of Taoist empiricism sketched by Needham in 1956. It does not change noticeably in later volumes (see, for instance, p. 35 above).[87]

Summary of findings. Let me sum up the results of this discursive inquiry into linkages of Taoist philosophy and scientific attitudes. Needham's case for empiricism is not built on the *Chuang-tzu* and *Lao-tzu*, but on works of a more eclectic nature. His evidence for the constituents of empiricism—recourse to experience of Nature rather than to authority, avoidance of preconceptions that interfere with experience—consistently refers instead to the responsive and creative stilling of the mind practiced by mystics in more than one religious tradition. Its aim is not science. More than that, it is not knowledge.

One must agree with Needham that the indigenous Chinese forms of mysticism did not rule out acceptance of the natural order, or curiosity about it. But early Chinese sources do not say that "to be able to practise *wu-wei* implied learning from Nature by observations essentially scientific." The most famous Taoist master of the T'ang period expressed quite the opposite view.

---

85. II, 89, 124–125; *Chuang-tzu, 12:* 55, one of the "outer chapters."

86. II, 449, my italics.

87. One finds an occasional reflection concerned less with philosophical attitudes and more with practice. For instance, on optics, "The Taoists might talk about the wonders and beauties of Nature, the Naturalists might bring forward their generalised explanations of her phenomena, the Logicians might argue about the proper way of discussing, but only the Mohists actually took mirrors and light-sources and carefully looked to see what happened" (IV. 1, 78).

VII

In 711 "the emperor summoned Ssu-ma Ch'eng-chen, the Taoist master of Mt. T'ien-t'ai, and asked him about yin-yang and the disciplines based on study of regularities *(shu-shu* 數術*)*." This term refers to both quantitative and qualitative methods of prediction, from computational astronomy to fortune-telling. Ssu-ma replied "One devoted to the Way does less and less until he reaches the stage of no purposive action *(wu wei)*. Why should he be willing to tax his mind with the study of regularities?" When the emperor went on to ask whether one can govern a state that way, he went on "The state is just like the self. If you go along with the spontaneity of phenomena, and you have nothing selfish in your mind, the state will become ordered." The story ends with Ssu-ma receiving the emperor's permission to return to his mountain.[88]

Three other components of the empiricism argument call for further thought:

1. Needham asserted in 1959 that Taoists failed to construct a systematic theoretical account of Nature because their concepts were inadequate. Since then, Needham and many others have demonstrated that there were systematic theories and that they used just those concepts. They are not concentrated in treatises on philosophy but in technical writings on medicine, alchemy, geomancy, etc. No one has shown that any of these theories originated in or were the particular property of a Taoist movement. One can only agree with Strickmann's profound study of alchemy in the origins of the Supreme Purity movement: "As for all the technology, I see no reason to call it 'Taoist' except where it occurs in an indubitably Taoist social context." Strickmann suggests that alchemy and other arts be "visualized as separate entities, weaving in and out of Taoist (and other) contexts in the course of history, rather than as somehow being integral parts of a 'Taoism' that depends on them for its definition yet lacks any social dimension."[89]

2. The generalization that Confucians (undefined) limited their rationalism to human society "and did not even admit that the world of Nature was worth theorising about at all" sets up a general Taoist/Confucian dichotomy. Confucius and his immediate successors were, to be sure, humanists, but the neo-Confucian orthodoxy of the second century B.C. was built around parallels between the state and the cosmos. It was grounded in theorizing about Nature. It so successfully absorbed the ideas of the *Chuang-tzu* and other unconventional books that, however many formal contradictions the undergraduate may easily identify, one can no longer speak of Taoist and Confucian philosophies engaged in conflict.

---

88. II, 7. II, 71. The anecdote is from the comprehensive history *Tzu chih t'ung chien* 資治通鑑, *200:* 6669–6670. Ssu-ma uses the language of *Lao-tzu* 48. For the context see Schipper 1982: 140 and Benn 1987. Ho Peng Yoke 1991 explains *shu*.

89. Strickmann 1979: 166.

3. The Baconian and Newtonian stress on avoiding preconceived theories has maintained a strong influence upon British philosophy, a field on which Needham made an enduring mark long before he began writing on Chinese science.[90] Historians of science in the last couple of decades, however, have tended to see this seventeenth-century notion not as an accurate reflection of the method that generated modern science, but as a specious claim meant to impress laymen with the superiority of the new science over scholasticism. Experimental science proceeded in early modern Europe, as it does now, by hunches, gambles, and premonitions, undesirable only if they are badly founded or uncritically maintained.

One should thus hardly be offended by preconceptions in the science of ancient China. Among the most common are the ideas that the cosmos is an organic whole, that a universal Tao comprises the tao of each individual phenomenon, and that the void state of meditation provides access to an underlying reality that abstract principles cannot encompass.[91] They were no more inimical to discovery than the Christian faith of the Oxford theologians who worked out the mathematical behavior of falling bodies.

The doubts conveyed above do not settle the question of Taoism and scientific attitudes. Empiricism is only one aspect of Needham's broad approach. When one evaluates his arguments that agnostic naturalism, mechanism, materialism, experimentalism, democracy, and collectivism were inherent in early Taoist philosophical writings, the problems I have raised are typical. In many instances, when the evidence is interpreted literally it does not support Needham's thesis. In others, the plausibility of the argument depends on how one defines vague terms. The reader will already have noticed that most of the isms just listed are notably vague.

If the case is unproven for a tight connection between Taoist philosophy and scientific attitudes, what can we now conclude about the role of Needham's threefold Taoism in the history of science?

1. Popular religion and orthodox Taoism overlapped considerably, but whether a given instance of liturgy, self-cultivation, magic, or divination was Taoist must be determined individually.

2. By almost any common stereotype of "Taoist" (naturalist thinker, dropout, priest, iconoclast, magician, official at home in the evening, etc.), a certain number of *fang-shih* (varying in each case) would undoubtedly qualify as Taoists. The only definition that makes all *fang-shih* Taoist is tautologous. It reveals merely that someone considered someone (about whom perhaps nothing more is known)

---

90. Nakayama 1973: 23–30.

91. On this last theme within science see Sivin 1989.

competent in marginal disciplines. Whether a given *fang-shih* was a Taoist by a given definition must be determined individually.

3. The authors of the *Lao-tzu* and similar books influenced other schools and inclinations. From the Han on, one can no more call this interaction "influence" than one would use that word for the role of vegetables in a stew. There is no doubt that the *Lao-tzu* etc. contributed to the Han neo-Confucian stew a spirit of openness toward Nature. At the same time Taoist philosophers were not responsible for the move of Confucianism away from humanism. One must also look within Han orthodoxy at the radiating influence of the Book of Changes as its commentaries reinterpreted it and made it part of the synthesis. The mountain of erudition in *Science and Civilisation in China*, in Needham's other writings, and in the publications of Sinologists who agree with it, has yet to prove, in concrete cases, that Taoist philosophers motivated concrete scientific explorations to an extent that other intellectual convictions did not. What seem to be instances of this motivation must be evaluated one by one.

4. It is impossible to settle the question of links between Taoism and science by confronting doctrinal or canonical texts on a high level of abstraction. After examining Buddhist scriptures it is natural to conclude, at least as a broad and preliminary generalization, that "in the last resort, Buddhism was a profound rejection of the world ... One of the pre-conditions absolutely necessary for the development of science is an acceptance of Nature, not a turning away from her." But then one is left unprepared to encounter individuals such as I-hsing (682/683–727), a major figure in both Tantrism and mathematical astronomy, or the T'ien-t'ai monk Tsan-ning 贊甯 (919–1001?), a devotee of natural history and physical studies who nevertheless earned the epithet "Tiger of Monastic Discipline."[92]

Taoism as a spirit. History is too subtle a matter to be neatly encompassed by definitions. As much may be lost by rigidly adhering to them as by not using them. It is necessary to consider a different aspect of Needham's Taoism.

Needham's Taoism is not only three sets of practices and convictions that affected Chinese civilization in particular ways. It is also a spirit, a cast of mind: curious, devoted to Nature, undistracted by convention, aware in a past-oriented society that the future can be shaped, convinced that when they strive for the social good all men and women stand on the same high level. This spirit shines through Needham's portrayals of individual scientists. It is the mind-set of modern science—not of what it is, but of what scientists who strive to be more than technicians tend to believe it should be. Most of these ideals were definitively

---

92. II, 430–431. On I-hsing see Ang 1979 and the table on pp. 51-54 below; on Tsan-ning, Franke 1976: 1040–1046.

VII

stated in the 1660's, and are often restated today. Needham held them high. They are, I think, largely responsible for his remarkable eloquence in portraying the science of all ages and places as part of an ecumenical and convergent enterprise.[93]

This most original and ample of all Needham's Taoisms subsumes the others, and ultimately determines the character of the linkage with science—or, to put it more exactly, the linkage from science. Needham's starting point was his view, formed by scientific and historical research, of what science has always been about.[94] His search in the scholarship on Taoism for this spirit led him toward the three aspects I have just discussed.

This is an ideal picture of science, not a description of Taoism. It does not correspond to what the documents of any Taoism reveal and to what its most penetrating students understand. But these objections are disabling only if Needham's work is misread as an attempt to write a definitive history of Chinese science. He made it clear at the outset that it is meant as "but a reconnaissance" addressed "not to sinologists, nor to the widest circles of the general public, but to all educated people, whether themselves scientists or not, who are interested in the history of science, scientific thought and technology, in relation to the general history of civilisation."[95] His picture of Taoism, as of much else that scientists, engineers, and physicians would find exotic, is heuristic. It has justified itself by holding the attention of readers who otherwise would never have given a moment's critical reflection to ancient Chinese thought and practice.

For two thousand years Taoism has been a religion, a matter of the spirit, as well as a social affiliation. No history that fails to consider both aspects can be fully adequate. But the flaw in Needham's account is confounding faith and collectivity, and reading modern ideals into both.

The old cliché has it that every gentleman was a Confucian at work and a Taoist at leisure. True; that is all that survived of the dissenting spirit of the *Chuang-tzu* once the book had been conventionalized and trivialized. The cliché tells us nothing about any class of Taoists except the one it defines: the class of all gentlemen at leisure. When these "Taoists" are mistaken for, say, the class of initiates who shaped the orthodox religion, or the class of artisans associated with the evolution of technology, the potential for misunderstanding is obvious. So long as we avoid this blurring of thought, even clichés can tell us something about the career of Taoism that we ignore at our peril.

---

93. For a cogent statement of this view see Needham 1973.
94. Nakayama 1973.

VII

TAOISM AND SCIENCE

Great Scientists

It should be possible to circumvent the shortcomings of philosophical and other doctrines as a guide to the attitudes of particular human beings by examining the affiliations of individual scientists. That is what I now propose to do. In this section I will examine the careers of a group of men and women consistently numbered among the greatest scientists, physicians, and practitioners of technology in traditional China. I will ask how many were Taoists, and in what sense.

The table on pp. 51–54 includes all those represented in the two best general collections of scientific biographies, supplemented by a few outstanding names chosen from two more specialized collections of scientific biographies, one on medicine and one on Sung and Yuan mathematics and astronomy.[96] The table thus includes those considered by specialists to be of the first rank in the history of science, and an assortment of others nearly as famous. There is bound to be a great deal of arbitrariness in the composition of any reasonably short list. Nevertheless, after a number of mental experiments in substituting names, I believe that a wide range of changes would not greatly affect the outcome.

The tastes of modern Chinese scholars are, of course, different from those of most historians of European science. This list is more catholic in its range of pursuits and social backgrounds than what one would expect to see in a corresponding list devoted to Western science, medicine, and technology for the past 2500 years, or for that matter in any list based on the preferences of Chinese historians before modern times. That breadth recommends this selection for my purpose.

The table provides data on a far from simple issue. When we ask which of a group of scientific figures were Taoist, what definition of "Taoist" makes sense? If we admit every sense habitually used by Sinologists, the interest in technical matters that all these people shared would obviate biographical study. A Buddhist patriarch and a career bureaucrat who belonged to an orthodox school of Confucianism would be equally Taoist.

The obvious alternative is the narrow criterion of initiation. It lets us know, at least, what we are dealing with. Still, the simplicity of the criterion may distract attention from less clear-cut but no less interesting characteristics shared by many scientists. I have adopted neither extreme, but have attempted in investigating the

---

95. I, 5, 8.

96. Li Kuang-pi 李光璧 & Ch'ien Chün-yeh 錢君曄 1955; Li Yen 李儼 1963; Jen An 靭庵 1963; Ch'ien Pao-ts'ung 钱宝琮 et al. 1966.

careers of these thirty-nine people to look for any obvious pattern.

In the brief summaries of careers and contributions, I have characterized each scientist by occupation or intellectual concern, and specified one or two well-known achievements. The table pays special attention to formal religious affiliations and to civil service careers. It notes biographies in the Standard Histories as indications of conventional success. To indicate stereotypes, it records the labels (such as technician, *fang-chi*) under which certain biographies are grouped. It notes accounts in treatises on immortals in the Taoist Canon, not as proof of initiation, but to signify that the individual has played a role in the ideology of immortality. What that role was can be determined only by studying each account.

The most striking result to emerge from the table is that most of the people listed were regular civil servants. At least fifteen of the thirty-nine had full careers in the bureaucracy. Other careers were interrupted by dynastic transition or personal choice. On the other hand, only four subjects are accorded biographies in the various "lives of the immortals." Three of these four, Ko Hung, T'ao Hung-ching, and probably Sun Ssu-mo as well, were Taoist initiates. One, Hua T'o, was a legendary physician and surgeon whose marvelous exploits the hagiographies merely retell from the Standard Histories.[97]

Ko, T'ao, and Sun are conspicuously connected with laboratory alchemy as well as medicine. The connection between the two fields is not surprising, since the materials, implements, and methods of alchemy were largely derived from pharmacy. Claims that alchemy made important contributions to medicine are generally expressions of faith, not based on documented comparisons.

Hardly less striking is the unimportance of links with more ambiguous senses of "Taoist." For few who were not initiates does some indication of an interest in Taoist philosophy or religion emerge from the biographical information. When it does, its significance is regularly ambiguous.

There is a famous anecdote from I-hsing's childhood about an encounter with the Taoist master Yin Ch'ung 尹崇 at the Abbey of the Dark Capital (Hsuan-tu Kuan 玄都觀) in Ch'ang-an. Yin was amazed at how quickly the boy memorized the Canon of Supreme Mystery (written, as Chap. III indicates, squarely in the interest of Confucian orthodoxy). Tsu Ch'ung-chih compiled commentaries on the *Lao-tzu* and *Chuang-tzu*, hardly eccentric even for a conventional literatus in

---

97. Of the three biographies of Hua in the *Tao tsang*, that in *Li shih chen hsien t'i tao t'ung chien* 歷世眞仙體道通鑑, 20: 10a–10b, and the second in *San tung ch'ün hsien lu*, 8: 4b, copy the *Hou Han shu*, cited in the table. The first biography in *San tung ch'ün hsien lu*, 1: 20b–21a, copies the *San kuo chih* version. For the Indian origins of the Hua T'o legend see Ch'en Yin-k'o 1930. For Sun, see Chap. VI, n. 18.

VII

FAMOUS CHINESE SCIENTISTS, PHYSICIANS, AND TECHNOLOGISTS

NAME	DATES	BIOGRAPHIES IN HISTORIES	TYPE	BIOS. IN TT	TAO IST?	REFERENCES	CAREER
Ch'in Yüeh-jen 秦越人 (Pien Ch'üeh 扁鵲)	fl. ca. 501 B.C.?	Shih chi, 105	shared	none	no	Li 7–9	Physician, deified; case records attributed to him survive
Kung-shu Pan 公輸般 or Lu Pan 魯班	born after 470?	none	none	none	no	Li 1–6	Semi-legendary artisan and inventor, deified
Li Ping 李冰	fl. ca. 250	none	none	none	no	IV. 3, 288–296; Li 11–14	Career official; he and son built Kuan-hsien (Szechwan) water transport and irrigation system; deified
Ts'ai Lun 蔡倫	ca. A.D. 57–ca. 121	Hou Han shu, 68	eunuchs	none	no	Li 15–17	Eunuch, career official, classicist, ennobled; invention of paper attributed to him
Chang Heng 張衡	78–139	Hou Han shu, 49	own	none	no	Sun 1935; Lai 1956	Career official, inventor, instrument-maker, astronomer, cosmologist, poet
Hua T'o 華佗	ca. 145–ca. 208	Hou Han shu, 72B; San kuo chih, Wei, 29	technicians (fang-shu 方術, fang-chi 方伎)	2	no	Li 37–41	Legendary physician; prodigious feats of surgery, diagnosis, use of anesthesia
Chang Chi 張機 (Chung-ching 仲景)	ca. 150–ca. 219	none	none	none	no	Li 31–36	Physician; systematic approach to therapy
Ma Chün 馬鈞	fl. 220/250	none	none	none	no	IV. 2, 39–42, 158; Li 42–47	Inventor; south-pointing chariot, ballistae, looms, water-raising machinery, mechanical puppets, etc.
P'ei Hsiu 裴秀	224–271	Chin shu, 35	shared	none	no	Li 48–53	Career official, cartographer, ennobled
Liu Hui 劉徽	active before 263	none	none	none	no	Ho in DSB; Wagner 1975: 4	Mathematician, career unknown

VII

Name	Dates	Source	Category	Number	Biography?	Reference	Description
Ko Hung 葛洪 (Pao-p'u-tzu 抱樸子)	283–343	Chin shu, 72	shared	7	yes	Li 55–59	Official, wrote books he called Taoist and Confucian, ennobled; medical author; amateur of alchemy, immortality
Tsu Ch'ung-chih 祖沖之	429–500	Nan Ch'i shu, 52; Nan shih, 72	litterateurs	none	no	Li 61–72	Official, mathematician, astronomer, mechanician, classicist
T'ao Hung ching 陶弘景	456–536	Liang shu, 51; Nan shih, 76	recluses (ch'u-shih 處士; yin-i 隱逸)	many	yes	Mugitani 1976; Strickmann 1978a	Official, polymath, medical scholar, creator of Mao Shan movement with imperial patronage
Li Tao-yuan 酈道元	465 or 472–527	Wei shu, 89; Pei shih, 27	harsh officials; shared	none	no	Li 73–78	Career official, geographer, classicist
Chia Ssu-hsieh 賈思勰	fl. 532/555	none	none	none	no	Amano 1978	Official, agriculturalist
Li Ch'un 李春	fl. ca. 605	none	none	none	no	IV. 3, 175–178; Li 97–99	Career unknown; planned and built famous Hopei bridge
Sun Ssu-mo 孫思邈	alive 673	Chiu T'ang shu, 191; Hsin T'ang shu, 196	technicians; recluses (yin-i)	6	probably	Sivin 1968: 81–144	Physician and alchemist, courtier, legendary figure in both Taoist and Buddhist writing
I-hsing 一行 (secular name Chang Sui 張遂)	682/683–727	Chiu T'ang shu, 191	technicians	none	no	Li 1965; Ang 1979	Ch'an monk, then Tantric patriarch, astronomical official; astronomy, mathematics, clock design, meridian survey
Wang Ping 王冰	ca. 710–805	none	none	none	no	Okanishi 1958: 5	Medical scholar, military and civil administrator
Pi Sheng 畢昇	fl. 1041/1048	none	none	none	no	Li 119–122	Commoner; movable-type printing
Su Sung 蘇頌	1020–1101	Sung shih, 340	shared	none	no	Needham, Wang & Price 1960: 5–9; Li 135–144	Career official, prime minister, polymath; planned astronomical clock, compiled materia medica
Shen Kua 沈括	1031–1095	Sung shih, 331	shared, appended	none	no	Holzman 1959; Sivin 1975	Career official, polymath; contributions to every field
Li Chieh 李誡	d. 1110	none	none	none	no	Yetts 1927; Yü 1959	Career official, polymath, artist; architectural standards

VII

TAOISM AND SCIENCE

Name	Dates	Source	Category	?	References	Taoist?	Description
Liu Wan-su 劉完素	b. ca. 1110	*Chin shih*, 131	technicians	none	Jen 75–80; Rall 1970: 38–52	no	Physician, deified
Chang Ts'ung-cheng 張從正	ca. 1156–ca. 1228	*Chin shih*, 131	technicians	none	Jen 81–85; Rall 1970: 52–58	no	Physician, briefly Imperial Physician
Li Kao 李杲 (Tung-yuan 東垣)	ca. 1180–1251	*Yuan shih*, 203	technicians	none	Jen 86–90; Rall 1970: 58–69	no	Physician, wealthy
Li Yeh 李冶 (or Li Chih 李治)	1192?–1279?	*Yuan shih*, 160; *Hsin Y. s.*, 171	shared	none	Ch'ien 1966: 104–148; Ho in DSB	no	Mathematician, man of letters, high exam rank, briefly official
Ch'in Chiu-shao 秦九韶	1202–1261?	none	none	none	Ch'ien 1966: 60–103; Libbrecht 1973: 22–30	no	Mathematician, military and then civil official, corrupt
Yang Hui 楊輝	fl. ca. 1261/1275	none	none	none	Ch'ien 1966: 149–165; Lam 1977	no	Mathematician, official
Kuo Shou-ching 郭守敬	1231–1316	*Yuan shih* 164; *Hsin Y.s.* 171	shared	none	Li 1966	no	Career official, astronomer, hydraulic engineer
Chu Shih-chieh 朱世傑	fl. ca. 1280–1303	*Hsin Yuan shih*, 171	shared, appended	none	Ch'ien 1966: 166–209; Ho in DSB	no	Mathematician, teacher of mathematics
Huang Tao-p'o 黃道婆	fl. ca. 1296	none	none	none	Li 165–170	no	Entrepreneur, deified; cotton manufacture
Wang Chen 王禎	fl. 1295/1300	*Ming shih*, 289	loyal officials	none	Li 151–155; Amano 1967	no	Model official, inventor, scholar of agriculture
Chu Chen-heng 朱震亨 (Tan-hsi 丹溪)	ca. 1281–1358	*Yuan shih*, 189	Confucians	none	Jen 91–96; Rall 1970: 69–95	no	Physician; initiate of Chu Hsi school
Li Shih-chen 李時珍	1518–1593	*Ming shih*, 299	technicians	none	Chang 1954; Sivin 1973	no	Physician, man of letters, briefly medical official
Hsu Kuang-ch'i 徐光啓	1562–1633	*Ming shih*, 251	shared	none	ECCP 316–319	no	High official, pious Christian, astronomer, translator, agriculturalist
Hsu Hung-tsu 徐宏祖 (Hsia-k'o 霞客)	1586–1641	none	none	none	Li Chi 1974: 13–22	no	Explorer, geographer

Sung Ying-hsing 宋應星	1587–ca. 1665	none	none	no	Career official until Ch'ing, technological encyclopedist, cosmologist
Minggantu 明安圖	ca. 1692–1763	none	none	no	Mongol, career astronomical official, mathematician, cartographic surveyor

Notes

1. Sources cited without dates: DSB = Gillispie 1970–1990, s.v.; ECCP = Hummel 1943–1944, s.v.; Li = Li Yen 1959; Jen = Jen An 1963.

2. To indicate type of biography in the Standard Histories, "shared" means that the biography takes up part of a *chüan* labelled by the names of the subjects rather than by type, an indication of prestige. "Appended" indicates that the subject's biography is appended to that of a more famous relative or teacher. The titles of biographies grouped by type are noted.

3. Only the number of biographical accounts in the *Tao tsang* is noted; for full references see Weng 1935.

4. Only one or two citations are given here, with preference for a scholarly monograph if available and Li 1959. A reference to the latter alone usually means that my characterization is based mainly on primary sources.

5. In dates, the form "1210/1215" means "some time between 1210 and 1215."

6. Persons listed here were chosen because they are generally known, and appear regularly in biographical collections. Sources for more detailed studies include Chuang Wei-feng 1989 (2300 astronomers), Li Ching-wei 1988 (6000 physicians), and Li Yun 1989 (10,500 physicians) or, for local studies, to mention only two on Kiangsu, Hsu Po-ch'un 1983 and Ch'en & Hsieh 1985.

the late fifth century. Su Sung produced a critical edition of the *Huai-nan-tzu*.

But these involvements must be weighed against other themes in each career. I-hsing was committed to the life of a Buddhist monk by the time he reached maturity. Tsu Ch'ung-chih also annotated the Confucian Analects and the Canon of Filial Piety. Su Sung's literary remains testify to his overwhelming activity on behalf of "state Confucianism" (i.e., the imperial cult), including its religious aspect. His prefaces to several Buddhist writings survive. Men of letters were seldom forced to choose between religious or philosophic enthusiasms. These data remind us that the two Taoist classics were an integral part of high culture, and that curious people tend to be curious about a great many matters.[98]

In this group every instance of a Taoist involvement short of initiation is counterbalanced by involvements that can be stereotyped as Confucian or Buddhist. It is surprising that no peripheral but clearly partisan Taoist affiliations have surfaced among these thirty-nine figures. No doubt less superficial research than mine and that of my secondary sources will reveal a few. At the moment practically nothing is known about the lives of several people on the list. But it will take more than a few such affiliations to justify concluding from "Taoist" quotations, encounters, or book titles that the people involved were empiricists, agnostic naturalists, democrats, etc.

The pattern of biographies in the Standard Histories supports the hypothesis about *fang-shih* advanced above (p. 30). Twenty-three of the thirty-nine scientists have biographies in the Standard Histories. Only one of these, Chang Heng, has a chapter of his own. Ten share an unlabeled chapter with others, which indicates uncomplicated elite status. Of the various labels under which the biographies of thirteen people fall, only "technicians" *(fang-chi, fang-shu)* and "recluses" recur. The rubrics that occur only once include eunuchs, litterateurs, harsh officials, loyal officials, and Confucian scholars.[99]

A bias in the biographic approach works to the detriment of those who did

---

98. For the texts that Tsu annotated see his two official biographies. The preface to Su Sung's ed. of the *Huai-nan-tzu,* compiled from recensions in the imperial library and the collection of his own family, still exists. It is entirely concerned with bibliographic detail, and expresses not a hint of interest in the content of the book *(Su Wei-kung wen chi* 蘇魏公文集, *66:* 7a-8b). For his prefaces to Buddhist writings see *67:* 7b-12b. His interest in the subject matter emerges clearly in these. On Su see also Weng Fu-ch'ing 1986.

99. For Chu Chen-heng's prominent position in the orthodox line descended from Chu Hsi 朱熹 (1130-1200), see *Sung Yuan hsueh an* 宋元學案, *82:* 39a. The biographical notice there asserts that Chu studied medicine as the best means to practice the ethical principles he learnt from his teacher Hsu Ch'ien 許謙 (1199-1266).

not follow conventional careers, including Taoist and Buddhist clerics. We are most reliably and usually most fully informed about the lives of the regular bureaucrats, the "Confucians" of current clichés. For instance, among the most important alchemical authors were the unknown author of the *Chou i ts'an t'ung ch'i* 周易參同契 (second century A.D. or later) and Ch'en Shao-wei 陳少微 (fl. ca. 712?). We know nothing about the life of either, so they do not appear in the biographical collections. Medicine is not badly covered, since it was practiced at every level of society, but mathematics and astronomy are no doubt underrepresented. Siting ("geomancy")[100] does not appear at all, due in part to the same paucity of biographical information, and in part to its modern official classification as superstition. The names of most inventive craftsmen and engineers are not recorded.

Even when due regard is paid to this bias, the conclusion is inescapable. The only regularity conspicuous among scientists is official position. With a very few exceptions, neither the texts of the Taoist philosophers nor the allegiances demanded by the Taoist religion played a dominant role in the lives of the best-known scientists, physicians, and technologists. The sources disclose the usual concomitants of office, aristocratic status or family traditions of prestige, the privileged upbringing that marked the highly educated. The Tantric patriarch and all three Taoists belonged to great families with long office-holding traditions. I-hsing, T'ao, and Sun were intimates of emperors, and Ko an ennobled military and civil officer. These were hardly resolute enemies of "feudal bureaucratism."

Conclusions

This biographical investigation supports the conclusion based earlier on a critical examination of Needham's hypotheses. There is no evidence for any regular and necessary link between Taoism and science that will let us predict, given an individual's affiliation to one Taoism or another, that we will find attitudes friendly to scientific investigation; nor, given an individual's involvement in science, technology, or medicine, that we will find Taoist motivations.

This is true whether we consider the philosophy or the religion. The other dozen or so "Taoisms" frequent in Sinological publications merely confuse the issue. Previous hypotheses have relied heavily on the "Taoism" of the *fang-shih*. As we have seen, that label and its synonyms tell us only that someone has mastered an art and that the writer does not consider him a peer. They tell us nothing about his alignment with any other Taoism.

---

100. Bennett 1978.

VII

This clearing of the air would have been a great deal more difficult had it not been for the scholarly care with which Needham built and documented, on his usual bold scale, the most substantial set of hypotheses about the relations of Taoism and science. This view, formed in the 1950's, reflected the conventions of the time. By his participation in the meetings that formed modern studies of Taoism, Needham also encouraged the emergence of a more adequate view.

What part of his argument is likely to survive for more than heuristic purposes? The part that is at the same time most subtle and most central, and perhaps most likely to be overlooked as studies of Taoism become more specialized.

I remarked earlier that Needham means most fundamentally by "Taoism" a spirit that infuses science, that defines a timeless fellowship, curious, skeptical, honoring experience above authority, valuing Nature in its own right, willing to unite the manual and the intellectual, and so on. Believing that this spirit is necessary for science to flourish, he discovered that the *Chuang-tzu* and the *Lao-tzu* expressed some of these themes for the first time. Not every aspect, to be sure, but more than are found in other classics before the Han synthesis. This perception is true. That is obvious when we find the great physicist Hideki Yukawa reminiscing that the *Chuang-tzu* taught him serendipity. But it is true only so long as, like Yukawa, we read the *Chuang-tzu* as a timeless classic.[101]

Why, then, are we unable to trace the spirit in the flesh? Why do particular arguments about "the Taoists" in the history of science after 250 B.C. turn out to be so equivocal? Why do we not find these Taoist notions playing an overwhelming role in the particular motivations and imaginations of most great scientists?

These philosophies represent a short and in some senses a transitory interval in the long history of China's encounter with the Way. The *Chuang-tzu* and *Lao-tzu*, two very different statements that began the tradition, unite popular beliefs and mystical striving with profound esthetic impulses. To Chinese readers of later centuries they were, rather than the inception, two culminations of an archaic view of man and Nature. All but a few critical scholars believed that Lieh-tzu, Kuan-yin-tzu, and their ilk were also archaic Taoist philosophers. Only in recent times are they generally understood to be late imitations and extensions of the two pre-Han books.

Just as the philosophers of the Tao drew on early popular traditions, well before the Later Han they had ceased to speak—if indeed they ever did—to a restricted audience of "Taoists." They were integrated—along with the Book of Changes and other "Confucian" writings—into a spectrum of comprehensive

---

101. Yukawa 1973: 64–69.

world-views to which every educated person had access. They were ceasing to be uniquely Taoist at just the moment when Ssu-ma T'an's effusion about a "Taoist school" was first suggesting to many generations of careless readers that there had been a single Taoist philosophy.

To the educated reader from the Han on, the *Lao-tzu* was not the spearhead of a movement that might well strike terror in the mind of a feudal bureaucrat. It was a text that spoke to everyone, not only to aspiring mystics but to careerists and indeed to emperors. To the Taoist initiate it was not a supreme canon, and did not proclaim opposition to the established order. It became one among many epochal divine revelations. It became a mighty spell when recited. Imperial cults used the book from the second century B.C. on to proclaim that the established order of society was in fact the promised millennium.

Needham has left us with some intriguing possibilities for the sciences that opened up during the philosophic moment of Taoism. But these possibilities have nothing to do with Taoism's eighteen centuries as a cluster of religious traditions the goal of which was communal or personal salvation. Nor do they concern Taoist initiates as persons joined in making history. How in studying the influence of Taoism on science can we ignore either the goal or the agents?

Needham's survey also does not prove that science influenced Taoism, in any historically significant sense of either word. He has clearly documented his claim, allowing the conclusion that if we mean by science more than general cosmological perspectives, such an influence rests unproven and will be difficult to prove.[102]

Prospects

The new history of Taoism as a specialized outgrowth of popular religion has begun to reveal concrete linkages to science in specific circumstances. Ultimately these will disclose general patterns. An accurate and thorough inventory of ignorance is an obvious first step. It is thus fitting to admit how poorly we understand both the early Taoist masterpieces and the many religious movements as forces in history. Once we are no longer distracted by questions about what this or that ism contributed to the evolution of this or that field in traditional China, we can proceed with less abstract and more open-ended research questions that will eventually lead to a third history.

---

102. If on the other hand we take "science" to mean nothing more precise than general cosmological perspectives, their role in shaping liturgy has already been proven. See, e.g., Strickmann 1979 and Schipper & Wang 1986.

References

Early Chinese Sources

Classics not listed here are cited from the texts in standard concordances. Standard Histories are cited in the Chung Hwa Book Co. 1974 series. Treatises in the Taoist Canon are cited from the Commercial Press ed. by vol. number (TT) and item number in Schipper 1975 (S).

Ch'üan Hou Han wen 全後漢文 (Complete Essays of the Later Han Period). In *Ch'üan shang ku san tai Ch'in Han San kuo Liu ch'ao wen* 全上古三代秦漢三國六朝文, compiled by Yen K'o-chün 嚴可均, 1879. Reprint of 1894.

Ch'un-chiu fan lu 春秋繁露 (Abundant Dew on the Spring and Autumn Annals). Tung Chung-shu 董仲舒, parts written 156/130 B.C. *Ch'un-chiu fan lu i cheng* 春秋繁露義證.

Huai-nan-tzu 淮南子 (Book of the King of Huai-nan). Compiled under patronage of Liu An 劉安, King of Huai-nan, presented to the throne 139 B.C. *Huai-nan hung lieh chi-chieh* 淮南鴻烈集解.

Huang-ti nei ching 黃帝內經 (Inner Canon of the Yellow Lord). Anonymous, probably first century B.C. The *Su wen* 素問 (Basic questions), ed. by Wang Ping 王冰, 762, is cited from Jen Ying-ch'iu 1986 by *p'ien*, *chang*, line, and page number, and the Wang commentary by *chüan* and page number from the 1931 Commercial Press typeset ed.; reprint, Shanghai, 1955.

I hsueh ju men 醫學入門 (Introduction to Medical Studies). Li Ch'an 李梴, 1575. 1636 ed.

Kuan-yin-tzu 關尹子 (The Writings of the Gatekeeper). Anonymous, probably Southern Sung. TT 347, S 667, under title *Wu shang miao tao Wen shih chen ching* 無上妙道文始眞經. Cited by *p'ien* and page. On date see Jen Chi-yü 1991: 475, item 662.

Li shih chen hsien t'i tao t'ung chien 歷世眞仙體道通鑑 (Comprehensive Mirror of the Embodiment of the Way by immortals through the ages). Chao Tao-i 趙道一, ca. 1300. TT 142, S 296.

Lin-ch'uan chi 臨川集 (Short Writings of Wang An-shih). Compiled by Chan Ta-ho 詹大和, 1140. Revised ed. entitled *Wang An-shih ch'üan chi* 王安石全集, Taipei, 1974.

Lü shih ch'un-ch'iu 呂氏春秋 (Springs and Autumns of Master Lü). Compiled under patronage of Lü Pu-wei 呂不韋, ca. 239 B.C. In Ch'en Ch'i-yu 1984. Cited by *chüan*, section, *p'ien*, and page.

ably A.D. 70/80. In Huang Hui 1938.

Mao shan chih 茅山志 (Gazetteer of Mt. Mao). Liu Ta-pin 劉大彬, by 1326. TT 153–158, S 304.

Pao-p'u-tzu nei p'ien 抱樸子內篇 (Inner Writings of the Master who Embraces Simplicity). Ko Hung 葛洪, completed ca. A.D. 320. In *P'ing-chin-kuan ts'ung-shu* 平津館叢書.

Pao-p'u-tzu wai p'ien 抱樸子外篇 (Outer writings of the Master who Embraces Simplicity). Ko Hung, completed ca. A.D. 335. *Idem*.

Pen-ts'ao ching chi chu 本草經集注 (Canon of Materia Medica with Collected Annotations). T'ao Hung-ching 陶弘景, ca. A.D. 492. Shanghai, 1955, reprint of incomplete Tunhuang MS from Six Dynasties or T'ang.

San tung ch'ün hsien lu 三洞群仙錄 (Record of the Immortals in the Tripartite Canon). Ch'en Pao-kuang 陳葆光, by 1154. TT 992–995, S 1248.

Su Shen liang fang 蘇沈良方 (Good Prescriptions by Su and Shen). Compiled anonymously, 1141/1151. Peking, 1956, photographic reprint of late eighteenth-century ed. Conflates *Liang fang* 良方 (1088/1095) by Shen Kua 沈括 and miscellaneous medical notes by Su Shih 蘇軾. See Vol. I, III 47. Dating based on unpublished study by Chin Cheng-yao 金正耀.

Su Wei-kung wen chi 蘇魏公文集 (Collected Literary Works of Su Sung). Su Sung 蘇頌, ed. Su Hsi 蘇攜, 1139. Critical ed., Chung Hwa Book Co., 1988.

Sung hui yao 宋會要 (Collected Essentials of Sung Institutions). Reconstituted fragments of portion by Li Hsin-chuan 李心傳 (1166–1243), date unknown. Taipei, 1957, entitled *Sung hui yao chi pen* 宋會要輯本.

Sung Yuan hsueh an 宋元學案 (Studies of Scholarly Lineages of the Sung and Yuan periods). Huang Tsung-hsi 黃宗羲 et al., 1755. *Tseng pu* 增補 *Sung Yuan hsueh an*, in *Ssu pu pei yao*.

T'ai-p'ing ching 太平經 (Canon of Great Peace). Anonymous. Most scholars argue that the extant version originated in a secular revelation meant to support the ruling dynasty, probably in the later Han; was adapted in the Six Dynasties to describe a Taoist theocratic community; and probably reached its current form late in the sixth century. Mansvelt Beck 1980 and Zürcher 1980: 145 see it as the original Han work. Critical ed. in Wang Ming 1960. Cited by *chüan*, *p'ien*, and page.

T'ai-p'ing yü lan 太平御覽. Compiled by Li Fang 李昉 et al., 983. Chung Hwa Book Co., 1960, facsimile of Sung ed.

Tao-chiao i shu 道教義樞 (Pivot of Meanings in the Teachings of the Way). Meng An-p'ai 孟安排, ca. 700. TT 762–763, S 1129.

Tao-chiao ling yen chi 道教靈驗記 (Record of Remarkable Confirmations of the Teachings of the Way). Tu Kuang-t'ing 杜光庭, after 905. TT 325–326, S 590.

T'ieh wei shan ts'ung t'an 鐵圍山叢談 (Jottings from the Iron Mountains that Surround the World). Ts'ai T'ao 蔡絛, ca. 1130. By the son of Ts'ai Ching 蔡京, with predictable biasses.

Tzu chih t'ung chien 資治通鑑 (Comprehensive Mirror [of History] for Aid in Government). Ssu-ma Kuang 司馬光, 1085/1086. Peking 1956 ed.

Wang An-shih ch'üan chi 王安石全集. See *Lin-ch'uan chi*.

Wu ching i i 五經異義 (Alternative Intepretations of the Five Classics). By Hsu Shen 許慎, ca. A.D. 100. In *Huang Ch'ing ching chieh* 皇清經解, vols. 1248-1250.

Modern Sources

Akizuki Kan'ei 秋月觀英, editor. 1986. *Dôkyô kenkyû no susume. Sono genzô to mondaiten o kangaeru* 道教研究のすすめ。その現狀と問題點を考える (Advances in Taoist Studies. Their State and their Issues). Tokyo: Hirakawa Shuppansha. Seven essays by experts, and a union list of Taoist texts outside the *Tao tsang* in Japanese libraries.

Amano Motonosuke 天野元之助. *Gen no Ô Tei Nôsho no kenkyû* 元の王禎農書の研究 (A Study of the Book of Agriculture of Wang Chen). In Yabuuchi 1967: 341–468.

Ang Tian Se. 1979. I-hsing (683–727 A.D.). His Life and Scientific Work. Ph. D. diss., Chinese, University of Malaya.

Anonymous. 1930. *Chiang-hu i-shu pi ch'uan* 江湖醫術秘傳 (Secretly Transmitted Methods of Itinerant Healers). Original probably Shanghai: Kuo-i Hsueh-she (?), ca. 1930; photographic reprints, Hong Kong: Li-li Ch'u-pan-she, n. d.; Taipei: Ta Fang Ch'u-pan-she, 1973, under title *Chiang-hu i-shu ch'üan shu* 江湖醫術全書.

Baldrian-Hussein, Farzeen. 1987. Taoism: An Overview. In *Encyclopedia of Religions*, XIV, 288–306.

Bell, Catherine. 1988. Ritualization of Texts and Textualization of Ritual in the Codification of Taoist Liturgy. *History of Religions*, 27. 4: 366–392.

Benn, Charles. 1987. Religious Aspects of Emperor Hsuan-tsung's Taoist Ideology. *Buddhist and Taoist Studies*, 127–145.

Bennett, Steven J. 1978. Patterns of the Sky and Earth: A Chinese Science of Applied Cosmology. *Chinese Science*, 3: 1–26.

VII

Bielenstein, Hans. 1976. Lo-yang in Later Han Times. *Bulletin of the Museum of Far Eastern Antiquities, 48:* 1–142.

Bokenkamp, Stephen R. 1983. Sources of the Ling-pao Scriptures. In Strickmann 1981–1985: II, 434–486.

Boltz, Judith M. 1989. *A Survey of Taoist Literature. Tenth to Seventeenth Centuries.* China Research Monographs, 7. Berkeley: Institute of East Asian Studies, University of California.

Chan, Wing-tsit. 1963. *A Source Book in Chinese Philosophy.* Princeton University Press.

Chang Hui-chien 張慧劍. 1954. *Li Shih-chen* 李時珍 (Biography of Li Shih-chen). Shanghai: Shang-hai Jen-min Ch'u-pan-she.

Ch'ang Pi-te 昌彼德 et al. 1976. *Sung-jen chuan-chi tzu-liao so-yin* 宋人傳記資料索引 (Index to Sung Biographical Materials). 6 vols., Taipei: Ting-wen Shu-chü.

Ch'en Ku-ying 陳鼓應, editor. 1992- . *Tao-chiao wen-hua yen-chiu* 道教文化研究 (Studies of Taoist Culture). 5 vols. to date, Shanghai: Shang-hai Ku-chi Ch'u-pan-she.

Ch'en Kuo-fu 陳國符. 1949. *Tao tsang yuan-liu k'ao* 道藏源流考 (Researches in the History of the Taoist Canons). Rev. ed., 2 vols., Peking: Chung Hwa Book Co., 1963. The 1963 ed., cited here, is expanded but expurgated.

Ch'en Tao-chin 陈道瑾; Hsueh Wei-t'ao 薛渭涛. 1985. *Chiang-su li-tai i-jen chih* 江苏历代医人志 (Kiangsu Doctors through History). Nanking: Chiang-su K'o-hsueh Chi-shu Ch'u-pan-she.

Ch'en Yin-k'o 陳寅恪. 1930. San-kuo chih Ts'ao Ch'ung Hua T'o chuan yü fo-chiao ku-shih 三國志曹沖華佗傳與佛教故事 (The Biographies of Ts'ao Ch'ung and Hua T'o in the History of the Three Kingdoms and their relation to Buddhist Legends). *Ch'ing-hua hsueh-pao* 清華學報, 6. 1: 17–20. On the Indian origin of a group of Chinese themata.

Ch'ien Pao-ts'ung 钱宝琮 et al. 1966. *Sung Yuan shu-hsueh-shih lun-wen-chi* 宋元数学史论文集 (Collected Essays on the History of Mathematics in the Sung and Yuan Periods). Peking: K'o-hsueh Ch'u-pan-she.

Chin Chung-shu 金中樞. 1974. Lun pei-Sung mo-nien chih ch'ung-shang tao-chiao 論北宋末年之崇尚道教 (On Faith in the Taoist Religion at the End of the Northern Sung Period). Part I. In *Sung shih yen-chiu chi* 宋史研究集 (Collected studies in Sung history), VII, 291–392. Taipei: Chung-hua Ts'ung-shu Pien-shen Wei-yuan-hui.

Chow Tse-tsung. 1978. The Childbirth Myth and Ancient Chinese Medicine: A

Study of Aspects of the *Wu* Tradition. In Roy & Tsien 1978: 43-89.
Chuang Wei-feng 庄威凤 et al., editors. 1989. *Chung-kuo t'ien-wen shih-liao hui pien* 中国天文史料汇编 (Collected Materials for the History of Chinese Astronomy). 3 vols., 1 vol. to date, Peking: K'o-hsueh Ch'u-pan-she. Biographical notices of ca. 2300 astronomers before 1911, with pinyin index.
Dean, Kenneth. 1993. *Taoist Ritual and Popular Cults of Southeast China*. Princeton University Press.
DeWoskin, Kenneth J. 1981. A Source Guide to the Lives and Techniques of Han and Six Dynasties *Fang-shih*. *Bulletin of the Society for Studies of Chinese Religions*, 9: 79-105. Discussion of the meaning of *"fang-shih,"* and an index of pertinent anecdotes and biographies.
DeWoskin, Kenneth J. 1983. *Doctors, Diviners, and Magicians of Ancient China: Biographies of Fang-shih*. Translations from the Oriental Classics. New York: Columbia University Press. From the Standard Histories of the Three Kingdoms through the Chin.
Fang Shih-ming 方诗铭. 1993. Huang-chin ch'i-i hsien-ch'ü yü wu chi yuan-shih tao-chiao te kuan-hsi 黄巾起义先驱与巫及原始道教的关系 (The Relationship between the Early Yellow Turban Rebels, Witchcraft, and Primitive Taoism). *Li-shih yen-chiu* 历史研究 (Historical researches), 3: 3-13.
Forke, Alfred, trans. 1907-1911. *Lun-Hêng*. I. *Philosophical Essays of Wang Ch'ung*. II. *Miscellaneous Essays of Wang Ch'ung*. Mitteilungen des Seminars für orientalische Sprachen, supplementary volumes, 10, 14. 2 vols., Shanghai: Kelly & Walsh; reprint, New York: Paragon Book Gallery, 1962.
Franke, Herbert, editor. 1976. *Sung Biographies*. Münchener ostasiatische Studien, 16.1-16.3. 3 vols., Wiesbaden: Franz Steiner.
Fraser, J. T.; N. Lawrence; F. C. Haber, editors. 1986. *Time, Science and Society in China and the West*. The Study of Time, 5. Amherst: The University of Massachusetts Press.
Fukui Fumimasa. 1995. The History of Taoist Studies in Japan and Some Related Issues. *Acta Asiatica*, 68: 1-18.
Fukui Kôjun 福井康順 et al., editors. 1983. *Dôkyô* 道教 (Taoism). 3 vols., Tokyo: Hirakawa Shuppansha.
Giles, Lionel. 1935-1937. Dated Manuscripts in the Stein Collection. *Bulletin of the School of Oriental and African Studies*, University of London, 8. 1-26.
Gillispie, Charles Coulston, editor. 1970-1990. *Dictionary of Scientific Biography*. 18 vols., New York: Charles Scribner's Sons.

Goodrich, L. Carrington, editor. 1976. *Dictionary of Ming Biography. 1368–1644.* 2 vols., New York: Columbia University Press.

de Groot, J. J. M. 1892–1910. *The Religious System of China. Its Ancient Forms, Evolution, History and Present Aspect, Manners, Customs and Social Institutions Connected Therewith.* 6 vols., Leiden: E. J. Brill.

Graham, A. C. 1981. *Chuang-tzu. The Seven Inner Chapters and Other Writings from the Book Chuang-tzu.* London: George Allen & Unwin.

Hang-chou Ta-hsueh Li-shih-hsi Sung Shih Yen-chiu-shih 杭州大学历史系宋史研究室 (Seminar on Sung History, Department of History, Hangchow University), editors. 1986. *Sung shih yen-chiu chi-k'an* 宋史研究集刊 (Collected Researches on Sung History). Hangzhou: Che-chiang Ku-chi Ch'u-pan-she.

Ho Peng Yoke. 1991. Chinese Science: the Traditional Chinese View. *Bulletin of the School of Oriental and African Studies,* 54. 3: 506–519.

Holzman, Donald. 1959. Shen Kua and his Meng-ch'i pi-t'an. *T'oung Pao* (Leiden), 46: 260–292.

Hsieh Kuan 謝觀 (Li-heng 利恆). 1935. *Chung-kuo i-hsueh yuan-liu lun* 中國醫學源流論 (On the History of Chinese Medicine). In *Hsieh Li-heng hsien-sheng ch'üan shu* 謝利恆先生全書, vol. I. Shanghai: Ch'eng Chai I She.

Hsu Po-ch'un 徐伯春. 1983. *Chiang-su ku-tai k'o-hsueh-chia* 江苏古代科学家 (Ancient Scientists of Kiangsu). Nanking: Chiang-su K'o-hsueh Chi-shu Ch'u-pan-she.

Huang Hui 黃暉. 1938. *Lun heng chiao shih* 論衡校釋 (Critical Edition of Discourses Weighed in the Balance). Reprint with marginalia by Hu Shih 胡適, Taipei: Commercial Press, 1964.

Hummel, Arthur W., editor. 1943–1944. *Eminent Chinese of the Ch'ing Period (1644–1912).* 2 vols., Washington: U.S. Government Printing Office.

Jen An 靭庵, editor. 1963. *Chung-kuo ku-tai i-hsueh-chia* 中國古代醫學家 (Ancient Chinese Physicians). Hong Kong: Shanghai Book Co.

Jen Chi-yü 任继愈, editor. 1990. *Chung-kuo tao-chiao shih* 中国道教史 (A History of Taoism in China). Shanghai: Shang-hai Jen-min Ch'u-pan-she.

Jen Chi-yü, editor. 1991. *Tao tsang t'i-yao* 道藏提要 (Abstracts from the Taoist Canon). Peking: Chung-kuo She-hui-k'o-hsueh Ch'u-pan-she. Notes on contents and composition of each text.

Jen Ying-ch'iu 任應秋, editor. 1986. *Huang-ti nei ching chang-chü so-yin* 黃帝內經章句索引 (Phrase Index to the Inner Canon of the Yellow Lord). Beijing: Jen-min Wei-sheng Ch'u-pan-she. Includes not only phrases but technical terms, with a good text in old-style characters divided logically into sections.

Kaltenmark, Max. 1979. The Ideology of the *T'ai-p'ing ching*. In Welch & Seidel 1979: 19–52.

Kleeman, Terry F. 1994. *A God's Own Tale. The Book of Transformations of Wenchang, the Divine Lord of Zitong*. State University of New York Series in Chinese Philosophy and Culture. Albany: State University of New York Press.

Knoblock, John. 1988–1994. *Xunzi. A Translation and Study of the Complete Works*. 3 vols., Stanford University Press. Study and translation of *Hsun-tzu*.

Kobayashi Masayoshi. 1995. The Establishment of the Taoist Religion *(Tao-chiao)* and its Structure. *Acta Asiatica, 68:* 19–36.

Kracke, Edward A., Jr. 1953. *Civil Service in Early Sung China. 960–1067. With Particular Emphasis on the Development of Controlled Sponsorship to Foster Administrative Responsibility*. Monograph Series, 13. Cambridge, MA: Harvard-Yenching Institute.

Lai Chia-tu 賴家度. 1956. *Chang Heng* 張衡 (Biography of Chang Heng). Shanghai: Shang-hai Jen-min Ch'u-pan-she.

Lam Lay Yong. 1977. *A Critical Study of the Yang Hui Suan Fa. A Thirteenth-century Chinese Mathematical Treatise*. Singapore University Press.

Lau, D. C., translator. 1970. *Mencius*. Harmondsworth: Penguin Books.

Legge, James. 1885. *The Texts of Confucianism*. II. *The Li Ki*. Sacred Books of the East, 27–28. 2 vols., Oxford University Press.

Li Chi. 1974. *The Travel Diaries of Hsü Hsia-k'o*. Hong Kong: The Chinese University Press.

Li Ching-wei 李经纬, editor. 1988. *Chung-i jen-wu tz'u-tien* 中医人物辞典 (Biographical Dictionary of Traditional Chinese Medicine). Shanghai: Shang-hai Tz'u-shu Ch'u-pan-she. Ca. 6000 notices, including ca. 800 after 1911. Indexes to names, including variants, and books.

Li Kuang-pi 李光璧; Ch'ien Chün-yeh 錢君曄, editors. 1955. *Chung-kuo k'o-hsueh chi-shu fa-ming ho k'o-hsueh chi-shu jen-wu lun-chi* 中國科學技術發明和科學技術人物論集 (Collected Essays on Chinese Scientific and Technological Invention and Scientific and Technological Personnel). Peking: San-lien Shu-chü.

Li Ti 李迪. 1965. *T'ang tai t'ien-wen-hsueh-chia Chang Sui (I-hsing)* 唐代天文学家张遂(一行) (The T'ang Astronomer Chang Sui [I-hsing]). Shanghai: Shang-hai Jen-min Ch'u-pan-she.

Li Ti. 1966. *Kuo Shou-ching* 郭守敬 (Biography of Kuo Shou-ching). Shanghai: Shang-hai Jen-min Ch'u-pan-she.

Li Ti. 1978. *Meng-ku-tsu k'o-hsueh-chia Ming-an-t'u* 蒙古科学家明安图 (The Mongol Scientist Minggantu). Hohhot: Nei-meng-ku Jen-min Ch'u-pan-she.
[Li Yen 李儼, editor.] 1959. *Chung-kuo ku-tai k'o-hsueh-chia* 中國古代科學家 (Ancient Chinese scientists). Peking: K'o-hsueh Ch'u-pan-she.
Li Yun 李云 et al., editors. 1988. *Chung-i jen-ming tz'u-tien* 中医人名辞典 (Biographical Dictionary of Traditional Chinese Physicians). Peking: Kuo-chi Wen-hua Ch'u-pan Kung-ssu. Notices of 10,500 physicians to 1911, in stroke order.
Libbrecht, Ulrich. 1973. *Chinese Mathematics in the Thirteenth Century. The Shu-shu chiu-chang of Ch'in Chiu-shao.* M.I.T. East Asian Science Series, 1. Cambridge, MA: The MIT Press.
Lin Fu-shih 林富士. 1988. *Han-tai te wu-che* 漢代的巫者 (Han Mediums). Taipei: T'ao-hsiang Ch'u-pan-she.
Liu, James T. C. 1959. *Reform in Sung China. Wang An-shih (1021–1086) and his New Policies.* Cambridge, MA: Harvard University Press.
Liu P'an-sui 劉盼遂. 1975. *Lun heng chi chieh* 論衡集解 (Collected Annotations to Discourses Weighed in the Balance). Peking: Ku-chi Ch'u-pan-she.
Liu, Wu-chi; Irving Lo, editors. 1975. *Sunflower Splendor. Three Thousand Years of Chinese Poetry.* Bloomington: University of Indiana Press.
Mansvelt Beck, B. J. 1980. The Date of the Taiping Jing. *T'oung Pao,* 66: 149–182.
Mather, Richard B. 1979. K'ou Ch'ien-chih and the Taoist Theocracy at the Northern Wei Court, 425–451. In Welch & Seidel 1979: 103–122.
Miyakawa Hisayuki 宮川尚志. 1975. Sô no Kisô to Dôkyô 宋の徽宗と道教 (Sung Emperor Hui-tsung and the Taoist Religion). *Tôkai Daigaku kiyo* (Bungakubu) 東海大學紀要(文學部), 23: 1–10.
Miyakawa Hisayuki. 1976. Rin Reiso to Sô no Kisô 林靈素と宋の徽宗 (Lin Ling-su and Sung Emperor Hui-tsung). *Ibid.,* 24: 1–8.
Mollier, Christine. 1990. *Une Apocalypse taoïste du Ve siècle. Le Livre des incantations divines des grottes abyssales.* Mémoires, 31. Paris: Collège de France, Institut des Hautes Études Chinoises.
Mugitani Kunio 麥谷邦夫. 1976. Tô Kôkei nempo kôryaku 陶弘景年譜考略 (Critical Study of the Chronology of T'ao Hung-ching). *Tôhô shûkyô* 東方宗教, 47: 30–61, 48: 56–83.
Nakayama, Shigeru. 1973. Joseph Needham, Organic Philosopher. In Nakayama & Sivin 1973: 23–43.
Nakayama, Shigeru; N. Sivin, editors. 1973. *Chinese Science. Explorations of an*

Ancient Tradition. M.I.T. East Asian Science Series, 2. Cambridge, MA: The MIT Press. *Festschrift* for Needham.

Needham, Joseph, et al. 1954– . *Science and Civilisation in China.* 16 vols. to date. Cambridge, England: At the University Press.

Needham, Joseph. 1958. *The Development of Iron and Steel Technology in China.* London: The Newcomen Society.

Needham, Joseph. 1969. *The Grand Titration. Science and Society in East and West.* London: George Allen & Unwin Ltd.

Needham, Joseph. 1970. *Clerks and Craftsmen in China and the West. Lectures and Addresses on the History of Science and Technology.* Cambridge, England: At the University Press.

Needham, Joseph. 1973. The Historian of Science as Ecumenical Man: A Meditation in the Shingon Temple of Kongôsammai-in 金剛三昧院 on Kôyasan. In Nakayama & Sivin 1973: 1–8.

Needham, Joseph; Lu Gwei-djen. 1963. China and the Origin of Examinations in Medicine. *Proceedings of the Royal Society of Medicine, 56:* 63–70. Reprinted in Needham 1970: 379–395 under the title "China and the Origin of Qualifying Examinations in Medicine."

Needham, Joseph; Wang Ling; Derek J. Price. 1960. *Heavenly Clockwork. The Great Astronomical Clocks of Medieval China.* Cambridge, England: At the University Press.

Ngo Van Xuyet. 1976. *Divination magie et politique dans la Chine ancienne. Essai suivi de la traduction des "Biographies des Magiciens" tirées de l'Histoire des Han postérieurs.* Bibliothèque de l'École des Hautes Études. Sciences Religieuses, 78. Paris: Presses Universitaires de France.

Okanishi Tameto 岡西爲人. 1958. *Sung i-ch'ien i chi k'ao* 宋以前醫籍考 (Studies of Medical Books through the Sung Period). Peking: Jen-min Wei-sheng Ch'u-pan-she.

Pan Chi-hsing 潘吉星. 1990. *Sung Ying-hsing p'ing chuan* 宋应星评传 (Critical Biography of Sung Ying-hsing). Chung-kuo ssu-hsiang-chia p'ing chuan ts'ung-shu 中国思想家评传丛书 ("Critical Biography Series of Chinese Thinkers"). Nanking: Nan-ching Ta-hsueh Ch'u-pan-she.

Pokora, Timotheus, trans. 1975. *Hsin-lun (New Treatise) and Other Writings by Huan T'an (43 B.C.–28 A.D.).* Michigan Papers in Chinese Studies, 20. Ann Arbor: University of Michigan, Center for Chinese Studies.

Rall, Jutta. 1970. *Die vier grossen Medizinschulen der Mongolzeit. Stand und Entwicklung der chinesischen Medizin in der Chin– und Yüan-Zeit.* Münchener ostasiatische Studien, 7. Wiesbaden: Franz Steiner.

Reischauer, Edwin O.; John K. Fairbank. 1958. *East Asia. The Great Tradition. A History of East Asian Civilization*, I. Boston: Houghton Mifflin Company.

Robinet, Isabelle. 1984. *La révélation du Shangqing dans l'histoire du Taoïsme*. Publications, 137. Paris: École Française d'Extrême-Orient.

Robinet, Isabelle. 1991. *Histoire du Taoisme. Des origines au XIVe siècle*. Paris: Éditions Cerf.

Roy, David T.; Tsuen-hsuin Tsien, editors. 1978. *Ancient China. Studies in Early Civilization*. Hong Kong: The Chinese University Press. *Festschrift* for H. G. Creel.

Sakade Yoshinobu 坂出祥伸. 1977. Dôkyô to kagaku gijutsu 道教と科學技術 (Science and Technology in Relation to Taoism). In Sakai 1977: 199-234.

Sakade Yoshinobu. 1978. Hôjutsu ten no seiritsu to sono seikaku 方術傳の成立とその性格 (On the Origin and Character of the Biographies of Technicians). In Yamada 1978: 627-641.

Sakai Tadao 酒井忠夫, editor. 1977. *Dôkyô no sôgôteki kenky-* 道教の綜合的研究 (General Studies of Taoism). Tokyo: Kokusho kankôkai. Papers from Second International Conference on Taoism. Includes a survey of Japanese Taoist studies.

Sakai Tadao; Fukui Fumimasa 福井文雅. Dôkyô to wa nanka? Dôkyô. Dôka. Dôjutsu. Dôshi. 道教とは何か。道教, 道家, 道術, 道士 (What is Taoism? *Tao-chiao, tao-chia, tao-shu, tao-shih*). In Sakai 1977: 429-449.

Schipper, Kristofer M. 1975. *Concordance du Tao-tsang. Titres des ouvrages*. Publications, 102. Paris: École Française d'Extrême-Orient.

Schipper, Kristofer M. 1982. *Le corps taoïste. Corps physique—corps social*. L'espace intérieur, 25. Paris: Fayard.

Schipper, Kristofer M.; Wang Hsiu-huei. 1986. Progressive and Regressive Time Cycles in Taoist Ritual. In *Time, Science and Society in China and the West*. The Study of Time, 5, ed. J. T. Fraser et al., pp. 185-205. Amherst, MA: University of Massachusetts Press.

Seidel, Anna K. 1969. *La divinisation de Lao Tseu dans le Taoisme des Han*. Publications, 71. Paris: École Française d'Extrême-Orient.

Seidel, Anna K. 1974. Taoism. In *Encyclopedia Britannica*, 15th ed., s.v.

Seidel, Anna K. 1989-1990. Chronicle of Taoist Studies in the West 1950-1990. *Cahiers d'Extrême-Asie*, 5: 223-347.

Seidel, Anna K. 1990. *Taoismus: Die inoffizielle Hochreligion Chinas*. Tokyo: Deutsche Gesellschaft für Natur- und Völkerkunde Ostasiens.

Sivin, Nathan. 1968. *Chinese Alchemy: Preliminary Studies.* Harvard Monographs in the History of Science, 1. Cambridge, MA: Harvard University Press.

Sivin, Nathan. 1973. Li Shih-chen. In Gillispie 1970–1990: VIII, 390–398.

Sivin, Nathan. 1975. Shen Kua. In *ibid.,* XII, 369–393.

Sivin, Nathan. 1976. Chinese Alchemy and the Manipulation of Time. *Isis, 67:* 513–527. Reprinted in Sivin (ed.), *Science and Technology in East Asia* (New York: History of Science Publications, 1976), pp. 108–122.

Sivin, Nathan. 1988. Science and Medicine in Imperial China. The State of the Field. *Journal of Asian Studies, 47.* 1: 41–90.

Sivin, Nathan. 1989. On the Limits of Empirical Knowledge in Chinese and Western Science. In *Rationality in Question. On Eastern and Western Views of Rationality,* ed. Shlomo Biderman & Ben-Ami Scharfstein, pp. 165–189. Leiden: E. J. Brill.

Sivin, Nathan. 1995. State, Cosmos, and Body in the Last Three Centuries B.C. *Harvard Journal of Asiatic Studies, 55.* 1: 5–37.

Stein, Rolf. 1963. Remarques sur les mouvements du Taoïsme politico-religieux au IIe siècle ap. J.-C. *T'oung Pao, 50:* 1–78.

Stein, Rolf. 1979. Religious Taoism and Popular Religion from the Second to Seventh Centuries. In Welch & Seidel 1979: 53–81.

Steininger, Hans. 1953. *Hauch– und Körperseele und der Dämon bei Kuan Yin-tze.* Sammlung orientalischer Arbeiten, 20. Leipzig: Otto Harrassowitz.

Strickmann, Michel. 1974. Taoist Literature. In *Encyclopedia Britannica,* 15th ed., s.v.

Strickmann, Michel. 1977. The Mao Shan Revelations: Taoism and the Aristocracy. *T'oung Pao, 63:* 1–64.

Strickmann, Michel. 1978. The Longest Taoist Scripture. *History of Religions, 17:* 331–354.

Strickmann, Michel. 1979. On the Alchemy of T'ao Hung-ching. In Welch & Seidel 1979: 123–192.

Strickmann, Michel. 1981. *Le Taoïsme du Mao Chan. Chronique d'une révélation.* Mémoires, 17. Paris: Collège de France, Institut des Hautes Études Chinoises.

Strickmann, Michel, editor. 1981–1985. *Tantric and Taoist Studies in Honour of R. A. Stein.* Mélanges chinois et bouddhiques, 20–22. Bruxelles: Institut Belge des Hautes Études Chinoises. See esp. the Introduction to Vol. I.

Sun Wen-ch'ing 孫文青. 1935. *Chang Heng nien-p'u* 張衡年譜 (Chronology of the Life of Chang Heng). Shanghai: Commercial Press; reprint, 1956.

Tambiah, S. J. 1968. The Magical Power of Words. *Man*, n. s., *3:* 175–208.
Teiser, Stephen F. 1995. Popular Religion. *Journal of Asian Studies, 54.* 2: 378–395.
Thompson, Paul M. 1979. *The Shen Tzu Fragments.* London Oriental Series, 29. Oxford University Press.
Thorndike, Lynn. 1923–1958. *A History of Magic and Experimental Science.* 8 vols., New York: Macmillan, Columbia University Press.
Uchiyama Toshihiko 內山俊彥. 1965. Hô Keigon: mukunron no itansei 鮑敬言。無君論の異端性 (Pao Ching-yen: Anarchy as Heterodoxy). *Chûgoku no bunka to shakai* 中國の文化と社會, *12:* 1–16.
Verellen, Franciscus. 1989. *Du Guangting (850–933). Taoïste de cour à la fin de la Chine médiévale.* Mémoires, 30. Paris: Collège de France, Institut des Hautes Études Chinoises.
Verellen, Franciscus. 1995. Taoism. *Journal of Asian Studies, 54.* 2: 322–346.
Wagner, Donald B. 1975. Proof in Ancient Chinese Mathematics. Liu Hui on the Volumes of Rectilinear Solids. M. A. thesis, Chinese, Copenhagen University.
Wang Ming 王明. 1960. *T'ai-p'ing ching ho chiao* 太平經合校 (Canon of Great Peace, Critical Edition). Beijing: Chung Hwa Book Co.
Ware, James R. 1933. The *Wei shu* and the *Sui shu* on Taoism. *Journal of the American Oriental Society, 53:* 215–250. Translations of *Wei shu, 114* and part of *Sui shu.*
Watson, Burton. 1968. *The Complete Works of Chuang Tzu.* Records of Civilization. Sources and Studies. New York: Columbia University Press.
Weber, Max. 1922. *The Religion of China. Confucianism and Taoism,* trans. Hans H. Gerth. New York: The Macmillan Company, 1964. Translation from the German (Tübingen, 1922).
Welch, Holmes; Anna Seidel, editors. 1979. *Facets of Taoism. Essays in Chinese Religion.* New Haven: Yale University Press.
Weng Fu-ch'ing 翁福清. 1986. Su Sung sheng-p'ing shih-chi yen-chiu 苏颂生平事迹研究 ("About Su Song's Life Story"). In Hang-chou Ta-hsueh 1986: 266–305.
Wieger, Léon. 1917. *Histoire des croyances religieuses et des opinions philosophiques en Chine depuis l'origine jusqu'à nos jours.* Hsienhsien: Mission Press.
Wu Yiyi. 1989. Auspicious Omens and their Consequences. Zhen-Ren (1006–66) Literati's Perception of Astral Anomalies. Ph.D. diss., East Asian Studies, Princeton University.
Yabuuchi Kiyoshi 藪內清, editor. 1967. *Sô Gen jidai no kagaku gijutsu shi* 宋元時代の科學技術史 (History of Science and Technology in the Sung and Yuan

Periods). Kyoto: Kyôtô Daigaku Jimbun Kagaku Kenyûsho.

Yamada Keiji 山田慶兒, editor. 1978. *Chûgoku no kagaku to kagakusha* 中國の科學と科學者 (Chinese Science and Scientists). Kyoto: Kyôtô Daigaku Jimbun Kagaku Kenyûsho.

Yamada Toshiaki. 1995. The Evolution of Taoist Ritual. K'ou Ch'ien-chih and Lu Hsiu-ching. *Acta Asiatica*, 68: 69–83.

Yates, Robin D.S. 1994. The Yin-Yang Texts from Yinqueshan: An Introduction and Partial Reconstruction, with Notes on their Significance in Relation to Huang-Lao Daoism. *Early China*, 19: 75–144.

Yetts, W. Perceval. 1927. A Chinese Treatise on Architecture. *Bulletin of the School of Oriental and African Studies*, 4: 473–492.

Yü Chia-hsi 余嘉錫. 1958. *Ssu k'u t'i yao pien cheng* 四庫提要辨證 (Critical Notes on the Ssu-k'u Catalogue). 1937; rev. ed., Peking: Ko-hsueh Ch'u-pan-she.

Yü Ming-ch'ien 余鳴謙. 1959. Li Chieh 李誡 (Biography of Li Chieh). In Li Yen 1959: 135–139.

Yukawa, Hideki. 1973. *Creativity and Intuition. A Physicist Looks at East and West*, trans. John Bester. Tokyo: Kodansha International.

Zürcher, Erik. 1959. *The Buddhist Conquest of China. The Spread and Adaptation of Buddhism in Early Medieval China*. Sinica leidensia, 11. 2 vols., Leiden: E. J. Brill, 1959.

Zürcher, Erik. 1980. Buddhist Influence on Early Taoism. A Survey of Scriptural Evidence. *T'oung Pao*, 66: 84–147.

VII

RETROSPECT

It may seem odd to be looking back at a newly published essay, but this one, as I noted in the Introduction, was drafted in 1979. In other words, the research and writing that led to it followed closely upon that for Chap. VI. Preparing this one for publication, however, has meant a thorough rethinking. It now incorporates, and the argument is built on, a new understanding of Taoism that was not nearly so clear to any of us sixteen years ago. The analysis is now, I believe, sounder, and perhaps more interesting. Some readers will find it instructive to compare this current viewpoint with that of 1978 reflected in the last chapter.

I was also eager to finish this paper so that it could provide a summary of changes in our understanding of Taoism over the last generation, and references to the scholarship that created this largely new field. The needs of specialists have been met to a large extent by the comprehensive bibliographical survey of Seidel in 1989-1990, supplemented by Verellen 1995 (which appeared when this paper was in the final stages of revision). My dilettante's view of the new history of Taoism, and this essay's very selective citations, will complement those in the two earlier publications.

Chaps. VI and VII differ in the issues that they address. The former points out the unreflective and confusing use of terminology in conventional writing about Chinese religion. It applies to Confucianism and Buddhism as well as to Taoism. Since it was published, the prevalent reliance on vague isms has abated not at all. Such matters are generational. Perhaps Chap. VI will play a part in persuading a new generation of Sinologists and students of religion to make distinctions that any historian of ideas or sociologist would consider elementary. My modest proposal that authors simply state what they mean when they use "Taoism" and other isms will, I believe, prove its worth once it is tried.

This chapter is about not usage but history, specifically the history of science, technology, and medicine. It asks how Taoism was actually related to these enterprises, scrutinizing numerous claims made over the past century based on a large number of undefined Taoisms. Some overlap cannot be avoided. To answer such a question requires attention to terminology, just as certain historical issues had to be taken up in Chap. VI. I have minimized repetition when preparing this chapter for publication. There is in fact more overlap with Chap. VIII. The emphasis of the latter is different, however, because of its narrower topic. I trust that in all three chapters I have shown that there is a broader horizon to be seen once we step outside the comfortable perimeter of our unexamined assumptions.

VIII

Research on the History of Chinese Alchemy

Since the 1950's it has been part of folklore that the development of technical knowledge falls naturally into 'normal' and 'revolutionary' phases, and that the latter justifies the sedulous and infinitely less challenging labor of the former. But when we look at the history of science itself, the division is not so simple.

In the humanities, to which I insist the history of science belongs, revolutions are scarcely perceptible, except perhaps in the emergence of new disciplines.[1] What we see in healthy fields, rather, is a constant interplay between 'normal' processing of sources - finding materials and assessing their relevance to the enterprise, using philology to authenticate them and explain what they say, annotating and translating them as necessary - and their use toward the two ends of history: deepening our understanding of the past, and recasting what is already known to help us understand the present. The results of this use are evolutionary, not revolutionary. The historian can only admire the verve of the scientist who says, as a very eminent one once said to me, that he would question the health of his field if his own work were not outmoded within five years after publication. Very little solid work in the humanities becomes obsolete. A work worth reading is often not bettered a century after publication.

When a field stops going anywhere, the work of processing sources tends to continue at a slow rate. Some of the editing and explicating of texts may be very good indeed. The pursuit of disinterested curiosity for its own sake and, if I may be candid, for the sake of livelihood, may lead to impressive personal contributions. But old arguments are not settled, understanding does not **deepen, new issues do not transform**

---

1. Nor is the history of science an autonomous discipline. The field is defined by the application of historical methods (and insights from other fields, such as philosophy, anthropology, and the sociology of knowledge) to scientific practice and thought.

the quest, and the present is abandoned to those who specialize in the present. Nothing flows into the larger structures of understanding that, according to the charter myth of the research universities, eventually are supposed to make specialized work, no matter how narrow, useful in the learned world. One hears often from practitioners in static fields, even in the fields longest established, that it is too soon to draw any conclusions because not enough texts have become the subjects of monographs or translations. In such fields it is always too soon. The seed continues to be sown, but the fruit withers on the vine. To make a long story short, fields die when work in them, no matter how technically sound, ceases to have significance.

The State of the History of Alchemy

The history of alchemy is, if not dead, at least moribund. It will take more than a gradual effusion of textual studies, and occasional new thoughts on old issues, to revive it.

The point at which I began to follow alchemical history, thirty years ago, marks the completion of its *problematique*. The questions that it set out to solve, as given more or less final definitions by Von Lippmann, Sherwood Taylor, Holmyard, and others, were questions about the evolution of chemical processes. This was even true of Read's studies of alchemical iconography. Jung's psychological studies culminated a very different line of exploration, begun a century ago by the American Civil War general Ethan Allen Hitchcock.[2] But Jungian attempts to interpret alchemy as a symbolic quest for individuation never took seriously the possibility that alchemists ever did anything outside their minds. Jung's work has been almost wholly ignored by historians, who have seen alchemy significant only as, in Read's phrase, a 'prelude to chemistry.'

There have been, to my knowledge, only two broadly informed attempts to define alchemy in a way that holds for all the great world traditions. Joseph Needham defines alchemy as a combination of 'aurifaction' - 'the belief that it is possible to make gold . . . indistinguishable from, and as good as . . . natural gold, from other quite different substances, notably the ignoble metals' - and 'macrobiotics' - 'the belief that it is possible to prepare, with the aid of botanical, zoological, mineralogical and above all chemical, knowledge, drugs or elixirs . . . which will prolong human life beyond old age . . . ' Needham opposes the alchemist's aurifaction (his metallurgical naïveté, if you will) to the sophistication, in both senses, of 'aurifiction' - 'the conscious imitation of gold' and other precious substances by artisans, 'often with specific intent to deceive.' H. J. Sheppard calls alchemy 'the art of liberating parts of the cosmos from temporal existence to achieve perfection,

---

2. Cohen (1951).

which for metals, was gold, and for man, longevity, immortality and, finally, redemption.' Needham's blend of chemical technique and self-deception contrasts with Sheppard's emphasis on a spiritual process. Two formidably learned scholars, both aiming to encompass the full range of alchemical phenomena in all times and places, differ more than they overlap. But so low is the vitality of the history of alchemy that no one has attempted since to overcome this dissonance and define more comprehensively the object of study.[3]

There are several reasons that scholarship on alchemy is static. Most of the important early work was done by chemists who had developed antiquarian interests. That accounts for the focus on chemical aspects. More recently, in most of the world the historical study of modern science has become respectable. Chemists who a generation ago might have become fascinated by the ancient origins of their profession are now more likely to record the achievements of their immediate forebears or the background of their own contributions. At the same time, scientists soundly trained in classical languages have recently become as rare in China and Japan as in Europe. In the United States, where the Ph.D. language requirement is nominal and Fortran or Turbo Pascal increasingly satisfies part of it, few chemists can consider studying the remote past.

Unlike China and Japan, however, in the West scientific illiteracy remains for most Orientalists a badge of pride. Most of those prepared to spend their lives contemplating the past find alchemy, which the literature tells them is early chemistry, abhorrent. Those attracted to it as a facet of the occult, Hermeticism, etc., seldom see any point in exploring its technical intricacies. This difficulty is even more serious with respect to Asian alchemy, for few historians anywhere combine

3. J. Needham (1954-), vol. V, part 2, pp. 9-12; Sheppard (1981), pp. 9-10. Sheppard's definition is largely congruent with the understanding of Mircea Eliade, who saw the alchemist as 'the aid and savior of Nature,' who fulfills her tendency toward perfection in the mineral sphere, perfecting himself in the process (Eliade (1956), p. 55 and id. (1968)). When offered at the Sixteenth International Congress of the History of Science, Bucarest, 1981, it did not draw substantial discussion from Needham or others present.

Both of these definitions are flawed, because in important Chinese sources the perfected material is not metallic; cinnabar is as important as gold. Needham's definition is particularly flawed because his dichotomy of alchemical naïveté and artisanal sophistication - intended to reflect what he considers a key social distinction - is not consistently reflected in Chinese sources.

Although not noteworthy for its definition of alchemy, Yoshida Mitsukuni 吉田光邦 (1962) is the only world history of alchemy by an author able to read Occidental and Oriental languages. For additional references to recent publications on alchemy see Sivin (1988).

mastery of the languages and their philology with ability to comprehend laboratory procedures. There are thus few potential investigators among scientists or Sinologists.

The world frontier of the history of science has moved away from the study of technical ideas for their own sake, toward a concern for science as a simultaneously intellectual and social activity, understanding of which is impoverished when these two aspects are forcibly separated. In doing so it has drawn increasingly on the insights of other disciplines, from religious hermeneutics to sociology. From this point of view the study of alchemy is retrograde. It has ceased on the whole to attract professional historians of science.

Historians of science have gone beyond summarizing ancient books and stitching excerpts into narratives of progress, and have insisted on a more systematic interdisciplinary concern with the reasons for change. Partly for this reason and partly because of great changes in the organization and economics of publication, contributions of antiquarian enthusiasts limited in their command of method and in their familiarity with the literature as a whole are less likely than a generation ago to find their way into print.

With all this in mind, it is not surprising that as academia grows the number of active experts in the history of alchemy has been shrinking.

Historians of Alchemy in China

In China publication in the history of alchemy is growing rather than shrinking. That is not because the field is in a healthy state, but because it is coextensive with the history of chemistry. That field continues to attract scientists seeking to expand the honor roll of ancient Chinese technical achievements. Although, as I have just explained, compiling such honor rolls is no longer an important goal of professional historians of science, in China it is encouraged as a matter of nationalistic priority.

In the generation before the Great Proletarian Cultural Revolution a few erudite chemists explored much of the literature, but this work came to a halt from the mid-sixties to the late seventies. Interest has spread recently among young chemists seeking to find the roots of modern science in their own culture.

The government's disdain (lately restrained for the use of religion, ritual, and other 'superstitious' activity, combined with its socialist faith in the inevitable progress of science, do not encourage research scholars to comprehend alchemy as anything but early science.[4] Non-chemists are little involved, which is not surprising in a country where most universities are in effect institutes of technology. Most historical research is done outside the educational system in specialist institutes, which skim off most of the humanistic talent. At the same time central institutes offer few openings to those who develop new interests in mid-career. The restricted interests of chemists, lack of incentive for others to study alchemy, and the political risks of historiography combine to make studies of alchemy, although more frequent than in the past, no more diverse in theme or approach than they were half a century ago.

The Study of Chinese Alchemy as Chemical History
Historical research that went beyond vague generalities began with the work of Wang Chin 王 璡, who from 1920 on published a series of papers on early chemical industry. He was soon joined by Chang Hung-chao 章鴻釗 and others. Monographic work on alchemy began with Ts'ao Yuan-yü's 曹元宇 1933 study of the apparatus and methods of early alchemists, seen as a contribution to the history of chemical experimentation. Wang was a metallurgist trained at Lehigh University. Chang, a geologist, and Ts'ao, a chemist, were both educated in Tokyo. They were doubtless unaware of attempts going on in Europe to examine the links between alchemy, mysticism, art and psychology, and their comparisons were extremely narrow in scope.[5] The issue was always who did or thought something earlier than in Europe. This was natural enough as a corrective to the European insistence that Chinese alchemy, if indeed anything in China fit a proper European definition of alchemy, was irrelevant to the prehistory of chemistry. Robert P. Multhauf in his excellent survey of *The Origins of Chemistry* found himself forced to conclude from the scholarly literature available in the mid-1960's that 'the continuity from which the science of chemistry was to emerge was a phenomenon peculiar to the West. Natural philosophy, practical chemistry, and alchemy were all pursued in the Far East from

---

4. During my most recent stay in China in the first half of 1987, historians of medicine from all over the country reported to me that no research on religious healing is under way. I met no scholar of alchemy who was approaching it from any but a chemical point of view. On the other hand, the Chinese Academy of Social Sciences had just organized a research group to document the community rituals of renewal now being revived in southeast China. It is now safe to take up what George Sarton once called 'the study of wretched subjects' if any Chinese scholar is so inclined, but those with long memories may be pardoned for hesitating on the ground that it may become retrospectively unsafe.
5. Shimao Eikō 島尾永康 (1986) gives an interesting chronological account of historical studies in Chinese chemistry.

a very early time, and with a high degree of independence. But it was surely the West which saw the origins of *our science* of chemistry.'[6] The ethnocentric bias in this typical judgment, obvious enough to anyone familiar with the rich Asian sources, has spurred the work of two generations of Chinese scholars. They find curious and annoying the notion that the continuity of such Asian novelties as distillation, gunpowder, and the elixir should be counted only from their arrival in the Occident.

A key figure for the broadening of alchemical studies was Ch'en Kuo-fu 陳國符, who took his doctorate in chemistry at Darmstadt in 1942. He became interested in sorting out the textual traditions contained in the Taoist canons, the voluminous fifteenth-century version of which was reprinted and thus became widely available in the nineteen twenties. These canons were unknown to the first generation of chemical historians, who relied entirely on sources that had been passed down in general literature. This is true not only of the Chinese scholars mentioned above but of pioneers who had done less focussed work elsewhere - Hiortdahl (1909) in Europe, for instance, Chikashige Masumi 近重真澄 in Japan (1918), and Johnson (1928) in the United States. Ch'en was aware that the canons included roughly a hundred alchemical books, and that an important division of their original organization was devoted to alchemical traditions. From about 1950 on, then, the number of alchemical treatises that every student was expected to consult had increased by more than an order of magnitude.[7]

Consciousness of the Chinese alchemical tradition in the West began with the publications of Tenney L. Davis, an M.I.T. professor of chemistry, who from 1930 to the mid-1940's collaborated with Chinese graduate students and others to translate a series of readily available alchemical classics. Some of these sources actually were concerned with internal alchemy (*nei tan* 內丹, a matter of meditative techniques rather than the chemical manipulations of external alchemy (*wai tan* 外丹), but all the books studied were understood, or misunderstood, as milestones on the road toward modern chemistry. Despite these limitations and the resulting problems in Davis' interpretations, his publications ensured at least passing mention of China in general writings on alchemy. In addition to their impact on alchemy, Davis' translations were among the more important sources of Mircea Eliade's pathbreaking *The Forge and the Crucible* of 1956. Eliade argued that themes of death,

---

6. Multhauf (1967), p. 15; my italics.
7. See Ch'en (1949 and 1983). Although Ch'en included reading notes on a few of the treatises, his work was preoccupied with filiations of texts. Only in his later work did he treat the alchemical writings extensively, but even there he largely restricted himself to dating them and explaining the meanings of technical terms in them.

resurrection, and marriage of substances in alchemy worldwide put its practitioners closer to the mythology of the archaic metal-worker than to the theory of the modern chemist. Davis' main influence was, of course, upon chemical historians. They in turn found Eliade's argument too inconsequential to cite, much less to rebut.

As in many fields of the history of Chinese science, a broader *problematique* that set the agenda for alchemy was provided by Joseph Needham in general writings on science from the 1940's on, in monographic essays on alchemy beginning in 1959, and in the four volumes devoted to this topic (three on external and one on internal alchemy), written in the 1960's and published between 1974 and 1983.[8]

Needham's studies of alchemy cannot be considered in isolation from the rest of his work, since they bear on a pervasive set of hypotheses: the importance of China and other non-European civilizations in the evolution of universal modern science, which has shed all ethnic characteristics; the early superiority of China in science and technology and its decline due to socioeconomic factors; the unitary development of science and technology through history, which makes it unnecessary to study them separately; the central role of 'the Taoists' in the development of objective science, and so on. Every one of these hypotheses is controversial, a point that need not detain us.[9]

In any case Needham and his collaborators have provided an overview of the entire literature, and have made it impossible for anyone to deny that alchemy and early chemistry - which he firmly identified - were world enterprises, to which Europeans were merely prominent contributors.

---

8. Since Davis did not read Chinese and his collaborators were not acquainted with philology, their publications tended to be closer to paraphrases than translations. For a bibliography and information on Davis, see Leicester & Klickstein (1950). In addition to Eliade (1956), see his interesting afterthoughts as well as the supplementary references in his essay of 1968. Although Needham is by far the most catholic of those scholars who have approached alchemy as the prehistory of chemistry, in three large volumes on external alchemy he does not discuss Eliade's interpretation (the discussion in vol. V, part 4, p. 246 is part of my contribution). The volume on internal alchemy mentions Eliade often, but, with the exception of one passing remark on his use of sources, is exclusively concerned with his writings on yoga.

9. The most comprehensive analysis of Needham's hypotheses is in Restivo (1979). See Needham (1954), for the alchemical volumes, Ho & Needham (1959) for his earliest essay on the topic, and Teich & Young (eds., 1973) and Li, Zhang & Cao (eds., 1982) for bibliographies of his publications. Generally speaking, Needham's studies of alchemy have drawn on research by Ho and ancillary work on medicine by Lu Gwei-djen. All their earlier publications have been worked into the argument of *Science and Civilisation in China*.

10

Most pertinent, however, is Needham's conception of Taoists as empiricists, experimenters, and implacable foes of feudalism. Although these attitudes are documented primarily out of two books written by 300 B.C., 'the Taoists' appear in Needham's story regularly as the primary driving force behind technical innovation. This achievement depends crucially on use of the term 'Taoist' in a multitude of interchangeable senses, generally vague and occasionally circular.[10]

Needham often writes 'Taoist alchemy' when he means 'alchemy,' a habit that has been widely copied by authors on whose part the history of Taoism arouses no curiosity whatever. This habit has deflected the attention of European and American scholars from the questions of who did alchemy and what its practitioners had in common besides their alchemical activity. In China since 1949 it has not had that effect, since Taoism, as feudal superstition, could not have been responsible for anything progressive. But the emphasis of Chinese historians on who discovered what first has deflected attention from questions of context.

We remain profoundly ignorant of alchemy's matrix in society. It is clear by now that members of Taoist religious movements did not invent alchemy. They did not provide its theory or dominate its practice. We find alchemy prominent in the schemes of classification of the Taoist libraries because of pre-Taoist alchemical traditions that were taken over by certain sects, notably the Mao Shan movement, for their own purposes. These purposes included cultivating an elite clientele and imperial sponsorship, hardly an anti-feudal aspiration.

The most penetrating study so far of alchemy in Taoist social history asserts of its subject that 'I see no reason to call it 'Taoist' except where it occurs in an indubitably Taoist social context . . . [its] technologies would also emerge more clearly . . . if visualized as separate entities . . . rather than as somehow being integral parts of a 'Taoism' that depends on them for its definition yet lacks any social dimension.' Alchemy was attractive to Taoist initiates in certain circumstances because it was a means (among many) to spiritual power or secular patronage, but it was attractive to non-Taoists for the same reason.

Equally to the point, it has turned out to be singularly difficult, after the close look at the Taoist canons that has been under way for two decades, to find evidence for the past two thousand years that Taoist masters performed experiments (or knew what they were), valued empirical observation as a source of knowledge, or for that matter included among their goals collecting knowledge about Nature or understanding it theoretically. They did not see themselves discovering nov-

---

10. On this habit, deeply ingrained in Sinology, see Sivin (1978).

elties, but rediscovering archaic revelations of timeless truth.[11]

Like the global hypotheses that have shaped his view of alchemy, Needham's analysis of it is an interesting blend of old and innovative themes. External alchemy is, as in the work of his predecessors, proto-chemistry, despite the deep concern in its literature for moral and spiritual matters. Needham's *problematique* is that of the chemical antiquarianism of the first half of this century.

Needham breaks new ground with his claim, developed at great length, that internal alchemy is in effect proto-biochemistry. By way of contrast, the only Sinologist to specialize in internal alchemy defines it in this way: 'Internal alchemy is a very complex syncretic system' meant, like external alchemy, to produce an elixir that will make the adept immortal. 'But it is from his own body that the adept makes his laboratory,' and everything else needed for the process. The practitioner adapts alchemical language to describe interior 'processes of purification aimed at a spiritual and corporeal transformation.'

But Needham's emphasis is on an 'essentially materialist' process, for, in his words, the practitioner 'did not seek psycho-analytic peace and integration directly': 'Physiological alchemists believed they could use . . . the control of respiration, the mastery of neuro-muscular coordination . . . sexual activity . . . bodily exposure to light, and the management of the mind in methods of meditation and mental concentration' to make a 'real chemical substance formed from the juices and *pneumata* of the body . . . .' that would confer 'material immortality, for no other was conceivable . . . .'

This approach requires that Needham redefine familiar terms, so that, for instance, *ch'i* 氣, the Chinese counterpart of the Stoic pneuma, becomes for the early internal alchemist 'the dissolved gases in his body-fluids.' But 'the truly proto-scientific character of their endeavour appears . . . in that *tour de force* of medieval achievement, the preparation of active hormones from urine, worked up in almost manufacturing quantities.' This achievement, a reader who pays attention to footnotes may notice, is chronicled not in alchemical or Taoist literature but medical books and memoirs of literary figures who according to Sinological convention would be classified as 'Confucian' rather than 'Taoist.'

It would take more space than this essay offers to determine from the primary sources whether internal alchemy is, as Needham claims, 'physiological through and through' and 'always infused with character-

---

11. The quotation is from Strickmann (1979), p. 166. Taoist libraries are taken up in Sivin (1978). On the non-Taoist origins of alchemy see Robinet (1984). I have discussed alchemy as one of many means to spiritual transcendence in a 1976 essay. I will return below to Taoists' goals in connection with alchemy.

istically Chinese sanity, sobriety, empiricism and rationality.' I will simply note in passing that the last four attributes are singularly given to meaning what one wants them to mean. My own reading finds empiricism as a philosophic stance related to modern science conspicuously rare in the documents of alchemy, external or internal. Rationality, as one would expect among mystics striving for spiritual perfection and exemption from death, is usually inseparable from its opposite.[12]

My point, rather, is the aftereffect of this line of argument. Rather than challenge the conventional wisdom by suggesting that psychoanalysis and chemistry are not the only options when trying to understand meditational traditions and symbolic systems of meaning, Needham's positivism has given the equation of alchemy and proto-chemistry a new lease on life. This larger point has been quite ignored by historians of chemistry in Asia, for they are not inclined to deemphasize the religious component of internal alchemy. It is precisely what fits alchemy neatly into the official category of feudal superstition.

Asian scholars over the past decade, rather, have made the question of whether early alchemists isolated active sex hormones from the urine of pubescent boys the most frequently explored topic in what they see as the prehistory not of biochemistry but of organic chemistry. Needham's claim was not based on experimental reconstruction of the ancient processes. Several attempts to repeat those processes in laboratories on the Chinese mainland and Taiwan have concluded that their product could not be purified active sex hormones. But none has taken internal alchemy seriously enough as a topic of study to question the reasoning that led Needham to this fragile interpretation.[13]

As this example indicates, despite the increasing number of scholars exploring the literature, the outcome continues to be claims about ancient anticipations of modern chemistry. Superficial claims are freely made, since the search for Chinese priorities encourages them, and freely voided, since closer examination tends to reveal that alchemists and modern chemists have little in common. In Europe, where the tra-

---

12. Baldrian-Hussein (1984), p. 14; Needham, et al. (1954-), vol. V, pp. xxviii-xxx, 23. Needham's argument here, unlike that about external alchemy, draws extensively on scholarship concerned with yoga and other religious themes, but still emphasizes their physiological aspects.
13. Needham (1954-), pp. 312-37. The point has been refuted, among others, in four papers by Liu Kuang-ting 劉廣定 published in 1981, Meng Nai-ch'ang 孟乃昌 (1982), and Chang Ping-lun 張秉倫 & Sun I-lin 孫毅林 (1988). A preliminary paper presented by H. T. Huang, et al., at a 1988 conference, suggested that under certain circumstances traces of hormones may be present, but not in effective concentrations. This is hardly a vindication of Needham's claim. At best, far from being augmented, the high natural concentration of sexual hormones in the urine has been markedly reduced.

jectories of technical history, religious studies, and philological Sinology manage to traverse the alchemical literature without encountering each other, the most frequent product still tends to be annotated translation. I detect no sign of new frameworks of understanding that will be adequate to the many dimensions of the subject.

New Issues in the Study of Alchemy

Ways to vivify the study of Chinese alchemy do not differ fundamentally from those that apply to other world traditions. They involve new research issues that will capture the imaginations of historians as well as of scientists: who did alchemy, how practitioners were associated and why, what their quest signified to them, and what their own goals were; how aspects of this picture changed, and for what reasons.

Means and Ends: This rethinking will be expedited if we stop assuming that the ends of alchemy were chemical, and ignoring evidence to the contrary. The Chinese documents suggest, as do those of Hellenistic alchemy, that chemical knowledge, usually in small concentrations, was only a means. The goals were more spiritual than cognitive or utilitarian.

The most obvious way to proceed involves attention to what alchemists themselves have to say about what they were doing and thinking. As Robert Halleux, the most redoubtable student of Hellenistic alchemy, puts it, the distinction between a technical formula and an alchemical procedure can be sought 'only at the level of the operative's consciousness.'[14] A close study of the alchemists' theories and motivations in the entire Chinese literature left me unable to affirm that they were striving for chemical knowledge, nor indeed that their goal was cognitive. I argued on the basis of their own assertions that external alchemy's characteristic activity was making chemical models of cosmic process, with the time dimension telescoped so that the great cycles of time could be witnessed by the adept. I concluded that 'the content, tone, and balance of the evidence strongly suggest that the dominant goal of Chinese alchemy was contemplative, and even ecstatic. . . . The alchemists . . . made the cycles of the cosmic process accessible, and undertook to contemplate them, because they believed that to encompass the *Tao* with their minds . . . would make them one with it.' Even those who sought power, or medicine for others, understood that success depended on spiritual self-cultivation.

I have suggested that attempts to excavate clear modern principles out of the teeming synonyms, metaphors, and secret names in alchem-

---

14. Halleux (1981), pp. 29-30. See also the detailed 1984 study by Ann C. Wilson of the basic Hellenistic text *Physika kai mystika* which, despite flamboyant technical flaws, makes a strong case for its ritual origins.

ical writing are misdirected. The purpose of alchemical obscurity was not 'mystifications to addle the uninitiated' but a maximum of symbolic allusion packed into every utterance. This is not an attempt 'to protect against danger sometimes by purposive obfuscation,' as Needham explains it, but a corollary of the larger effort 'to build rich and complicated structures of meaning that would encompass the many levels of human experience and cosmic process that their chemical operations reflected and modelled.'[15]

There is no doubt that, in pursuit of their goals, alchemists were consumers of chemical knowledge, and that the knowledge that they used increased and became more sophisticated as time passed. But evidence that they originated this knowledge is rare indeed. Claims that alchemists invented various processes or types of apparatus are generally assumptions rather than conclusions. An alternative hypothesis, namely that alchemists on the whole took their techniques and apparatus from craftsmen and physicians, deserves to be explored. That entails, of course, looking beyond the alchemical literature.

This reconnaissance indicates that an adequate understanding of alchemy is unlikely to come from either the history of chemistry or the history of religion pursued separately as disciplinary specialties. That alchemy encompassed both in intimate juncture is a matter of the greatest possible interest for the history of human consciousness, all the more so because this intimacy is also perceptible in the documents of Western, Islamic, and Indian alchemy.

Mental and Social Activity: So far scholarship has been concerned primarily with alchemical ideas and the activity behind them. We are moderately well informed about who did astronomy or medicine in traditional China, but there is little save legends to satisfy curiosity about who the alchemists were. Why should they so consistently be surrounded by marvels and miracles - except for those, rarely studied, who turn out to be unremarkable literati? Until this mystery is unravelled, the history of alchemy will remain afloat, unrooted in the web of activity that we call Chinese society.

A clue can be found in the legends surrounding an alchemist about whose life we are tolerably well informed. T'ao Hung-ching, 'T'ao the Hermit' (436-536), appears in his hagiographies in the dynastic histories and elsewhere as a wonderworking recluse. The Chinese popular imagination usually pictures him deep in meditation on the second floor of

---

15. Sivin (1976), pp. 523-4; id. (1981), pp. 231-4. Evidence from the primary sources for the arguments in the 1976 essay is given *in extenso* in my contribution to Needham, *et al.* (1954-), vol. V, part 4, pp. 210-305, completed 1970 but published 1980. The quotation about obfuscation is from Needham, *et al.* (1954-), vol. V, part 3, p. 74.

the hermitage built for him on Mt. Hua-yang by a properly respectful emperor, where T'ao was insulated once and for all from contact with the profane world by solicitous disciples on the floor below, much as professors today are protected by the secretaries in their outer offices.[16]

But in the remarkable studies of Michel Strickmann, based in the main on close reading of T'ao's voluminous writings, he is an accomplished and worldly patrician. His scholarship, alchemical operations (chancy but in no sense experimental), pilgrimages, literary detective work, swordmaking, and excellent connections among the powerful led to extensive personal patronage by two emperors, one of whom was simultaneously repressing Taoism through the rest of his realm. T'ao's pursuit of alchemical elixirs was unsuccessful. But it was part of a conspicuously successful effort to construct a symbolic, liturgical, and technological foundation for a new Taoism oriented away from the grass-roots, community organization of the old movements and toward private self-cultivation by aristocrats like himself - and not least by monarchs.

The implication is that the character of alchemical practice is shaped by time, place, circumstances and interests, and that it is possible in given cases to learn what they were. Other recent studies by historians of religion have succeeded by similar means in seeing past the veil of legend to the actual careers of 'legendary' Taoists.[17] More such breakthroughs can be anticipated following publication of a massive bibliography of the Ming Taoist canons, which links many of the fifteen hundred treatises to particular Taoist movements, the histories of which are yielding to multi-disciplinary study.

A breakthrough has been made in several studies which demonstrate once and for all that hagiographies 'correspond to cults, and express the ideas of communities of believers,' and that their central theme is typically the genesis of holy places. In this sense the theme of the mirabilia of T'ao Hung-ching is the status of Mt. Hua-yang as a devotional center. Among studies of hagiography, three recent dissertations

---

16. *Liang shu* 梁書 (Beijing 1974 ed.), *ch. 51, pp.* 742-3; *Nan shih* 南史 (idem), *ch.* 76 pp. 1897-1901. For hagiographies in the Taoist canons see Weng Tu-chien 翁獨健 (1935), p. 208.

17. Strickmann (1979) and (1981). A parallel instance of iconoclastic biography which does not involve an alchemist is Verellen's 1986 study of Tu Kuang-t'ing 杜光庭 (850-933). Sivin (1968), p. 81-144 is a critical study of sources for the biography of a great alchemist, Sun Ssu-mo 孫思邈 (or Ssu-miao, alive 673).

have demonstrated how such legends change as they diffuse from one region and one community to others. There remains no warrant for letting legends stand in the stead of an understanding of the situation, the organization, even the economics of alchemical work.[18]

Conclusion

These nascent lines of inquiry, which did not originate in the history of chemistry, show some potential for shedding new light on the evolution of alchemy. I have mentioned other questions that will cast light on the matrix out of which alchemy evolved over the centuries: who did alchemy, how and why they were associated, what their quest signified to them, and what their own goals were, and how and why aspects of this picture changed.

But it will take more than a list of new issues to reinvigorate the study of alchemy. What has held it back has been the disciplinary blinders that have led scholars looking at it from different perspectives to ignore what they might have learned by combining viewpoints. Until the chemists and specialists in religion are willing to learn from each other, and the philologists and intellectual historians from both, there is little prospect of resolving old or new issues. With that in mind, this conference is a most auspicious occasion, but what matters is whether it inspires lasting change.

It may be that a broader, better-integrated approach will yield a more critically informed appreciation of the role of alchemy in the growth of chemical skills and knowledge. On the other hand, I find it conceivable that when the returns are in we will have to conclude that the Chinese alchemists' contribution to chemistry was negligible, and that what we have to learn from them mainly concerns the imaginative potentialities of the human spirit. This is not an outcome to be dreaded. One result, if this turns out to be the case, will be a salutary challenge to our preconceptions about the early history of science.

Traditions so richly endowed by the thought and imaginations of many cultures and ages deserve to be impartially understood. Regardless of the outcome, the field needs to be subjected, after a long hiatus, to the broad questioning and contention that is the surest sign of life in scholarship.

---

18. Notable on the general issue is Schipper (1985). Outstanding studies of change and variation are Kleeman (1985), Hansen (1987), and Duara (1988).

RESEARCH ON THE HISTORY OF CHINESE ALCHEMY

REFERENCES TO SECONDARY SOURCES

This list includes only works cited directly and typical works of persons cited. For full lists of works on alchemy see the voluminous bibliographies of Needham, et al. (1954-), vol. V, parts 2-5, supplemented for recent publications by Sivin (1988). An asterisk (*) marks titles which are recommended by the author as important studies of Chinese alchemy.

*Baldrian-Hussein, Farzeen, Procédés secrets du joyau magique. Traité d'Alchimie taoïste du XIe sicle, Paris 1984
Excellent short introduction on internal alchemy, the best on the topic.

Bynum, W. F., Browne, E. J., & Porter, R. (eds.), Dictionary of the History of Science, Princeton 1981
Actually a concise encyclopedia.

Chang Chueh-jen 张觉人, Chung-kuo lien-tan-shu yü tan yao 中国炼丹本与丹药 (Alchemy and elixir medicines in China), Chengdu: Chung-kuo k'o-hsueh chi-shu ch'u-pan-she, 1985
On drug formulas, mostly inorganic, which the author believes are descended from alchemical elixirs (both are called tan).

Chang Hung-chao, 'Chung-kuo yung hsin te ch'i -yuan' 中國用鋅的起源 (Origins of the use of zinc in China), K'o-hsueh 8 (1923), pp. 233-43. See Wang Chin, et al. (1955).

Chang Ping-lun & Sun I-lin, '"Ch'iu shih fang" mo-ni shih-yen chi ch'i yen-chiu 秋石方模拟实验及其研究 ('Three typical simulated tests and a physicochemical examination of autumn mineral'), Tzu-jan k'o-hsueh-shih yen-chiu, 7.2 (1987), pp. 170-183

*Chao K'uang-hua 赵匡华, Chung-kuo ku-tai hua-hsueh-shih yen-chiu 中国古代化学史研究 (Researches in the history of ancient chemistry in China), Beijing: Pei-ching ta-hsueh ch'u-pan-she, 1985
Reprints 54 articles published from 1977 on.

*Ch'en Kuo-fu, Tao tsang yuan-liu k'ao 道藏源流考 (Studies in the history of the Taoist canons), 1949, sec. ed., rev. and enl., Beijing: Chung-hua shu-chü, 1963
Primarily a history of scriptural collections, but contains pioneering studies of alchemical traditions as well. The second edition is much fuller on alchemy, but the first is less affected by ideological pressures.

*Id., Tao tsang yuan-liu hsu k'ao 道藏源流續考 (Supplementary studies in the history of the Taoist canons), Taipei: Ming wen shu-chü, 1983
Mainly on the dating of treatises - largely from internal evidence - and the identification of chemical constituents and apparatus.

Chikashige Masumi, 'Tōyō ko tōki no kagakuteki kenkyū 東洋古銅器の化學的研究' (Chemical researches on Oriental copper-alloy artifacts), Shirin 3.2 (1918), pp. 177 ff.
Chikashige's publications are summarized in the next item.

Id., Alchemy and Other Chemical Achievements of the Ancient Orient. The Civilisation of Japan and China in Early Times as Seen from the Chemical Point of View, Tokyo: Rokakuho Uchida, 1936

Cohen, I. B., 'Ethan Allen Hitchcock. Soldier-Humanitarian-Scholar. Discoverer of the "True Subject" of the Hermetic Art', *Proceedings of the American Antiquarian Society* 61 (1951), pp. 29-136

Diergart, P., *Beiträge aus der Geschichte der Chemie dem Gedächtnis von Georg W. A. Kahlbaum gewidmet*, Leipzig 1909

Duara, Prasenjit, 'Superscribing Symbols: The Myth of Guandi, Chinese God of War', *Journal of Asian Studies* 47.4 (1988), pp. 778-95

Eliade, M., *Forgerons et alchemistes*, Paris 1956; English translation, *The Forge and the Crucible*, New York 1962

Id., 'The Forge and the Crucible: A Postscript', *History of Religions* 8.1 (1968), pp. 74-88

Halleux, R., *Les alchimistes grecs. I: Papyrus de Leyde. Papyrus de Stockholm. Fragments et recettes*, Paris 1981
First of ca. 12 vols. in a complete translation of Greek alchemical texts.

Hansen, Valerie, 'Popular Deities and Social Change in the Southern Song Period (1127-1276)', Ph.D. dissertation, History, University of Pennsylvania, 1987

Hiortdahl, Th., 'Chinesische Alchimie', in Diergart ed. 1909, pp. 214-26
Generalities based on secondary sources and one ill-understood primary source.

Ho Ping-yü & Needham, J., 'The Laboratory Equipment of the Early Chinese Alchemists', *Ambix* 7 (1959), pp. 57-115.

Huang, H. T., *et al.*, 'Preliminary Experiments on the Identity of *Chiu shi* (Autumn Mineral) in Medieval Chinese Pharmacopoeias', Paper presented at 5th International Conference on the History of Science in China, San Diego, 9 August 1988

Johnson, Obed Simon, *A Study of Chinese Alchemy*, Shanghai: Commercial Press, 1928
A very unsatisfactory doctoral dissertation based on a narrow range of sources.

Kleeman, T. F., 'Wenchang and the Viper. The Creation of a Chinese National God', Ph.D. dissertation, Oriental Languages, University of California, Berkeley, 1985

Leicester, H. M., & Klickstein, H. S., 'Tenney Lombard Davis and the History of Chemistry', *Chymia* 5 (1950), pp. 1-16.
See the bibliography of Davis' writings on Chinese alchemy, pp. 6-10.

Li Guohao, Zhang Mengwen & Cao Tianqin (eds.), *Explorations in the History of Science and Technology in China. Compiled in Honour of the Eightieth Birthday of Dr. Joseph Needham, FRS, FBA*, Shanghai: Shanghai Chinese Classics Publishing House, 1982
Biographical sketches, bibliographies, and 28 papers by leading Chinese and foreign scholars.

Liu Kuang-ting (= 1981a), 'Ts'ung Pei-Sung jen t'i-lien hsing-chi-su shuo t'an k'o-hsueh tui k'o-chi-shih yen-chiu te chung-yao-hsing 從北宋人提煉性激素説談科學對科技史研究的重要性' (On the importance of science to research in the history of science and technology, with reference to the claim that Northern Sung people extracted and purified sex hormones), *Kuo-li T'ai-wan Ta-hsueh wen-shi-che hsueh-pao* 30 (1981), pp. 363-76

Id. (= 1981b), 'Jen niao chung so te ch'iu-shih wei hsing-chi-su chih chien-t'ao 人尿中所得秋石爲性激素之檢討' (Critical examination of the assertion that the Autumn Mineral derived from human urine was sex hormones), *K'o-hsueh yueh-k'an* 5 (1931); repr. with the following two articles in Chao K'uang-hua ed. 1985, pp. 502-12

Id. (= 1981c), 'Pu t'an ch'iu-shih yü jen niao 補談秋石與人尿' (Supplementary discussion of Autumn Stone and human urine). Ibid., 6 (1981)

Id. (= 1981d), 'San t'an ch'iu-shih yü jen niao 三談秋石與人尿', (A third discussion of Autumn Stone and human urine), Ibid., 8 (1981)

RESEARCH ON THE HISTORY OF CHINESE ALCHEMY

Meng Nai-ch'ang, 'Ch'iu shih shih i 秋石试议' ('A tentative discussion on the Qiu Shi [Autumn Mineral]'), *Tzu-jan k'o-hsueh-shih yen-chiu* 1.4 (1982), pp. 289-99; repr. in Chao K'uang-hua (ed.), 1985, pp. 483-501
Multhauf, R. P., *The Origins of Chemistry*, New York 1967
  Deals with chemical processes and alchemy to the eighteenth century.
*Needham, J., et al., *Science and Civilisation in China*, Cambridge 1954-
  Vol. V, parts 2-4, are devoted to external alchemy, part 5 to internal alchemy, and part 7 considers the role of alchemists in the origin of gunpowder.
Restivo, S. P., 'Joseph Needham and the Comparative Sociology of Chinese and Modern Science', *Research in Sociology of Knowledge, Sciences and Art* 2 (1979), pp. 25-51
*Robinet, I., *La révélation du Shangqing dans l'histoire du Taoisme* (Publications de l'École Française d'Extrême-orient, 87), Paris 1984
Schipper, K. M., 'Taoist Ritual and Local Cults of the T'ang Dynasty', in Strickmann (ed.), 1985, pp. 812-34
Sheppard, H. J., 'Alchemy', in Bynum, et al. (eds.), 1981, pp. 9-10
*Shimao, Eikō, 'Chūgoku kagakushi kenkyū no tembō 中国化学研究の展望 ' ('Review of researches into the history of Chinese alchemy'), *Kagakushi* 2 (1986), pp. 72-81
  Conspectus of recent studies, with attention to Japanese, Chinese, and European contributions.
*Sivin, N., *Chinese Alchemy: Preliminary Studies* (Harvard Monographs in the History of Science, 1), Cambridge, Mass. 1968
  Includes bibliographical essays on the identification of substances and medical disorders mentioned in alchemical texts.
Id., 'Chinese Alchemy and the Manipulation of Time', *Isis* 67 (1976), pp. 513-26
Id., 'On the Word "Taoism" as a Source of Perplexity. With Special Reference to the Relations of Science and Religion in Traditional China', *History of Religions* 17 (1976), pp. 303-30
*Id., 'The Theoretical Background of Elixir Alchemy', in Needham, et al., 1954- , vol. V, part 4, pp. 21-305
  On the aims and theoretical perspectives of Chinese alchemists.
Id., 'Discovery of Spagyrical Invention', *Harvard Journal of Asiatic Studies* 41.1 (1981), pp. 219-235
  Essay review of Needham, et al. (1954-), vol. V, part 3.
Id., 'Science and Medicine in Traditional China. The State of the Field', *Journal of Asian Studies* 47.1 (1988), pp. 41-90
  Mainly covers important publications since 1980.
*Strickmann, M., 'On the Alchemy of T'ao Hung-ching', in Welch & Seidel (ed.), (1979), pp. 123-92
Id., *Le Taoïsme du Mao Chan. Chronique d'une révélation* (Mémoires de l'Institut des Hautes Études Chinoises, 17), Paris 1981
Id. (ed.) (= 1985), *Tantric and Taoist Studies in Honour of R. A. Stein*, 3 vols., Brussels: Institut Belge des Hautes Études Chinoises, 1981-1985
Teich, M., & Young, R. (eds.), *Changing Perspectives in the History of Science. Essays in Honour of Joseph Needham*, London 1973
  Includes a biography and bibliography.
Ts'ao Yuan-yü, 'Chung-kuo ku-tai chin-tan-chia te she-pei ho fang-fa 中國古代金丹家的設備及方法' (Equipment and methods of ancient Chinese alchemists), *K'o-hsueh* 17 (1933), pp. 31-54
  See Wang Chin, et al. (1956).

Verellen, F., 'Du Guangting (850-933). Un taoïste de cour et son temps', Dissertation, Docteur ès lettres, Université de Paris VII, 1986

*Wang Chin, et al., *Chung-kuo ku-tai chin-shu hua-hsueh chi chin-tan-shu* 中國古代金屬化學及金丹朮 (Ancient Chinese metallurgical chemistry and alchemy), Shanghai: Chung-kuo k'o-hsueh t'u-shu i-ch'i kung-ssu, 1955
Reprints articles by Wang, Chang Hung-chao, Ts'ao Yuan-yü, et al. originally published from 1920 on.

Welch, H. H., & Seidel, A. (eds.), *Facets of Taoism. Essays in Chinese Religion*, New Haven 1979

Weng Tu-chien, *Tao tsang tzu-mu yin-te* 道藏子目引得 ('Combined indices to the authors and titles of books in two collections of Taoist literature'; Harvard-Yenching Institute Sinological Index Series, 25), Beijing, Harvard-Yenching Institute, 1935

Wilson, C. A., *Philosophers, Iosis, and Water of Life* (Proceedings of the Leeds Philosophical and Literary Society. Literary and Historical Section, 19.5), Leeds Philosophical and Literary Society, 1984

Yoshida Mitsukuni, *Renkinjitsu. Senjitsu to kagaku no aida* 煉金术. 仙术と科學の間: Between art of immortality and science), Tokyo: Chüö kõron, 1962
A world history of alchemy.

IX

AN INTRODUCTORY BIBLIOGRAPHY OF TRADITIONAL CHINESE MEDICINE:

BOOKS AND ARTICLES IN WESTERN LANGUAGES

This annotated bibliography covers scholarship in Western languages on medicine in traditional and modern China. A similar list for science appears in *Science in Ancient China*. Publications are included mostly because of their quality and usefulness, a few in order to warn readers that the promise of their titles is specious.

History of Medicine in Imperial China 1
    Reference Works 1
    Studies Useful for Orientation 3
    Medicine and Related Topics 4
    Materia Medica 11
History of Medicine in Twentieth-Century China 11
    Reference Works 11
    Medicine and Related Topics 12

The obvious differences in the size of sections reflect the varying character and extent of the literature in each category. Historians have conspicuously neglected recent medicine. Most publications on the topic are concerned with policy rather than the work and the people who did it. Much of the work on policy published more than a decade ago is already obsolete.

History of Medicine in Imperial China

Reference Works
Bensky, Dan, et al. 1986. *Chinese Herbal Medicine. Materia Medica.* Seattle: Eastland Press. Discusses and illustrates over 400 plants, with data from Chinese textbooks. Not scholarly, but relatively full and clear, with useful appendices.
Bretschneider, E. 1881–1895. Botanicon sinicum. Notes on Chinese Botany from Native and Western Sources. *Journal of the North China Branch, Royal Asiatic Society*, 16 (Bibliographical sources), 25 (Botanical identifications), 29 (Uses in medicine). Fascimile reprints, Tokyo, 1937; Nendeln, Lichtenstein, 1967.

Chen Chan-yuen (Ch'en Ts'un-jen 陳存仁). 1968. *Chung-kuo i-hsueh-shih t'u chien* 中國醫學史圖鑑 ("History of Chinese Medical Science Illustrated with Pictures"). Hong Kong: Shang-hai Yin-shu-kuan. Pictorial archive with condensed English versions of the captions. Nothing in the sloppy text should be used without checking; the author has missed no opportunity to represent legends as historical facts and late imaginative depictions as portraits. The section on "Spread of Chinese Medical Science into Great Britain," France, the United States, Japan, etc., contains mostly snapshots of the author's travels.

Ferreyrolles, Paul. 1953. *L'acupuncture chinoise.* Lille: Éditions S.L.E.L. Recommended solely for the bibliography of European writing on Chinese medicine from 1671 to 1950 (pp. 177–191).

Ghani, A. R. 1965. *Chinese Medicine and Indigenous Medical Plants.* PANSDOC Bibliographies, 396. Karachi: Pakistan National Scientific & Technical Documentation Centre. An unannotated list of about 300 books and papers in European languages and Chinese, included because of availability to the compiler rather than coverage or value. Most are either nontechnical or concerned with individual plants. The majority of important Chinese publications in medicine and pharmacology since 1949 are omitted. Almost all titles are given in English only.

Hu, Shiu-ying. 1980. *An Enumeration of Chinese Materia Medica.* Hong Kong: Chinese University Press. A concise, authoritative reference list by Chinese names; provides pharmaceutical and botanical identifications. The most up-to-date handbook on botanical and pharmaceutic nomenclature in English; Read, Stuart, etc., are largely obsolete for Latin names of plants.

Huard, Pierre. 1968. *Chinese Medicine,* trans. Bernard Fielding. World University Library. New York: McGraw-Hill. This well-illustrated paperback deserves praise for its concern with the connections of other Asiatic traditions to that of China, and European knowledge of Chinese medicine and vice versa in the last three centuries. At the same time, it is perfunctory, carelessly thought through, and full of elementary errors of fact, interpretation, translation, and transliteration.

Huard, Pierre; Ming Wong (Wang Ming). 1956. Bio-bibliographie de la médecine chinoise. *Bulletin de la Société des Études Indo-chinoises,* n.s., *31:* 181–246. Index of principal Chinese medical writers, and brief bibliography of Western (including Russian) and Chinese works.

Keys, John D. 1976. *Chinese Herbs. Their Botany, Chemistry, and Pharmacodynamics.* Rutland, VT: C. E. Tuttle Company. Uninformative despite the rich sources that the author claims to have drawn on.

BIBLIOGRAPHY 3

Read, Bernard E. 1931-1941. *Chinese Materia Medica.* Peking Natural History Bulletin. Standard references for animal drugs. For detailed list of fascicules, see Needham, *Science and Civilisation in China,* III, 784-785. Although Read's publications are the best of their sort in English, recent publications in Chinese have rendered them obsolete.

Read, Bernard E.; C. Pak. 1936. *A Compendium of Minerals and Stones Used in Chinese Medicine From the Pen Ts'ao Kang Mu. . . [of] Li Shih Chen . . . 1597* A.D. 2d ed., Peking Natural History Bulletin. Unlike *Chinese Materia Medica,* this is not a translation but an ill-digested mass of information from hither and yon. The order of accuracy is nevertheless high.

Sivin, Nathan. 1988. Science and Medicine in Imperial China. The State of the Field. *Journal of Asian Studies, 47.* 1: 41-90. Survey with bibliography of important recent research.

Sivin, Nathan. 1989. A Cornucopia of Reference Works for the History of Chinese Medicine. *Chinese Science,* 9: 29-52. Reports on a large number of reference sources published in China and Japan since ca. 1985.

Smith, F. Porter. 1911. *Chinese Materia Medica. Vegetable Kingdom.* Revised edition, ed. G. A. Stuart. Shanghai: American Presbyterian Mission Press. Reprint (ed. Ph. Daven Wei), 1969, Taipei: Ku T'ing Book House. The fullest of the older monographs on Chinese materia medica. Still worth consulting, particularly in view of its indexes to Chinese names of herbs. There is also valuable translated material on identities and characteristics of medicinal plants in Bretschneider 1881-1895: Part III.

Yuan, Tung-li. 1958. *China in Western Literature. A Continuation of Cordier's Bibliotheca Sinica.* New Haven: Yale University, Far Eastern Publications. Pp. 549-561 list standard reference works on botany, zoology, etc.

Studies Useful for Orientation

Brieger, Gert. 1980. History of Medicine. In *A Guide to the Culture of Science, Technology, and Medicine,* ed. Paul T. Durbin, pp. 121-194. New York: Free Press. A discussion of issues and trends in the study of European medicine.

Chang, Kwang Chih, editor. 1977. *Food in Chinese Culture. Anthropological and Historical Perspectives.* New Haven, CT: Yale University Press. Essays on the culinary arts, dynasty by dynasty, mostly by leading historians. Little on nutrition, but the book is the best of its kind.

Kuriyama, Shigehisa. 1992. Between Mind and Eye: Japanese Anatomy in the Eighteenth Century. In Leslie & Young 1992: 21-43 (p. 7 below). A sophisticated study of the interaction between Western representations of anatomy and traditional ideas of spirituality.

Lock, Margaret. 1980. *East Asian Medicine in Urban Japan. Varieties of Medical Experience.* Comparative Studies of Health Systems and Medical Care, 4. Berkeley: University of California Press. Excellent field study of Chinese-style medicine and its social matrix.

Lock, Margaret. 1993. *Encounters with Aging: Mythologies of Menopause in Japan and North America.* Berkeley: University of California Press.

McNeill, William H. 1977. *Plagues and Peoples.* Garden City: Anchor Press. On the role of micro-organisms and infectious disease in world history. Weak on China but important for epidemiology applied to history.

Ohnuki-Tierney, Emiko. 1984. *Illness and Culture in Contemporary Japan. An Anthropological View.* Cambridge University Press. A penetrating analysis of lay conceptions of body, illness and health care. Like Lock, a model for studies of China.

Pomeranz, Bruce; Gabriel Stux, editors. 1989. *Scientific Bases of Acupuncture.* New York: Springer Verlag. Papers from a 1987 Dusseldorf conference.

Schipper, Kristofer. 1982. *The Taoist Body,* tr. Karen C. Duval. Berkeley: University of California Press, 1993. Trans. of *Le corps Taoïste* (Paris: Fayard, 1982). Penetrating study of conceptions of the body in Taoist religion.

Medicine and Related Topics

Ågren, Hans. 1982. The Conceptual History of Psychiatric Terms in Traditional Chinese Medicine. In Li Guohao 1982: 573–581.

Anonymous. 1975. *Herbal Pharmacology in the People's Republic of China. A Trip Report of the American Herbal Pharmacology Delegation.* Washington: National Academy of Sciences. In addition to first-hand reports on clinical use of herbal drugs and on the cultivation of herbs, includes overly brief but informative notes on 248 drug substances.

Chiu, Martha Li. 1981. Insanity in Imperial China. A Legal Case Study. In Kleinman & Lin 1981: 75–94. On *k'uang* 狂, a technical term for manic behavior or mania as a syndrome.

Chiu, Martha Li. 1986. Mind, Body, and Illness in a Chinese Medical Tradition. Ph.D. diss., History and East Asian Languages, Harvard University. Excellent critical study of the Inner Canon of the Yellow Lord *(Huang ti nei ching t'ai su* 黃帝內經太素*)* that finds exceptions to the predominant holistic viewpoint.

Demiéville, Paul. 1937. Byô 病. In *Hôbôgirin. Dictionnaire encyclopédique du Bouddhisme d'après les sources chinoises et japonaises,* ed. Demiéville, III, 224–265. Paris: Adrien Maisonneuve. Magisterial survey of the connections between Buddhism, sickness, and medicine. Trans. Mark Tatz, *Buddhism and Healing: Demiéville's Article "Byô" from Hôbôgirin* (Lanham, MI: University Press of America, 1985).

Despeux, Catherine. 1985. *Shanghan lun. Traité des coups de froid.* Paris: de la Tisserande. Translation of the Treatise on Cold Damage Disorders *(Shang han lun* 傷寒論, between 196 and 220).

Despeux, Catherine. 1987. *Prescriptions d'acuponcture valant mille onces d'or. Traité d'acuponcture de Sun Simiao du VIIe siècle.* Paris: Guy Trédaniel. Translation of part of Prescriptions Worth a Thousand *(Ch'ien chin fang* 千金方), with an introduction and scholarly notes.

Epler, Dean C. 1980. Blood-letting in Early Chinese Medicine and its Relation to the Origin of Acupuncture. *Bulletin of the History of Medicine,* 54: 337–367.

Furth, Charlotte. 1986. Blood, Body and Gender: Medical Images of the Female Condition in China, 1600–1850. *Chinese Science,* 7: 43–66. Mainly on menstruation and women's disorders in Ming and Ch'ing medical sources.

Furth, Charlotte. 1988. Androgynous Males and Deficient Females. Biology and Gender Boundaries in Sixteenth- and Seventeenth-Century China. *Late Imperial China,* 9. 2: 1–31.

Harper, Donald J. 1982. The 'Wu Shih Erh Ping Fang': Translation and Prolegomena. Ph.D. diss., Oriental Languages, University of California at Berkeley. Translation of and excellent commentary on a fragmentary medical formulary excavated in 1973 at Mawangdui, Hunan.

Hillier, Sheila M.; John A. Jewell. 1983. *Health Care and Traditional Medicine in China, 1800–1982.* London: Routledge & Kegan Paul. Compiled from secondary sources of greatly varying quality.

Hoeppli, R. 1959. *Parasites and Parasitic Infections in Early Medicine and Science.* Singapore: University of Malaya Press. A large part of this collection of previously published essays is devoted to China and Southeast Asia.

Hoizey, Dominique. 1988. *Histoire de la médecine chinoise. Des origines à nos jours.* Médecine et sociétés, 12. Paris: Éditions Payot. Based not on research but on recent Chinese textbooks. Primarily bibliographical, unreliable on such matters as translation of technical terms.

Hsu Ta-ch'un. 1989. *I Hsueh Yuan Liu Lun. Forgotten Traditions in Ancient Chinese Medicine,* trans. Paul U. Unschuld. Brookline, MA: Paradigm Publications. Unreliable, wooden translation of an idiosyncratic, witty book of 1757.

Huard, Pierre; Ming Wong. 1971. *Oriental Methods of Mental and Physical Fitness: The Complete Book of Meditation, Kinesiotherapy & Martial Arts in China, India & Japan.* New York: Funk & Wagnalls. The authors' usual chaotic more-or-less-historical once-over-lightly. Some of the material on calisthentics is novel, but only the illustrations (many of them excellent) can be trusted.

Hymes, Robert P. 1987. 'Not Quite Gentlemen.' Physicians in the Sung and Yuan. *Chinese Science*, 8: 9–76. Meticulous, important study of changing patterns of recruitment to medical careers.

Institute of Acupuncture and Moxibustion, Chinese Academy of Traditional Chinese Medicine. 1990. *State Standard of the People's Republic of China. The Location of Acupoints*. Beijing: Foreign Languages Press. Profusely illustrated. Includes substantial material from early sources with discussions to resolve contradictions, and data on related anatomical, incl. nerve, structures. Does not discuss therapy. Makes previous publications on loci obsolete.

Kaptchuk, Ted J. 1983. *The Web That Has No Weaver. Understanding Chinese Medicine*. New York: Congdon & Weed. An insightful introduction for laymen. The author has had some training in a school of traditional medicine in Macao.

Keegan, David. 1988. *Huang-ti nei-ching*. The Structure of the Compilation, the Significance of the Structure. Ph.D. diss., History, University of California, Berkeley. A pathbreaking technical study of how the Inner Canon (and by implication other Han classics) came together.

Kleinman, Arthur M., et al., editors. 1975. *Medicine in Chinese Cultures: Comparative Studies of Health Care in Chinese and other Societies*. Bethesda; Washington, DC: National Institutes of Health. This conference volume includes a couple of good historical papers. A number of the anthropological contributions are useful, but no attempt is made to test the applicability of field work in today's Taiwan and Hong Kong to questions about traditional China.

Kleinman, Arthur M.; Tsung-yi Lin, editors. 1981. *Normal and Abnormal Behavior in Chinese Culture*. Culture, Illness, and Healing. Studies in Comparative Cross-Cultural Research, 2. Dordrecht: D. Reidel.

Kuriyama, Shigehisa. 1986. Varieties of Haptic Experience. A Comparative Study of Greek and Chinese Pulse Diagnostics. Ph.D. diss., History of Science, Harvard University.

Larre, Claude. 1987. *La voie du Ciel. Huangdi, l'Empereur Jaune, disait... La médecine chinoise traditionnelle*. Paris: Desclée de Brouwer. A series of meditations on classical doctrine, with special attention to *p'ien* 1–2 of the *Inner Canon of the Yellow Lord*.

Larre, Claude, and Elisabeth Rochat de la Vallée, translators. N. d. *Toux. Présentation, traduction et commentaire du Suwen, chapitre 38 (Kelun) et autres textes se rapportant au sujet*. Paris: Institut Ricci.

Larre, Claude, and Elisabeth Rochat de la Vallée, translators. 1983. Plein ciel. Les authentiques de haute antiquité. Texte, présentation, traduction et commentaire du "Su Wen," chap. I. *Méridiens, 61–62:* 13–67. Translation of *p'ien* 1.

Larre, Claude, and Elisabeth Rochat de la Vallée, translators. 1983–1986. Vif. Texte, présentation, traduction et commentaire du "Su Wen," chap. III. *Méridiens,* 1983, 69–70: 39–90; 1985, 71–72: 15–71; 1986, 73–74: 13–78.

Larre, Claude, and Elisabeth Rochat de la Vallée, translators. 1984. Assaisonner les esprits. Texte, présentation, traduction et commentaire du "Su Wen," chap. II. Part 2. *Méridiens,* 67–68: 13–54.

Larre, Claude, and Elisabeth Rochat de la Vallée, translators. 1985. *Par cinq. Discours méthodique sur les phénomènes et comment ils répondent au yin/yang. Texte, présentation, traduction et commentaire de Suwen chapitre* 5. Paris: Institut Ricci. Translation of *p'ien* 5.

Larre, Claude, and Elisabeth Rochat de la Vallée, translators. 1986–87. Fil. Texte, présentation, traduction et commentaire du "Suwen," chap. VIII. *Méridiens,* 1986, 75–76: 13–44; 1987, 77: 13–44.

Larre, Claude, and Elisabeth Rochat de la Vallée, translators. 1988–1989. Cascade. Texte, présentation, traduction et commentaire du *Lingshu,* chap. 8. *Méridiens,* 1988, 81: 25–44; 83: 13–43; 1989, 85: 17–41, 87: 17–37.

Larre, Claude, and Elisabeth Rochat de la Vallée. 1992. *Les mouvements du coeur. Psychologie des Chinois.* Variétés sinologiques, 76. Paris: Desclée de Brouwer. Translation and explication of *Huang ti nei ching ling shu, p'ien* 18.

Leslie, Charles, and Allan Young, editors. 1992. *Paths to Asian Medical Knowledge.* Comparative Studies of Health Systems and Medical Care. Berkeley: University of California Press. One paper on Japan and four on China.

Li Guohao et al., editors. 1982. *Explorations in the History of Science and Technology in China.* Shanghai Chinese Classics Publishing House. *Festschrift* for Needham. See Ågren 1982 and Porkert 1982.

Lu Gwei-Djen; Joseph Needham. 1980. *Celestial Lancets. A History and Rationale of Acupuncture and Moxa.* Cambridge University Press. The first scholarly history of acupuncture. Esp. valuable material on its transmission outside China and its influence in Europe from the sixteenth century on.

Lu Gwei-Djen; Joseph Needham. 1988. A History of Forensic Medicine in China. *Medical History, 32:* 357–400.

Miyasita Saburo [Miyashita Saburô 宮下三郎]. 1976. A Historical Study of Chinese Drugs for the Treatment of Jaundice. *American Journal of Chinese Medicine,* 4. 3: 239–243. This important series of essays is mainly concerned with historical changes in drugs of choice for various remedies. This and Miyasita

1980 are confusing because of careless editing.

Miyasita Saburo. 1977. A Historical Analysis of Chinese Formularies and Prescriptions: Three Examples. *Nihon ishigaku zasshi,* 23. 2: 283–300.

Miyasita Saburo. 1979. Malaria *(yao)* in Chinese Medicine during the Chin and Yuan Periods. *Acta Asiatica,* 36: 90–112.

Miyasita Saburo. 1980. An Historical Analysis of Chinese Drugs in the Treatment of Hormonal Diseases, Goitre and Diabetes Mellitus. *American Journal of Chinese Medicine,* 8. 1: 17–25.

Needham, Joseph. 1980. *China and the Origins of Immunology.* Occasional Papers and Monographs, 41. University of Hong Kong, Centre of Asian Studies.

Needham, Joseph; Lu Gwei-djen. 1970. Medicine and Chinese Culture. In Needham, *Clerks and Craftsmen in China and the West.* Cambridge University Press. Pp. 263–293. Good overviews of traditional medicine, although some interpretations are forced. Contains three other articles of general interest: "Proto-endocrinology in Medieval China" (pp. 294–315), "Hygiene and Preventive Medicine in Ancient China" (pp. 340–78), and "China and the Origin of Qualifying Examinations in Medicine" (pp. 379–395). Revised versions will appear in *Science and Civilisation in China,* vol. VI, part 7.

Needham, Joseph; Lu Gwei-djen. 1984. Medicine in Traditional China. In *Proceedings,* Symposium 9, 17th International Congress of Internal Medicine, Kyoto, pp. 3–11. Brief, informal survey of issues.

Ng, Vivien. 1990. *Madness in Late Imperial China. From Illness to Deviance.* Norman: University of Oklahoma Press. The main emphasis of the book is on law. A chapter on medicine contains interesting material, but suffers from failure to consult original sources. Cf. the more acute Chiu 1981.

Otsuka, Keisetsu. 1976. *Kanpo. Geschichte, Theorie und Prax● der Chinesisch-Japanischen traditionellen Medizin.* Tokyo: Tsumura Juntendo. Translation of the standard introduction to Chinese-style medicine as practiced in Japan, by its leading practitioner. Different in many fundamental ways from Chinese practice.

Porkert, Manfred. 1974. *The Theoretical Foundations of Chinese Medicine. Systems of Correspondence.* MIT East Asian Science Series, 3. Cambridge, MA: The MIT Press. The deepest available analysis of the basic concepts of medicine; restricted to those concerning the body and its functions. For a heated comparison of Porkert's and Needham's approaches to translating technical terms, see Needham's review of this book in *Annals of Science,* 1975, 32: 491–502.

Porkert, Manfred. 1978. *Klinische chinesische pharmakologie.* Heidelberg: Verlag für Medizin Dr. Ewald Fischer. An introduction of about a hundred pages on the functions and therapeutic applications of Chinese drugs, and articles on individual drugs classified by functions.

Porkert, Manfred. 1982. The Difficult Task of Blending Chinese and Western Science: The Case of Modern Interpretations of Traditional Chinese Medicine. In Li Guohao 1982: 553–572. On basic differences between traditional Chinese and modern medicine.

Porkert, Manfred. 1988. *Chinese Medicine. Its History, Philosophy and Practice, and Why it May One Day Dominate the Medicine of the West.* New York: William Morrow. A general textbook and synthesis of Porkert's previous writings. Somewhat more simply argued than 1974.

Sivin, Nathan. 1986. Traditional Chinese Medicine and the United States: Past, Present, and Future. *Bulletin,* The American Academy of Arts and Sciences, May, *39.* 8: 15–26. Informal survey of Western knowledge of Chinese medicine since the seventeenth century.

Sivin, Nathan. 1987. See below, p. 15. Contains a long introduction on classical medicine.

Sivin, Nathan. 1995. Text and Experience in Classical Chinese Medicine. In *Knowledge and the Scholarly Medical Traditions,* ed. Don G. Batess, pp. 177–204. Cambridge University Press. A study of the role that transmission of written texts played in Han-dynasty medicine.

Sung Tz'u. 1981. *The Washing Away of Wrongs (Hsi yuan chi lu): Forensic Medicine in Thirteenth-Century China,* trans. Brian E. McKnight. Science, Medicine, and Technology in East Asia, 1. Ann Arbor, MI: University of Michigan, Center for Chinese Studies. Translation of the oldest extant book on forensic medicine (ca. 1247), which discusses how to determine whether a death is an accident, suicide, or homicide, and who was responsible. Detailed, comparative introduction. Contains much of medical interest.

Unschuld, Paul Ulrich. 1979. *Medical Ethics in Imperial China. A Study in Historical Anthropology,* trans. M. Sullivan. Berkeley: University of California Press. English version of *Medizin und Ethik. Sozialkonflikte im China der Kaiserzeit* (Wiesbaden: Steiner, 1975). Contains much material previously unavailable in Western languages, but analytically uncritical and sinologically unreliable.

Unschuld, Paul Ulrich. 1982. Ma-wang-tui *Materia Medica.* A Comparative Analysis of Early Chinese Pharmaceutical Knowledge. *Zinbun,* 18: 11–63. Summarizes the pharmaceutical import of 1973 discoveries.

Unschuld, Paul Ulrich. 1985. *Medicine in China. A History of Ideas.* Comparative Studies of Health Systems and Medical Care, 13. Berkeley: University of California Press. Innovative in its attention to medical pluralism and in the range of issues discussed, but use with caution. For this and the next two titles see the essay review in *Isis*, 1990, *81:* 722-731.

Unschuld, Paul Ulrich. 1986. *Medicine in China. A History of Pharmaceutics.* Idem, 14. Idem. Extremely detailed descriptive bibliography of the main books in the *pen-ts'ao* tradition. Ignores pharmaceutics as reflected in other genres.

Unschuld, Paul Ulrich. 1987. *Medicine in China. Nan-ching. The Classic of Difficult Issues.* Idem, 18. Idem. First integral translation of a medical classic. Also translates a number of its commentaries. Perceptive introduction, innovative interpretation, but the Chinese ed. is sloppy and the translation unreliable.

Unschuld, Paul Ulrich, editor. 1988. *Approaches to Traditional Chinese Medical Literature. Proceedings of an International Symposium on Translation Methodologies and Terminologies.* Dordrecht: Kluwer Academic Publishers. Essays on problems of translation vary greatly in competence.

Unschuld, Ulrike. 1972. *Das T'ang-yeh pen-ts'ao und die Übertragung der klassischen chinesischen Medizintheorie auf die Praxis der Drogenanwendung.* München: Privately published. Provides an overview of a therapeutic treatise by a leading doctor of the mid 13th century. The argument that the *T'ang-yeh pen-ts'ao* marks the first integration of medical theory and pharmaceutical practice should be approached critically. For a summary see the next item.

Unschuld, Ulrike. 1977. Traditional Chinese Pharmacology: An Analysis of its Development in the Thirteenth Century. *Isis*, 68: 224-248.

Veith, Ilza, trans. 1949. *Huang Ti Nei Ching Su Wen, The Yellow Emperor's Classic of Internal Medicine.* Baltimore: Williams & Wilkins. Incompetent partial translation. The University of California Press 1966 "revised edition" revises only a few sentences.

Yamada Keiji. 1979. The Formation of the *Huang-ti Nei-ching. Acta Asiatica*, 36: 67-89. Brilliant reconstruction of the theoretical classic by Yamada and Akahori Akira. Yamada's identification of different interlocutors in *Huang ti nei ching* with different schools is an assumption, not a conclusion.

Yamada Keiji. 1991. Anatometrics in Ancient China. *Chinese Science*, 10: 39-52. A study of ideas in early medicine about dimensions and proportions of the human body.

Zhang Zhongjing. 1986. *Treatise on Febrile Diseases Caused by Cold (Shanghan lun),* trans. Luo Xiwen. Beijing: New World Press. This and the next item are loose translations, based on superficial understanding, of early therapeutic classics.

Zhang Zhongjing. 1987. *Synopsis of Prescriptions of the Golden Chamber (Jinkui yaolue fanglun),* trans. Luo Xiwen. Idem.

Materia Medica

Bray, Francesca. 1989. Essence and Utility. The Classification of Crop Plants in China. *Chinese Science*, 9: 1-13. Although studies of Chinese taxonomy are generally based exclusively on materia medica, Bray shows this is insufficient.

Bretschneider, E. 1870. The Study and Value of Chinese Botanical Works. *Chinese Recorder*, 3: 157-163, and in following issues. The author, a famous physician and botanist, studied most of the Chinese writings specifically devoted to botany. An introductory essay, better informed than most writings on the subject a century later.

Li, C. P. 1974. *Chinese Herbal Medicine*. Bethesda; Washington, DC: National Institutes of Health. Mainly an account of recent Chinese experimental work and clinical applications, with information on 44 important plants.

Li Hui-lin, translator. 1979. *Nan-fang ts'ao-mu chuang. A Fourth Century Flora of Southeast Asia*. Hong Kong: Chinese University Press. Translation of the classic on southern plants of A.D. 304, with copious notes and good introduction by a highly competent botanist. The authenticity of the book translated is a matter of debate; see Ma Tai-loi 1978.

Ma Tai-loi. 1978. The Authenticity of the *Nan-fang ts'ao-mu chuang. T'oung Pao* (Leiden), *64*: 218-252. Cf. Li Hui-lin 1979.

Needham, Joseph. 1968. The Development of Botanical Taxonomy in Chinese Culture. In *Actes du douzième congrés international d'histoire des sciences*. Paris, pp. 127-133. Tentative but suggestive.

Nguyen Tran Huan. 1957. Esquisse d'une histoire de la biologie chinoise des origines jusqu'au IVe siècle. *Revue d'histoire des sciences*, *10:* 31-37. Summary account of early developments.

History of Medicine in Twentieth-century China

Reference Works

Akhtar, Shahid. 1975. *Health Care in the People's Republic of China. A Bibliography with Abstracts*. IDRC-038e. Ottawa: International Development Research Center. Summaries of ca. 600 articles and books, mostly in English, ranging in quality from excellent to useless. The compiler provides no help in determining which is which.

Anonymous. 1973. *A Bibliography of Chinese Sources on Medicine and Public Health in the People's Republic of China: 1960-1970*. Bethesda: National Institutes of Health. A massive compilation; most of the articles and books have been abstracted and translated into English through Joint Publications Research Service.

Medicine and Related Topics

Anonymous. 1974. *A Barefoot Doctor's Manual.* Bethesda: National Institutes of Health. Reprint, Philadelphia: Running Press, 1977. Translates a compilation by the Institute of Traditional Chinese Medicine, Hunan Province, 1970. Invaluable guide to the integrated Western-Chinese primary therapy of the Cultural Revolution period, with information on plant drugs in wide use. Detailed and practically oriented.

Anonymous. 1975. *Plant Studies in the People's Republic of China: A Trip Report of the American Plant Studies Delegation.* Washington: National Academy of Sciences. First-hand reports on agriculture and the biological sciences.

Anonymous. 1977. *Creating a New Chinese Medicine and Pharmacology.* Peking: Foreign Languages Press. Published for the information of foreigners. In addition to the title essay this short book contains essays on innovations in treatment of fractures, acute abdomen, and cataract.

Bowers, John Z.; Elizabeth F. Purcell, editors. 1974. *Medicine and Society in China.* New York: Josiah Macy, Jr. Foundation. Conference papers on scattered subjects; those on current medicine are useful.

Bullock, Mary Brown. 1980. *An American Transplant. The Rockefeller Foundation and Peking Union Medical College.* Berkeley: University of California Press. Good critical study of the most important Western medical teaching institution in Republican China.

Chen, C. C.; Frederica M. Bunge. 1989. *Medicine in Rural China: A Personal Account.* Berkeley: University of California Press. Chen was the most eminent public-health reformer of the Republican period. His work continued until he was purged in 1957. Valuable for his experiences and for his critique of public health policy and education since 1949.

Croizier, Ralph. 1968. *Traditional Medicine in Modern China. Science, Nationalism, and the Tensions of Cultural Change.* Harvard East Asian Series, 34. Cambridge, MA: Harvard University Press. On the struggle of traditional medicine to survive before 1949. Cf. Zhao 1991, which uses a wider range of sources but is not as well informed about certain aspects of Republican politics.

Farquhar, Judith Brooke. 1987. Problems of Knowledge in Contemporary Chinese Medical Discourse. *Social Science and Medicine,* 24. 12: 1013–1021. Argues against imposing theory/practice, reality/symbol dichotomies on contemporary Chinese medicine.

Farquhar, Judith Brooke. 1994. *Knowing Practice. The Clinical Encounter of Chinese Medicine.* Boulder: Westview Press. Original, extremely perceptive analytic study based on field work in a traditional medical school. Important for understanding classical medicine as well.

Fox, Renée C.; Judith P. Swazey. 1982. Critical Care at Tianjin's First Central Hospital and the Fourth Modernization. *Science,* 20 August, *217:* 700–705. A report of field work, with special attention to ethical issues.

Henderson, Gail E. 1989. Issues in the Modernization of Medicine in China. In *Science and Technology in Post-Mao China,* ed. Denis Fred Simon & Merle Goldman, pp. 199–221. Cambridge, MA: Harvard University Council on East Asian Studies.

Henderson, Gail E.; Myron S. Cohen. 1984. *The Chinese Hospital. A Socialist Work Unit.* New Haven: Yale University Press. A sociological field study of the effects of the work organization on the individual, with valuable observations on recent health care.

Henderson, Gail E., et al. 1988. High-technology Medicine in China. The Case of Chronic Renal Failure and Hemodialysis. *New England Journal of Medicine,* 14 April, *318:* 1000–1004. Discusses the effects of economic reforms on health care, and the role of occupation and insurance in access to expensive therapy.

Horn, Joshua. 1969. *"Away with all pests..." An English Surgeon in People's China.* London: Hamlyn. Mainly on public health, a sympathetic and informed account by a surgeon who practiced in China 1954–1965. Ignores traditional medicine.

Hu Teh-Wei. 1975. *An Economic Analysis of the Cooperative Medical Services in the People's Republic of China.* Bethesda: National Institutes of Health. Concludes that the PRC health care system ca. 1970 was economically successful, but because it depends on China's unique political system, its direct applicability to other countries is limited.

Jamison, Dean T., et al. 1984. *The Health Sector in China.* Population, Health and Nutrition Department, Reports, 4664-CHA. Washington: World Bank. An exceptionally well-informed, systematic digest of information.

Jennerick, Howard P. 1973. *Proceedings. NIH Acupuncture Research Conference. February 28 and March 1, 1973.* Bethesda: National Institutes of Health. Short reports on a cross-section of acupuncture research in the U. S. show contradictory results due to lack attention to standards for professional competence.

Kleinman, Arthur M. 1980. *Patients and Healers in the Context of Culture. An Exploration of the Borderland between Anthropology, Medicine, and Psychiatry.* Comparative Studies of Health Systems and Medical Care, 3. Berkeley: University of California Press. Based on field work in Taiwan; perceptive on medical pluralism.

Kleinman, Arthur M. 1986. *Social Origins of Distress and Disease. Depression, Neurasthenia, and Pain in Modern China.* New Haven: Yale University Press. Based on field work in a Hunan psychiatric facility; comparative in interpretation. This and the next item are important for general implications.

Kleinman, Arthur M.; Byron Good, editors. 1985. *Culture and Depression. Studies in the Anthropology and Cross-Cultural Psychiatry of Affect and Disorder.* Berkeley: University of California Press.

Lampton, David M. 1974. *Health, Conflict, and the Chinese Political System.* Michigan Papers in Chinese Studies, 18. Ann Arbor: University of Michigan, Center for Chinese Studies. Explores health policy to see how a range of social interests has been incorporated in Chinese bureaucracy.

Lisowski, F. P. 1979. The Emergence and Development of the Barefoot Doctor in China. *Nihon ishigaku zasshi, 25*: 339-392. Reprinted in *Eastern Horizon*, April 1980, *19.* 4: 6-20.

Liu Xingzhu; Junle Wang. 1991. An Introduction to China's Health Care System. *Journal of Public Health Policy, 12.* 1: 104-116. Brief but informative survey of system ca. 1990.

Lucas, AnElissa. 1982. *Chinese Medical Modernization. Comparative Policy Continuities, 1930s-1980s.* New York: Praeger Publishers.

Mann, Felix. 1963. *Acupuncture. The Ancient Art of Healing.* New York: Random House. Useful only as a description of modern European acupuncture.

Quinn, Joseph R., editor. 1975. *China [sic] Medicine as We Saw It.* Bethesda: National Institutes of Health. An anthology of reports by visitors to China on health administration, research, and practice, with three papers on epidemiology.

Risse, Guenter B., editor. 1973. *Modern China and Traditional Chinese Medicine. A Symposium Held at the University of Wisconsin, Madison.* Springfield, IL: Thomas. Little of scholarly value, except for an article by Ralph Croizier updating his work on traditional medicine in modern China.

Rosenthal, Marilynn M. 1987. *Health Care in the People's Republic of China. Moving toward Modernization.* Westview Special Studies on China. Boulder: Westview Press. Not very informative, badly edited essays largely based on 1979 and 1981 visits.

Sidel, Ruth; Victor W. Sidel. 1974. *Serve the People. Observations on Medicine in the People's Republic of China.* Boston: Beacon Press. Informative eyewitness book on health care in the PRC.

Sidel, Ruth; Victor W. Sidel. 1982. *The Health of China*. Boston: Beacon Press. Popular overview, not as well informed as the earlier book.

Sivin, Nathan. 1987. *Traditional Medicine in Contemporary China. A Partial Translation of Revised Outline of Chinese Medicine (1972), with an Introductory Study on Change in Present-day and Early Medicine.* Science, Medicine and Technology in East Asia, 2. Ann Arbor: Center for Chinese Studies, University of Michigan. Translates discussions of theory and its application to diagnosis and the planning of therapy from *Hsin pien Chung-i-hsueh kai yao* 新編中醫學概要, with an extensive introduction on change in classical and recent medicine.

Wegman, Myron E., et al. 1973. *Public Health in the People's Republic of China*. New York: Josiah Macy, Jr. Foundation. Conference papers, several of them useful, but nothing of value on history. Annotated bibliography mostly of sources in English, good summaries but uncritical.

Wong, K. Chimin [Wang Chi-min]; Wu Lien-teh. 1932. *History of Chinese Medicine. Being a Chronicle of Medical Happenings in China from Ancient Times to the Present Period.* Tientsin: Tientsin Press. 2d ed., Shanghai: National Quarantine Service, 1936. Only about a quarter is devoted to traditional medicine, and is hopeless as history. The rest is about the impact of European medicine; since it ignores Chinese sources, it is almost entirely an institutional history.

Wu Lien-teh. 1959. *Plague Fighter. The Autobiography of a Modern Chinese Physician.* Cambridge, England: W. Heffer & Sons Ltd. Includes a famous account of the 1910–1911 Manchurian plague epidemic, which the author was in charge of relieving.

Zhao Hongjun. 1991. Chinese versus Western Medicine. A History of their Relations in the Twentieth Century. *Chinese Science*, 10: 21–37. Summarizes a remarkable study in Chinese of competition and shifting legal status to 1949. Cf. Croizier 1968.

INDEX

Chinese words, except for book titles, are listed under the English translations used here. Cross-references are provided for Pinyin items in Chap. V.

Academy, Chi-hsia, putative: IV 19; evidence: IV 24; meaning: IV 27
Accuracy, and precision: V 179
Alchemy, defined cross-culturally: VIII 4; historians of, in China: VIII 6; internal and external: VIII 8; new methods for study: VIII 13
Astronomy, limitations of: V 166, 168

Book of Changes. See *Chou i*
Brahe, Tycho: V 177
Brooks, E. Bruce: IV 3
Buddhism: VII 25, 26; and science: VII 47
Bureaucracy, within human body: I 8

Cai Yong. See Ts'ai Yung
Cai Yuanding. See Ts'ai Yuan-ting
Celestial bureaucracy: I 8
Celestial Masters movement: VI 321; VII 7
Chang Chi: VII 51
Chang Chieh-pin: V 181
Chang Chung-ching. See Chang Chi
Chang Heng: VII 51
Chang Hung-chao: VIII 7
Chang Sui. See I-hsing
Chang Ts'ung-cheng: VII 53
Ch'ao, Master (physician): II 8
Chao Chih-tse: II 8
Chao Hsiu-wu: II 12
Chavannes, Edouard: VII 4
Chemistry, history of: VIII 6
Chen Ch'üan: II 9
Chen-jen (realized immortal): VII 32
Ch'en Kuo-fu: VII 4; VII 27; VIII 8
Chen Li-yen: II 9
Ch'en Shao-wei: VII 56
Chen-tzu hsin-fa: V 182

Cheng Ho: VII 17
Cheng Hsuan: IV 12
Cheng ming (correction of names) : I 3
Cheng Po-ch'iao: IV 15
Cheng-t'ung tao tsang: VII 3
Cheng Yin: VI 323
Ch'i, capital city of: IV 21
Chi-hsia (section of Ch'i capital): IV 19
Ch'i ti chi: IV 25
Ch'i-wu Huai-wen: VII 33
Chia Jang: VII 31
Chia K'uei: V 169
Chia Ssu-hsieh: VII 52
Chiao Hsun: VI 320
Chiao Kan: III 7
Chieh ch'ao: III 14
Chieh Yü: IV 22
Ch'ien hsu: III 8n10
Ch'ien Mu: IV 20
Chikashige Masumi: VIII 8
Ch'in Chiu-shao: VII 53
Ch'in Yueh-jen: VII 51
Ching-chou school: III 13n15
Ching Fang: III 5
Ching-yueh Ch'üan shu: V 181
Chiu T'ang shu: VII 29n52
Chou i, dating: IV 2; Hsi tz'u ta chuan: III 1; scholarship on: III 3n4, 6n6
Chou i ts'an t'ung ch'i: VII 56
Chou li, K'ao kung chi: VII 17
Chou Lü-ch'ing: VII 35
Chu Chen-heng: II 5, 10; VII 53
Chu Hsi: V 176, 178; and Yang Hsiung: III 14
Chu ping yuan hou lun: II 5
Chu Shih-chieh: VII 53
Chu Tan-hsi. See Chu Chen-heng

Chu Tan-hsi i an: II 5, 10n10
Ch'ü Yuan: VII 37
Chu yun: IV 11
Chuang Tsun: III 14
Chuang-tzu: VI 305, 313; VII 5, 26; opposition to technology: VII 43
Ch'un-ch'iu fan lu: I 6; IV 6; V 168; VII 26n46
Ch'un-yü K'un: IV 22
Chung lun: IV 25
Ch'ung Shang: IV 15
Chung shih. See Tsou-tzu chung shih
Chunqiu fan lu. See *Ch'un-ch'iu fan lu*
Circulation, in body and in politics: I 6
Clock, astronomical: VII 21
Collaboration, principles: I 2
Complete Perfection movement: VII 7
Confucian, meanings: VI 317
Confucius: IV 4
Correspondences, in medicine: II 6

Dai Faxing. See Tai Fa-hsing
Dai Zhen. See Tai Chen
Davis, Tenney L.: VIII 8
De Woskin, Kenneth J.: VII 30
Debate, in Greece and China: I 8
Dissection (divination): VII 32
Divination: III 21, 24; V 175
Divine Empyrean movement: VII 7
Divine Spell movement: VII 7
Documents, Book of. See *Shang shu*

Eclipse prediction: V 170
Education, late Warring States: IV 26
Eliade, Mircea: VIII 8
Emotion, and illness: II 1; as medical category: II 4; Chinese terminology: II 7
Emperor, Chen-tsung, of N. Sung: VII 18; Hui-tsung, of N. Sung: VII 19; Kao-tsu, of E. Wei: VII 33; Kuang-wu, of E. Han: VII 32; Shih huang-ti, of Ch'in: IV 15; T'ai-tsu, of N. Wei: VII 18; Wu-ti, of Han: VII 18
Empiricism: VII 38, 46

Eno, Robert: IV 19
Epistemology, and medicine: V 182

Fairbank, John K.: VI 311
Fang-chi (technician): VII 27
Fang I-chih: V 180
Fang-shih (technician): IV 15; VII 27, 55; as epithet: VII 29, 34
Fang-shu (technique): VII 27
Fang Yizhi. See Fang I-chih
Feng-yuan li. See Oblatory Epoch system
First Emperor. See Emperor, Shih huang-ti, of Ch'in
Forke, Alfred: VII 33
Fu ch'i fa (breath disciplines): VII 17
Fu Shan: II 2
Fukui Fumimasa: VII 27
Fung, Yu-lan: IV 1

Giles, Lionel: VII 2
Girardot, Norman: VI 312
Graham, A. C., on Tsou Yen: IV 1, 7
Grand Inception system: V 169
Great Expansion system: V 172
Greece, comparisons with Chia: I 8
Groot, J. J. M. de: VII 34

Halleux, Robert: VIII 13
Han Fei-tzu: IV 7; VII 6
Han kuan i: VII 32
Han Shih-liang: II 11
Han shu: IV 12; VII 24
Han Yü: III 13; VII 36
Hao Yun. See Ho Yun
Head texts: III 20; structure of: III 29
Henderson, John: V 179, 182
Hilary of Poitiers, St.: V 184
Hiortdahl, Th.: VIII 8
Hipparchus of Nicaea: V 177
Ho Chieh: II 12
Ho-nan Shao shih wen-chien hou lu: II 8
Ho Yun: II 8
Hou Han shu: II 8

Hou Pa: III 3n2
Hsi chung: III 25n37
Hsi tz'u ta chuan. *See Chou i*
Hsiang Hsiu: VII 15
Hsiao-hsi kua: III 30
Hsien ching (canons of immortality): VII 35
Hsien-men Kao: IV 15
Hsin lun: III 12n15
Hsin T'ang shu: V 172
Hsu Ch'ien: VII 55n99
Hsu Hsia-k'o. *See* Hsu Hung-tsu
Hsu Hung-tsu: VII 53
Hsu Kuang-ch'i: VII 53
Hsuan-hsueh: VII 5
Hsuan-tu Kuan: VII 50
Hsun-tzu: IV 3; VII 26; as client of Ch'i king: IV 23
Hua shu: VII 13
Hua T'o: II 8; VII 51
Huai-nan-tzu: VII 40, 43
Huan T'an: III 12
Huan Yuan: IV 22
Huang-Lao: VII 9
Huang Tao-p'o: VII 53
Huang-ti nei ching: I 5; III 2; VII 27n48; cosmological synthesis: IV 6; on emotion and illness: II 1, 6; *Su wen* on state and body: I 7
Humanism: III 1

I fang k'ao: II 4
I-hsing: V 172; VII 30, 47, 52, 55
I hsueh ju men: VII 29n52
I jen (remarkable person): VI 308
I lin: III 7
I-shu (technical art): VII 27
Immortality, techniques of: IV 15; VI 319
Indeterminacy, terms for: V 167; role in history of science: V 182
Inequalities, in astronomy: V 177
Ixing. *See* I-hsing

Jealousy as medical disorder: II 2

Jia Kui. *See* Chia K'uei
Jih-che (diviner of propitious days) : VII 27
Jih chung: III 25n37
Jingyue Quanshu. See Ching-yueh Ch'üan shu
Johnson, Obed Simon: VIII 8

Kaltenmark, Max: IV 9n13
Kao Lien: VII 35
Kimura Eiichi: VI 314
King Chien-p'ing, of Liu Sung: VII 36
Knoblock, John: IV 1, 19
K'o (guest): IV 20
Ko Hsuan: VI 323
Ko Hung: VI 323; VII 13, 25, 52
K'ou Ch'ien-chih: VII 18
Kua ch'i: III 30
Kuan-yin-tzu: VII 13, 42
Kung-shu Pan: VII 51
Kuo Hsiang: VII 15
Kuo Shou-ching: VII 53
Kuo yü, dating: IV 3

Lai, Whalen W.: VI 312
Lao-tzu: III 16; V 167; VI 305, 313; VII 5; in Han thought: VII 47
Laozi. See Lao-tzu
Legge, James: VII 17n25
Li Chieh: VII 52
Li Chih. *See* Li Yeh
Li Ch'un: VII 52
Li Kao: VII 53
Li Ping: VII 51
Li Shih-chen: VII 53
Li shih chen hsien t'i tao t'ung chien: VII 50n97
Li Sung-ch'ing: II 8
Li Tao-yuan: VII 52
Li Tung-yuan. *See* Li Kao
Li Yeh: VII 53
Liang fang: VII 22n39
Lieh-tzu: VI 313
Lin-ch'uan chi: VII 23n40
Lin Fu-shih: VII 31
Lin Ling-su: VII 22

Liu An, King of Huai-nan; Tsou Yen as predecessor of: IV 15
Liu Hsiang: III 13
Liu Hsin: III 9, 13
Liu Hui: VII 51
Liu Hun-k'ang: VII 22
Liu Tsung-yuan: VII 36
Liu Wan-su: VII 53
Lloyd, G.E.R.: I 2
Lo Chih-t'i: II 15
Lu Hsiu-ching: VI 321
Lu Pan. See Kung-shu Pan
Lü Pu-wei: I 5; Tsou Yen as predecessor of: IV 14
Lü shih ch'un-ch'iu: I 5, II 7n7; on limits of knowledge: VII 40; proposals for government: IV 5
Lun heng: VII 26
Lun li: V 169

Ma Chün: VII 51
Ma-wang-tui texts: VII 16
Makeham, John: IV 19
Mao shan chih: VII 22n39
Mao Shan movement: VI 319
Maspero, Henri: VII 4
Medicine, correspondences in: II 6; indeterminacy in: V 181
Mencius: V 178
Meng Hsi: III 5
Mengqi bitan. See Meng-ch'i pi-t'an
Ming-chia and semantics: I 3
Minggantu: VII 54
Mo-tzu: VI 305, 320; VII 25
Mohist Canons: IV 7
Mu t'ien-tzu chuan: VI 321
Multhauf, Robert P.: VIII 7

Needham, Joseph, on alchemy: VIII 4, 9; on internal alchemy: VIII 10; on naturalists: IV 1; on Taoism: VI 309; VII 10; sources for discussion of Taoism: VII 13
Neo-Taoism: VII 9, 15

New Policies, scholarship on: VII 23
Ngo Van Xuyet: VII 29
Nü k'o hsien fang: II 2
Numinous Treasure movement: VII 7
Nylan, Michael: III 42

Oblatory Epoch system: V 173
Ôfuchi Ninji: VI 315
Optics: VII 33

Pai shu (Atractylis): VII 17
Pan Ku: III 9
P'ang P'u: IV 3
Pao Ching: VII 13n14
Pao Ching-yen: VII 13
Pao-p'u-tzu. See Ko Hung
Pao-p'u-tzu nei p'ien: VI 325
Pao-p'u-tzu wai p'ien: VII 13n14
Patronage: IV 19
P'ei Hsiu: VII 51
Pelliot, Paul: VII 4
Pen-ts'ao ching chi chu: VII 29n53, 35
Philosophy, Chinese rubrics analogous to: IV 8
Pi Sheng: VII 52
Pieh lu: IV 24
Pien Ch'ueh. See Ch'in Yueh-jen
Pien the Wheelwright: VII 36
Precision, and accuracy: V 179
Preconceptions, in Taoism: VII 41, 46
Psychiatry and Chinese medicine: II 1
Ptolemy of Alexandria: V 177

Reforms, of N. Sung: VII 19
Register, initiation: VII 8
Religion, popular: VI 318

Sakai Tadao: VII 27
San kuo chih: II 8
San shih (three corpseworms): VII 16
San tung ch'ün hsien lu: VII 50n97
San-t'ung li. See Triple Concordance system
Schwartz, Benjamin: IV 1

Seidel, Anna K.: VII 2, 25
Sexual techniques: VII 16
Shang shu, dating: IV 2
Shao Po-wen: III 34n47
Shao shih wen-chien hou lu. See *Ho-nan Shao shih wen-chien hou lu*
Shao Yung: III 34n47
Shen Gua. See Shen Kua
Shen Kua: V 173, 180; VII 52
Shen Kuo. See Shen Kua
Shen Tao: IV 22; VII 39
Sheng chi ching: I 7
Sheppard, H. J.: VIII 4
Shih chi, on celestial bureaucracy: I 8; on Hsun-tzu: IV 23; on Tsou Yen: IV 9, 11, 15
Shih i: III 3
Shih shan i an: II 11n11
Shu-shu (study of regularities): VII 45
Shui ching chu: IV 12
Song shi. See *Sung shih*
Ssu-ma Ch'eng-chen: VII 13, 45
Ssu-ma Kuang, on *T'ai hsuan:* III 8
Ssu-ma T'an: IV 17; VII 24
Ssu tuan: IV 4
Stein, Rolf: VI 322; VII 16
Stoics: V 184
Strickmann, Michel: VII 45; VIII 15
Su Hsun: III 10
Su Shen liang fang: VII 22n39
Su Shih: III 13
Su Sung: VII 20, 52, 55
Su Wei-kung wen chi: VII 55n98
Sun I-jang: VII 27n47
Sun Ssu-mo: VII 30, 52
Sung Chung: III 13n15
Sung shih: V 175; VII 22n37, 22n39
Sung shu: V 171
Sung Wu-chi: IV 15
Sung Ying-hsing: VII 54
Sung Yuan hsueh an: VII 55n99
Superstition: VII 6
Supreme Purity movement: VII 7

Ta-yen li. See Great Expansion system
Tai Chen: V 166
T'ai-ch'u li. See Grand Inception system
Tai Fa-hsing: V 171
T'ai hsuan ching, commentaries in: III 24; commentaries on: III 7n8; cyclic character: III 31; influence of *Lao-tzu:* III 16; meaning of title: III 3, 15; rarity in Sung: III 9; status as canon: III 3n2; 3-base notation in: III 35
Tai Nien-jen: II 11
T'ai-p'ing ching: VII 30
T'ai-p'ing Yü lan: VII 34n63, 37n71
T'ai-shang Lao Chün: VII 5
T'an Ch'iao: VII 13
T'an Chih: II 12
T'an Huang: II 12
Tao, as divinity: VI 306
Tao-chia (masters of the Way): VI 305; VII 24
Tao-chiao (teachings of the Way): VI 305; VII 25
T'ao Hung-ching: VI 319, 321; VII 30, 35, 52; VIII 14
Tao-jen: VII 26; as term for Buddhists: VII 34
Tao-shih: VII 26
Tao-shu: VII 26
Tao te ching. See *Lao-tzu*
Tao tsang. See *Cheng-t'ung tao tsang*
T'ieh wei shan ts'ung-t'an: VII 22n39
T'ien Ch'ang: IV 21
T'ien Pien: IV 22
T'ien-yin-tzu: VII 13
Taoism, and scientists: VII 49; conflict with Confucianism, putative: VII 45; international conferences on: VII 4; scholarship on: VII 4n4; terminology for: VII 24
Taoism, religious: VI 306; and popular religion: VI 318, 322
Taoist, meanings in English: VI 316
Taoist Canon. See *Cheng-t'ung tao tsang*
Ten Wings. See *Shih i*
Tetragrams, sequence of: III 30

Thomas Aquinas, St.: V 184
Thorndike, Lynn: VII 11, 38n72
Triple Concordance system: V 169
Ts'ai Ching: VII 19, 24
Ts'ai Lun: VII 51
Ts'ai Yuan-ting: V 176
Ts'ai Yung: V 166
Tsan-ning: VII 47
Tsao hua (shaping and transforming process): VII 29
Ts'ao Yuan-yü: VIII 7
Tso chuan, dating: IV 3
Tso Yu-hsin: II 9
Tsou-tzu: IV 9n13
Tsou-tzu chung shih: IV 9n13
Tsou Yen, career: IV 9, 22; goals of writing: IV 13
Tsu Ch'ung-chih: V 171; VII 52, 55
Tung chung: III 25n37
Tung Chung-shu: III 2; IV 6; Tsou Yen as predecessor of: IV 14
Tzu chih t'ung chien: VII 45
Tzu-kung: IV 4
Tzu-ssu: IV 3

Vermilion, as drug: II 9

Wang An-shih: III 13; VII 19, 23
Wang Chen: VII 53
Wang Ch'eng-fu: VII 36
Wang Chin: VIII 7
Wang Ch'ung: III 12
Wang Fu: VII 21
Wang Mang, and Yang Hsiung: III 10; coinage: VII 32
Wang Pi: VII 13
Wang Ping: VII 27, 52
Wang Tzu-hsi: VII 22
Wang Ya: III 28
Watts, Alan: VI 326
Weber, Max: VII 3n2
Welch, Holmes: VI 311
Wen Chih: II 7

Wen hsuan: IV 11
Whitehead, Alfred North: V 175
Wind Disorders: II 5
Wu (spirit medium): VII 31, 34
Wu ch'ang: IV 5
Wu ching i i: IV 25
Wu hsing, in mature synthesis: IV 6; temporal associations: V 174
Wu K'un: II 4, 16
Wu te (five virtues): IV 10
Wu te chung shih: IV 11. See also *Tsou-tzu chung shih*
Wu-wei (non-interference): VII 32, 38, 45

Xin Tang shu. See *Hsin T'ang shu*

Yang Hsiung, and astronomy: VII 37; posthumous reputation: III 12
Yang Hui: VII 53
Yen Chung-p'ing. See Chuang Tsun
Yin Chih-i: V 181
Yin Ch'ung: VII 50
Yin Chung-ch'un: V 181
Yin-i (recluse): VII 27
Yin-yang, in mature synthesis: IV 6; school of, putative: IV 1
Yin Zhiyi. See Yin Chih-i
Yin Zhongchun. See Yin Chung-ch'un
Ying Shao: III 41; VII 42
Yixing. See I-hsing
Yoshida Mitsukuni: VIII 5n3
Yoshioka Yoshitoyo: VI 313
Yü Chia-hsi: VII 27
Yü han shan fang chi i shu: IV 11n15
Yü Hsi: IV 24
Yuan yu: VI 318
Yueh Kuang: II 12

Zhang Jiebin. See Chang Chieh-pin
Zhenzi xinfa. See *Chen-tzu hsin-fa*
Zhu Xi. See Chu Hsi
Zu Chongzhi. See Tsu Ch'ung-chih
Zürcher, Erik: VII 7, 9

R
601
.S58
1995